HARTMUT WARM,
engineer and independer
omy, geometry, the hist
musical aesthetics. He ha
published widely on these and other subjects, and in particular on his discoveries relating to the solar system and its inherent order. He lives in Hamburg, Germany.

# Signature of the Celestial Spheres

*Discovering Order in the Solar System*

Hartmut Warm

Translated by J. Collis

Sophia Books

Sophia Books
Hillside House, The Square
Forest Row, East Sussex RH18 5ES

www.rudolfsteinerpress.com

Published by Sophia Books 2010
An imprint of Rudolf Steiner Press

Originally published in German under the title *Die Signatur der Sphären* in 2001 by Keplerstern Verlag, Hamburg. This edition is based on the 2004 German edition, specially revised and edited for the English by the author



A catalogue record for this book is available from the British Library

ISBN 978 1 85584 235 9

Cover by Andrew Morgan Design
Typeset by DP Photosetting, Neath, West Glamorgan
Printed and bound in Malta by Gutenberg Press Ltd.

# Contents

# Preface

The splendid word 'cosmos' originally signified 'regularity' or also 'adornment' in ancient Greek, but about two and a half thousand years ago it began to acquire its present meaning of 'universe as an ordered whole'. The night sky was adorned with stars which people began to see as images, gathering them together in seemingly never-changing representations stemming from mythology or the animal kingdom. A handful of stars, however, differed from all the others in that they did change; they moved through the constellations along paths, some more complex, some less so, which they repeated over longer periods of time. The movements perceived led to ideas about the regularity of the wandering stars, the planets, and also about the sphere of the fixed stars which was seen as a coherent whole. The majority of Greek thinkers did not doubt that there was good order in the cosmos and that it was constructed meaningfully in a way that could to some extent be comprehended by the human mind. Some of them, beginning with Pythagoras of Samos, concluded in addition that the order must be musical, resounding in the heavens as a kind of harmony of the spheres sent forth by the planets' orderly progression.

No one would nowadays dream of taking seriously those early models of the universe showing the Earth in the centre surrounded by circles or spheres. But as to the general order of the cosmos the matter is not quite so simple. On the one hand we now know about events such as supernova explosions or monstrous eruptions of gamma rays; we speculate about the existence of so-called black holes that devour whatever might stray too close to them; and we wonder about many other extraordinary phenomena as well. Such things do not appear to speak in favour of there being a harmonious order in the universe. On the other hand when we look at photographs of galaxies, those shining starry islands adorning the immeasurable cosmos, we sense that whatever might underlie its structure must surely be comparable in magnitude to its huge dimensions. Furthermore, scientific successes have confirmed what the Greek philosophers were beginning to understand, namely that at least in part the world is indeed governed by meaningful laws which we now call the laws of nature and are able to comprehend with our intellect. This is the basis on which modern human beings have founded their image of the cosmos. It is a view in which there is no place for a harmony of the spheres or for any specific order of our planetary system.

Some years ago I began to ask myself what we have actually grasped with the help of our scientific methods. By this I don't mean the discovery of laws of nature that enable us to force substances to enter into chemical reactions with one another or that tell us how to build technical apparatuses. What I wanted to know was how much we have discovered about the origin of the universe, about the structure of entities such as our solar system, about how life originated and then developed right up to the emergence of human consciousness. As time went on I arrived at a preliminary and personal estimation of certain theories arising from the discoveries of researchers in various fields. One of my basic assumptions grew from my appreciation of music and from my previous research into the history of music theory. This is where I initially discovered the ancient idea of a harmonious regularity in the heavens, i.e. in the movements of the planets. I took it as a given fact until it dawned on me that actually I knew nothing at all about it. So I began to search the literature to find out what, if anything, was indeed known. I found advocates for the harmony of the spheres chiefly among those whose leanings included the arts and humanities while others, who regarded such a thing as pure fantasy, belonged on the whole to the scientific camp. But neither the one side nor the other yielded any thorough treatment of the subject that included the astronomical and the mathematical points of view.

People either believe in the idea or they do not. And the things in which they do or do not believe are in many cases those that are encompassed by Johannes Kepler's work on this subject. Kepler was one of the first scientists in the modern sense. He discovered the fundamental laws of planetary movements early in the seventeenth century. Among other things he found that planets travelled in ellipses, each in a slightly different way that can be determined by the degree of its eccentricity. At the same time Kepler was also one of the last astronomers to be convinced that the order in the cosmos involved more than merely the effects of physical forces. The various eccentricities must be as they are, so he said, in order to take care of the harmonies between the movements of the celestial bodies. On the basis of the planetary laws he had discovered, Kepler succeeded in calculating the spatial relationships, velocities and other parameters with greater accuracy than had ever previously been possible. And in one of these instances, by comparing the various planets, he found what he considered to be a very exact correlation with the intervals of musical notes. He thus brought the idea of a harmonious order in the solar system right into the scientific age, and to this day his ideas on the subject are better known than any others.

My investigation of Kepler's harmony of the spheres became one of the central themes of the present work, so that my struggle with his ideas proved to be the starting point for what became a journey into a largely unknown territory to which I was led by my own subsequent researches concerning the hidden order of our cosmic home. This mysterious realm that lies, you might say, on our very doorstep revealed its treasures on a variety of different levels: in the spatial structures, in the ratios of velocities, in the coordination of long-term movements of the individual planets around the Sun, and also in their inherent movements, i.e. their individual rotations. Most astonishing of all, however, was the way a continuous theme of geometric interrelatedness began to emerge, uniting the various parameters with one another and telling a kind of story. With hindsight I have come to realize that in ferreting out and describing what the solar system is trying to tell us I was frequently accompanied by Johannes Kepler's profound thoughts, his philosophical convictions, his delight in discovery and his human qualities which continue to speak directly to us even after 400 years. In this sense this book has come into being through him and is dedicated to him.

Centuries after Kepler's discovery of the planetary laws, and also of Sir Isaac Newton's discovery of the laws of gravity, basic structures in the planetary system appear, surprisingly, to have remained undetected. There are various reasons for this, some of them psychological: put simply, one can say that one only finds what one is looking for. But there is also a powerful mathematical and astronomical reason. Only in the last 15 or 20 years has it become possible to carry out many of the calculations on which the results I shall be describing are based. Theoretically, of course, they could have been done long ago, but only with an effort so huge and time-consuming as to be entirely impractical on the scale called for by the task. For this reason, although the principles elaborated by Newton at the end of the seventeenth century have not changed, even the best mathematicians have continued to work at optimizing and simplifying the relevant methods of calculation. 'Not until' 1982 and 1987 did, among others, P. Bretagnon of the Bureau des Longitudes in Paris publish procedures that make it possible to calculate by relatively simple means (with inconceivable precision and, thanks to computers, with breathtaking speed) the positions and paths of the planets for at least 10,000 years (5000 years back into the past and 5000 on into the future). I offer my profoundest gratitude to all those who have contributed to these developments over the centuries.

The methods of calculation are now so accurate that one can accept their results as being scientifically secure within the period of time specified. The solar system is, in this sense, up to now the only realm within the universe surrounding us about which we can make truly precise statements regarding its ordering. Since humanity's view of the world is always closely bound up with what is known about the universe, this circumstance is profoundly significant. The signature of the celestial spheres revealed by the interplay between the planets and the central star around which they circulate furthermore points far beyond what is merely calculable: like a musical composition, it leaves room for retrospectively creative interpretation, indeed it positively challenges us to embark on such an exercise. I have taken up this challenge with the greatest delight, hoping, however, that in my descriptions I have clearly shown the dividing line between my own personal view, the mathematical and astronomical foundations, and my elaborations based on these.

The text is arranged in a way which will make it possible for those who lack any specific knowledge of the subject to follow it (sometimes perhaps with a little patience). In the main only basic arithmetic and some elementary knowledge of geometry (supplied in the Appendix) will be needed. But I do realize that the descriptions 'basic' and 'elementary' are always relative in the realm of mathematics and that entry into the world of numbers and geometrical figures is not necessarily equally fascinating for all and sundry. The more complicated calculations not absolutely essential for an understanding have therefore been banished to the Appendix which, since its purpose is to provide a more detailed access and make it possible for the separate steps to be checked, is comparatively voluminous.

Finally I wish to thank Kerstin Kreft for struggling through the 'relatively basic' calculations, while reading the manuscript, even though some of them appeared to her to be 'relatively complicated'. I also thank her warmly for her scrupulous care, her delicate sense of the German language, her pertinent questions and the many conversations we have had on the themes in this book. I thank Manfred Lellek for the seriousness and enthusiasm with which he has acknowledged these celestial discoveries and for the confirmation he offered me chapter by chapter. May the final result throw a new light on the beauty and perhaps also the uniqueness that adorns our cosmic home.

Hartmut Warm
Hamburg, 8 April 2000

# Note to the English edition

The 'signature of the celestial spheres' became clear to me about 10 years ago, and I am now delighted to be composing a note to an English edition of my book *Die Signatur der Sphären* in which I endeavoured to describe it. With English now being the number one language in the world, this will mean that the discoveries presented here, which breathe new life into the ideas of Pythagoras and Johannes Kepler concerning the music of the spheres, will become accessible to all and sundry.

Ten years is a not inconsiderable span of time, so one might be tempted to ask how a book based on exact astronomical data can still be relevant. Before expanding on this, let me first give a few brief hints to assist the reader by summing up the particulars presented in the list of contents.

Chapters 1 and 2 describe astronomical and then geometrical and harmonic basics, while 3 and 4 consider the music of the spheres understood as a correspondence between musical and planetary ratios. Chapters 5 and 11 give an account of current scientific thinking regarding the origin and future of the planetary system and, respectively, of the universe. Chapters 6 to 10 and also 12 deal with the movements and rotations of the celestial bodies, and Chapter 13 with their ordering in space. Each of these blocks can also be considered separately. The book in English is a slightly shortened version of the German.

Ten years is indeed a long span of time, but in another sense it is also an extremely brief period that is of virtually no significance at all in relation to the longer-term interactions of the planets which are the subject of this book. A great many new astronomical details have of course been discovered over the past decade, so that several statements and astronomical values, such as Hubble's constant, can now be more precisely defined than was possible at the end of the second millennium. But the overall modern scientific image of the cosmos has remained the same. So the descriptions in Chapters 5 and 11, although no longer entirely up to date, are by no means obsolete, while above all the philosophical conclusions drawn from the astronomical discoveries and theories have lost none of their significance.

With regard to Pluto, which has meanwhile been demoted from its status as a planet in a not undisputed declaration by the International Astronomical Union (IAU), the following remains to be said. As is unfortunately often the case in scientific thinking, this judgement, too, is based solely on consideration of external details such as for example the size of the body in question. However, the investigations presented here show that Pluto belongs without any doubt to the solar system as a whole or, one could say, to the community of the planets. So whether the IAU or any other organization declares Pluto or some other body to be a planet, a minor planet, or anything else is, as far as I am concerned, of secondary relevance.

In conclusion I wish to thank Sevak Gulbekian most warmly for taking the initiative and so resolutely carrying through the publication in English of this rather specialized work. Likewise I thank Johanna Collis for her sensitive, conscientious and pleasing translation and also Nick Thomas who supported her in matters of technical terminology and usage.*

And my wish for readers of this book is that they may experience profound joy and spiritual benefit as their familiarity with the celestial harmonies grows. Since it may not always be an easy read, perhaps they will be helped along the way by a quotation from a book, *Der hörende Mensch*, by Hans Kayser, who has re-founded Pythagorean harmonics in the twentieth century: 'What is described would be merely superficial data losing any significance if the reader did not know how to cause the plucked strings to resound. A book on harmonics cannot be "read" like an ordinary work; it has to be creatively assimilated.'

I hope my book will in this way bestow upon its readers (to speak once again with Hans Kayser): '. . . a full flow of that inner certainty, that meaningful order, with which the Creation has been endowed from the beginning.'

Hartmut Warm
July 2009

---

*TRANSLATOR'S NOTE: Without the help of Nick Thomas on the astronomy, geometry and mathematics in this book I would have been unable to complete this translation. His effort has been tireless and his support wonderfully reassuring. The author has been similarly patient in answering what must have appeared to be a never-ending stream of questions.

# 1. The Initial Approach

The conception humanity once had of the world began to change about four hundred years ago, perhaps more profoundly than ever before. Although the transformation had begun somewhat earlier, what set the ball rolling in earnest were certain events that took place around the year 1600. In 1572 a new star appeared in the sky—a supernova, as we now know. The Danish astronomer Tycho Brahe worked out its exact position. He had constructed instruments for observing the skies that were among the best in the world at the time, and in addition he approached his work more systematically than almost any other astronomer. He succeeded in proving that the new arrival, which faded away again after about 15 months, must belong in the sphere of the fixed stars—the first painful blow to the heavens which had until that moment been regarded as perfect and immutable. Then, five years later, a comet came into view. This apparition, too, was accurately surveyed by Brahe. He showed it to be far more distant than the Moon and that it was likely to sweep across the heavens, traversing the planetary orbits. This meant that there could be no crystal spheres to which the celestial bodies were attached, as had been thought until that moment. Tycho Brahe had discovered space, the wide-open distances of the universe.[1]

The image of the sky and its division into spheres had until then often been linked to the idea of a music of the spheres, an unearthly, harmonious ordering of the cosmos. Around the year 1600, these ideas, soon to be regarded even more definitely as outdated, were once more cast by William Shakespeare (1564–1616) into wonderful, indeed immortal, lines:

> *Sit, Jessica. Look how the floor of heaven*
> *Is thick inlaid with patines of bright gold;*
> *There's not the smallest orb which thou behold'st*
> *But in his motion like an angel sings,*
> *Still quiring to the young-ey'd cherubims:*
> *Such harmony is in immortal souls;*
> *But, whilst this muddy vesture of decay*
> *Doth grossly close it in, we cannot hear it.*[2]

The telescope was invented in 1608. In the following year Galileo built one too, and pointed it towards the heavens. He discovered that the Moon was no smooth globe but had a pitted surface quite unfitting for a perfect celestial body. And his telescope also showed him that Venus underwent phases just like those of the Moon. For him, this was a convincing indication of the rightness of the theory published by Nicolaus Copernicus almost 70 years earlier, for the changes to the shape of Venus can be most convincingly explained by placing the Sun at the centre of the planetary orbits. But perhaps most important of all was the discovery that the sky contained at least ten times more stars than could be seen with the naked eye. In 1610 Galileo published a book about his discoveries some of which were shocking, since they called into question not only the outdated view of the structure of the cosmos but also the position of the human being in it. In the same year the Englishman John Donne (1572–1631) wrote in his poem 'An Anatomy of the World':

*The Sun is lost, and th'earth, and no mans wit*
*Can well direct him, where to looke for it.*
*And freely men confesse, that this world's spent,*
*When in the Planets, and the Firmament*
*They see so many new; they see that this*
*Is crumbled out againe to'his Atomis.*
*'Tis all in pieces, all cohærence gone;*
*All just supply, and all Relation . . .*[3]

Within a short space of time human beings had been ejected from a comprehensible world which they believed they could explain to some extent—one with which they felt united—and had entered instead into a state of cosmic homelessness. They are still in that state today, though of course this generalization cannot be applied to every individual. Nevertheless, Donne's words do describe a world view which is still valid four hundred years on, or is perhaps even stronger than at the time of its inception. We now know that there are not ten times but billions of times more lights in the sky and that not only supernovae but also all suns eventually collapse. An idea has developed according to which the universe arose out of an unimaginably huge burst of energy which people call the big bang. Are there any modern individuals who can claim to have been granted some modicum of spiritual stability by the discoveries of astronomy, however fascinating these may be? Whatever order which might be found in the way conglomerations of matter coexist, they argue, must be explicable solely on the basis of the laws of gravity. At best, the music of the spheres can be seen as nothing more than a poetic fancy. Jacques Monod, the molecular biologist, put his finger on the world view held by very many people towards the end of the twentieth century: *'The ancient covenant is in pieces; man at last knows that he is alone in the unfeeling immensity of the universe, out of which he emerged only by chance.'*[4]

This sense of being alone derives not so much from the as yet unanswerable question as to whether other living creatures might exist on other planets but rather from the presumed non-existence of any spiritual, creative power that would give meaning to the existence of the universe as such, as well as to that of the human beings who inhabit it. Monod's world view is seen as being scientific, but what does science actually tell us? One thing is certain, and that is that the cosmos is more enormous than we can imagine. But, we ask, is it entirely indifferent to us? After all, it brought us into being and has thus far kept us alive. There is nothing at all obvious in this, since countless very specific circumstances are needed for it to happen. Astronomers are faced with huge riddles by the very fact that our planetary system has proved sufficiently stable over billions of years to provide conditions which guarantee the continued existence and further evolution of living creatures. (This will be considered in greater detail in Chapter 5.)

The changes in the way people thought which began to arise around 400 years ago commenced when it became possible to observe phenomena in the heavens in greater detail, especially the structure of the solar system. Sir Isaac Newton's discovery of the law of gravity, which he published in 1687, broadly showed what it was that kept the planets in their orbits and provided the regular movements on which the order of the universe was founded. Whereas Newton himself was convinced that it was the Creator who had set the celestial clock in motion, about a hundred years later people began to conclude that the planets must have emerged from some kind of primordial mist. Out of this mist what had originally been diffuse matter was formed, by the effects of physical forces, into a few solid bodies circling around a centre. This idea of a merely accidental play of 'chance and necessity' made it unnecessary to seek further for the causes of any order in the way the planets related to one another.

The changes that led to our scientifically oriented concept of the world were thus founded on exact measurements which left no doubt as to how they should be interpreted. We shall be considering in Chapter 11 the extent to which this foundation is valid for a view of the universe as a whole. As far as the planetary system is concerned we certainly have knowledge that enables us to make exceedingly accurate calculations about its structure as a whole as well as about the motions of the individual celestial bodies. The methods needed for this have been continuously improved right up to the present day, as described here in the Preface.

And the tools we now have at our disposal enable us to make a new approach to the long-forgotten question as to whether a harmonious order, a kind of music of the spheres, lies hidden within the solar system. Harmony is a term that describes mathematically the relationship between various aggregates belonging to a coherent whole. For example, harmonious musical intervals can be understood in terms of frequency ratios such as 2:1 (the octave) or 3:2 (the fifth). Other ratios, e.g. those of the golden section, are also often described as being harmonious. Therefore the matter of harmony in the planetary system calls for a rigorous investigation of such ratios, for example those of the periods of rotation or the distances from the Sun. Perhaps by once more taking up this theme so long forgotten we shall discover concrete evidence of the cosmos being more than a conglomeration of various clumps of matter relating to one another only on the basis of purposeless, fortuitous laws. Our knowledge about the solar system, our cosmic vicinity, is now so precise that it is possible for us to reach clear results with regard to its structures. So after four hundred years, at least for this portion of the universe, we have a new and very solid foundation for deciding which of the two views quoted above is the more fitting: whether *'there's not the smallest orb but in his motion like an angel sings'*, or whether *'. . . all [is] in pieces, all cohærence gone; All just supply, and all Relation . . .'*

**Fig. 1.1** *Saturn. In August 1995 its rings were visible laterally; on the left is its large moon Titan casting a shadow; on the right other, smaller moons. Picture by the Hubble Space Telescope*

So let us now get to know our cosmic home a little more closely. We shall look first at the relative dimensions in the planetary system in which the nine wandering stars, the planets, travel around their common centre. People speak of wandering stars because they change their position as seen from Earth whereas all the other stars stay put in the firmament and are therefore known as fixed stars. They are fixed in relation to one another, of course, for the view presented to us by the apparently turning sky changes during the course of a night or through the seasons of the year. Today we know that the fixed stars also move, or that their recognizable positions as seen from Earth or any other viewpoint shift over very long periods of time, so that the constellations as we see them today do not remain the same for ever. Noticeable changes here only occur over tens of thousands of years, however, so that it is very understandable that for a long time the fixed stars were seen as being absolutely static.[5]

To give a clear picture of the relative sizes of the orbits of the planets it is useful to show the inner and the outer regions separately, since the average distance of Pluto, the farthest from the Sun, is about 102 times as great as that of Mercury, the closest to the centre.

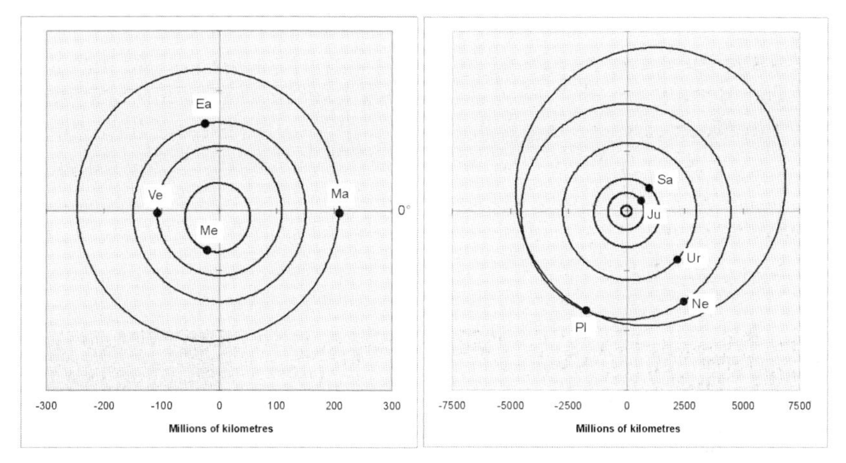


**Fig. 1.2** *Depiction to scale of the planetary orbits round the Sun (situated where the coordinates intersect). Left: the inner region with Mercury, Venus, Earth and Mars. Right: the outer region with Jupiter, Saturn, Uranus, Neptune and Pluto. These positions applied on 1 January 2000. (Planet sizes not to scale.) All the planets move anticlockwise*

In order to indicate the actual dimensions, the Mars orbit is also shown as the inner ring of the outer region. The distance between Jupiter and Mars is divided about halfway by the asteroidal belt. As far as is known today, this consists of about 50,000 (some say more) variously shaped small bodies also known as planetoids measuring more than one kilometre across, and probably countless smaller ones. The one with the greatest mass, Ceres, which was discovered in the night of the New Year 1801, is almost spherical and has a diameter of about 1000 kilometres. Beyond Pluto's orbit there is a further region consisting of as many rocky blocks of varying sizes known, after the eponymous astronomer, as the Kuiper Belt which is said to be the source of some of the short period comets (i.e. those with an orbital period of up to 200 years). Even farther out the Oort Cloud is presumed to exist. This was postulated by Jan Hendrik Oort as the possible source of other comets.

The presumed division of the planetary system into two parts by the natural barrier of the asteroid belt also shows in the size and consistency of the planets themselves. The inner planets are composed of rock and at their equator measure between 4879 km (Mercury) and 12,756 km (Earth), while most of the outer ones consist of gas surrounding a presumed core of iron or rock. At their equator they measure between 49,528 km (Neptune) and 142,984 km (Jupiter). Another characteristic of the four gas giants is their very rapid rotation. They turn once round their own axis in 9.9 (Jupiter) to 17.2 (Uranus) hours, whereas the rotation periods of the inner system vary greatly, as we shall see in detail. They, i.e. Jupiter, Saturn, Uranus and Neptune, also have their own satellite systems consisting of many moons, and they have ring systems amongst which that of Saturn is the best known, the most striking and the most beautiful. Among the moons of the inner planets our own is the only one that has a (very) significant part to play, as we shall show later. Mars has two tiny satellites, while Mercury and Venus are without moons.

Pluto is an exception among the relative homogeneity of the members of the outer planetary system. Having a diameter of only 2390 km it is the smallest planet in the solar system. Owing to its specific density derived from certain observational data[6] it is presumed to consist chiefly of rock. On the basis of its exceptional status, which includes some parallels with the moons of the outer planets, some astronomers even want to deny its planetary status* and declare it to be an inhabitant of the Kuiper Belt. To do this one would have to give an accurate definition of what the nature of a planet is, which would probably not be easy and which will not be attempted here. Whatever the case may be, we shall see that Pluto, at least in the matter of order in the planetary system, is undoubtedly a part of it and indeed must be a part of it on the basis of the logic inherent in that system. (I beg the reader to have patience with me when, at the beginning of this book, I keep referring to what is yet to come. Our overall theme is, after all, somewhat complex!)

Consideration of the data for Pluto and Mercury also reveals other similarities. The innermost and the outermost planet of our cosmic home are the two smallest. The eccentricities of their elliptical orbits are also by

---

* In the present work Pluto is counted as one of the nine planets regardless of the fact that the IAU downgraded it in 2006—a decision that has not gone unchallenged.

far the most pronounced (Mercury 0.206; Pluto 0.249. See Appendix 1.1 regarding the significance of these numbers). This shows not in the orbit itself, since even Pluto's orbit is still approximately circular, but in the distance of the orbit's centre from the Sun. The inclinations from the ecliptic (the plane in which all the other planetary orbits lie with maximal deviations of 3.39°) are noticeably higher, with Mercury at 7.0° and Pluto at 17.15°.

The conformities mentioned and a fundamental degree of structure in the planetary system become very obvious when the different elements are summarized as follows:

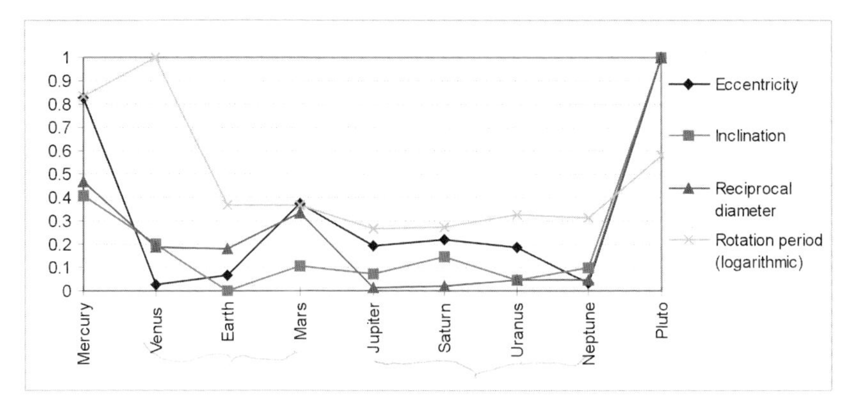


**Fig. 1.3** *Composite view of various data on the nine planets. The lines relate to the maximal value (= 1), or in the case of the diameter to the minimum. For reasons of clarity a logarithmic scale has been applied to the rotation period, since the range is greatest here*

This depiction offers us a meaningful division of the planets into three categories:

- the innermost and the outermost taken together as marking the boundaries;
- Venus, Earth and Mars (the exceedingly complex attunement of which will be discussed in detail in Chapter 7);
- the four great gaseous planets (which are also finely attuned to one another, as we shall see).

Let us now develop a general view of the ratios in our planetary system. We have already seen that the mean distances of Mercury and Pluto to the Sun increase by roughly a factor of 100, so that the total ratio of the radii from the innermost to the outermost orbit comes to almost 1:100. The corresponding ratio of the orbital periods is, perhaps surprisingly, *c.* 1:1000 and that of the mean velocities 10:1 (the velocity decreases the farther away the planets are from the Sun, thus reversing the ratio). The exact ratios of the three parameters are 1:102, 1:1029 and 10.1:1 (as derived from the data shown in Appendix 6.1).

Thus in the overall ratios of all the data relevant to us at this stage for an investigation of order in the solar system (we shall be meeting others later on), the number ten or its square and its cube, so greatly prized by the Pythagoreans,* makes its appearance. As we continue our investigations we shall discover good reasons why it is not exactly 10, 100 and 1000 and why—as Johannes Kepler would probably have pointed out—this is indeed not possible.

I have already said that the frequent appearance of the number ten or its powers might initially seem surprising. However, on the basis of the mathematical relations between the parameters mentioned, once the number ten has made its appearance (in connection with the velocities) it is bound to appear also in connection with all the parameters mentioned. Let us look first at the relationship between the mean distance and the orbital period. This is governed by the simple yet marvellous law Kepler discovered in 1618 and which was later named in his honour Kepler's Third Law. It was published in his book *The Harmony of the World* (*De harmonice mundi*). Somewhat abbreviated it goes as follows:

$$T^2 = a^3$$

(The square of the orbital periods equals the cube of the mean distances. More in Appendix 3.2.)

*'If you want to know the exact moment in time, it was conceived mentally on the 8th March in this year 1618 but submitted to calculation in an unlucky way, and therefore rejected as false, and finally returning on 15th of May and adopting a new line of attack, stormed the darkness of my mind. So strong was the support from the combination of my labor of seventeen years on the observations of Brahe and the present study, which conspired together, that at first I believed I was dreaming, and assuming my conclusion among my basic premises. But it is absolutely certain and exact . . .'*[7]

* The Pythagoreans were an order of scientific and religious thinkers headed by the Greek philosopher and mathematician Pythagoras of Samos in southern Italy during the sixth century BC. Many people unfortunately know Pythagoras only as the discoverer of the eponymous theorem about the square on the hypotenuse of a right-angled triangle, though this in itself cannot be too highly esteemed. One reason for the significance attached to the number ten was the fact that the sum of the first four numbers is 10. In the coming chapters I shall be speaking about further Pythagorean discoveries and ideas.

In purely physical terms Kepler's great discovery later came to be treated as a special case of Newton's more general Law of Gravity, namely where the mass of a body is insignificantly small in comparison with that of another, as is the case in the relationship between a planet and the Sun for almost all purposes. However, entering more deeply into the geometric significance of the equation one comes without fail to the conclusion that this has to do with the relationship between space (cf. the cube of the mean distance) and planar area, and it would be a very special area the length of whose sides corresponded to time. The area itself then has to be regarded as quadratic time (see Glossary), a form of time in which the linear flow of our normal time is no longer valid, in which past and future exist side by side as though in a plane. In other words: it appears that the ratios in our planetary system are regulated in a dimension that is not accessible to human beings. A philosophical and religious interpretation thus strongly suggests itself.

As to the ratios between planets, by applying $T^2 = a^3$ we find that where the ratio of the mean distances of two planets is 100:1, then the ratio of the orbital periods is 1000:1 and of the velocity 1:10 (see Appendices 3.2 and 3.3 respectively for the mathematical derivation). One might be inclined to regard the fact that the innermost and outermost planets are related in this way as mere chance, so the next question concerns the seven planets that lie in between, since to my knowledge no model exists that can depict in a convincing way the harmonious order so frequently claimed to exist in the planetary system. Kepler's ideas and the Titius-Bode law are probably the best known efforts in this direction. I shall discuss these and other attempts in Chapter 3. There is however one exception, or actually several exceptions to the above statement which therefore, after all, call for it to be modified. During August 1998 I gained an increasingly clear picture of the basic aspects of the spatial ordering of the nine planets in our solar system which, it turns out, accord with very simple and very beautiful geometrical forms. As this book progresses we shall be investigating this 'signature of the celestial spheres'—among much else, since the investigations I embarked on in consequence brought a good many other astonishing phenomena to light as well.

But first we must complete some preparatory work. The point of departure for all those who have hitherto struggled with this subject has been the premise that order in the heavens reveals itself in the ratios of the mean distances or of the orbital periods, or in the ratios of the points on the elliptical orbits closest to or farthest from the Sun, or in the velocities at those points. Johannes Kepler also considered the angles, as seen from the Sun, subtended by the planets in the same periods of time on its apparent path. Some also deal partially with the periods between sequential positions of the planets in conjunctions or oppositions (see Fig. 2.4 regarding these terms). As far as I know, no one has so far paid attention in this connection to the scarcely mentioned semi-minor axis of the ellipse.

For this reason we shall now take a closer look at the ellipse. (Appendix 1.1 offers a more detailed description of this geometrical form and its significance as a planetary orbit.) That the planets move in elliptical orbits and not in circles, as had previously been thought, was one of Kepler's earlier discoveries. During the year 1604, after copious calculations in connection with the orbit of Mars, he became aware of its elliptical form. Kepler's First Planetary Law states: The orbits of the planets are ellipses.

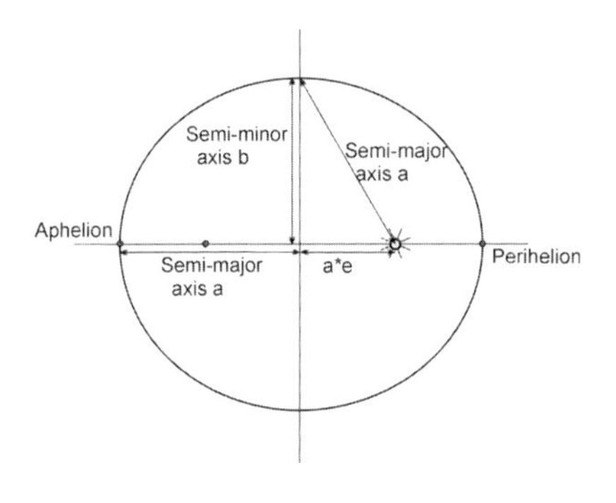


**Fig. 1.4** *The ellipse*

*'It can be said of Kepler, as of few great scientists, that what he accomplished would never have been done had he himself not done it. The discovery from the examination of naked-eye observational reports that planets move on ellipses, and according to the area law,* is so exceedingly improbable—and Kepler's manner of arriving at it was so decidedly personal—that it lies outside the course of any inevitable development.'*[8]

The figure above speaks for itself. The ellipse has two foci in one of which the Sun is located. In accordance with the famous theorem of Pythagoras ($a^2 = b^2 + e^2$) the two parameters normally specified in order to characterize a planetary orbit enter into the equation: the eccentricity and the mean distance, which is equal to the semi-major axis. Table 1.1 shows the semi-minor and the semi-major axis of all the planets and the ratios arising from these:

* The area law refers to Kepler's Second Law. The velocity of the planet in its orbit is at its fastest at perihelion, the point closest to the Sun, and correspondingly at its slowest at aphelion, the most distant from the Sun. The law governing the change of velocity states that the imaginary line joining the Sun and the planet sweeps out equal areas in equal times.

**Table 1.1** *The semi-major and semi-minor axes and their ratios. Interval from planet to planet signifies that e.g. the semi-minor axes of Mars and Earth are 1.5172:1.*

| | Semi-major axis a (km*10⁶) | a relating to Mercury | 1/a relating to Pluto | Interval from planet to planet | Semi-minor axis b (km*10⁶) | b relating to Mercury | 1/b relating to Pluto | Interval from planet to planet |
|---|---|---|---|---|---|---|---|---|
| Mercury | 57.9092 | 1 | 102.002 | | 56.6717 | **1** | 100.946 | |
| Venus | 108.2089 | 1.8686 | 54.5874 | 1.8686 | 108.2064 | 1.9094 | 52.8694 | 1.9094 |
| Earth | 149.5980 | 2.5833 | 39.4848 | 1.3825 | 149.5772 | 2.6394 | 38.2465 | 1.3823 |
| Mars | 227.9406 | 3.9362 | 25.9140 | 1.5237 | 226.9447 | **4.0045** | **25.2079** | 1.5172 |
| Jupiter | 778.3221 | 13.4404 | 7.5892 | 3.4146 | 777.4160 | 13.7179 | 7.3587 | 3.4256 |
| Saturn | 1427.607 | 24.6525 | 4.1376 | 1.8342 | 1425.402 | **25.1519** | **4.0135** | 1.8335 |
| Uranus | 2871.068 | 49.5788 | 2.0574 | 2.0111 | 2868.004 | 50.6073 | 1.9947 | 2.0121 |
| Neptune | 4498.187 | 77.6765 | 1.3132 | 1.5667 | 4498.000 | 79.3694 | 1.2719 | 1.5683 |
| Pluto | 5906.846 | 102.002 | 1 | 1.3132 | 5720.803 | 100.946 | **1** | 1.2719 |

The veil now begins to lift. And the small trick of ascertaining the ratios both outwards from within and vice versa does the rest. In the case of the semi-minor axis the overall ratio, to within one per cent, approaches the number $10^2$. There is significantly greater conformity (divergences of 0.11 and 0.34%) in the case of the first and fourth planet—also reckoned from inside and outside—which in each case relate in the ratio of the double octave 4:1.* With less than one per cent difference the first and sixth planet, counted both from inside and from outside (Mercury-Saturn and Mars-Pluto), form an interval of 25:1. This of course means that the ratio of Saturn to Mars is 25:4 or, to be exact, 6.281:1 instead of 6.25, a difference of merely 0.49%. (And let us not hide the fact that the ratios 25:1 and 25:4 lie almost exactly between two neighbouring musical intervals, the fifth 3:2 and the minor sixth 8:5.) Uranus divides the outer region of Pluto to Saturn almost exactly in the octave ratio of 2:1. Mars and Earth (although not quite so precisely, the deviation being 1.15%, the reasons for which will be seen later) form the interval of the fifth 3:2, which means (with about the same deviation) that the ratio Earth to Mars is 8:3, i.e. a fourth raised by one octave (2*4:3). Well, well, whoever would have expected such a thing!

Let us illustrate these findings. The following diagram does not represent the orbits of the planets around their centre but rather their interrelationships amongst each other, symbolized by circles:

* Those who are not familiar with musical terminology need not be put off here and I ask them to be patient. The analogy is simply too cogent to be ignored. In brief, a double octave is nothing other than the ratio of frequencies of 4:1 between two different musical notes. These concepts will be explained more fully in Chapter 2.

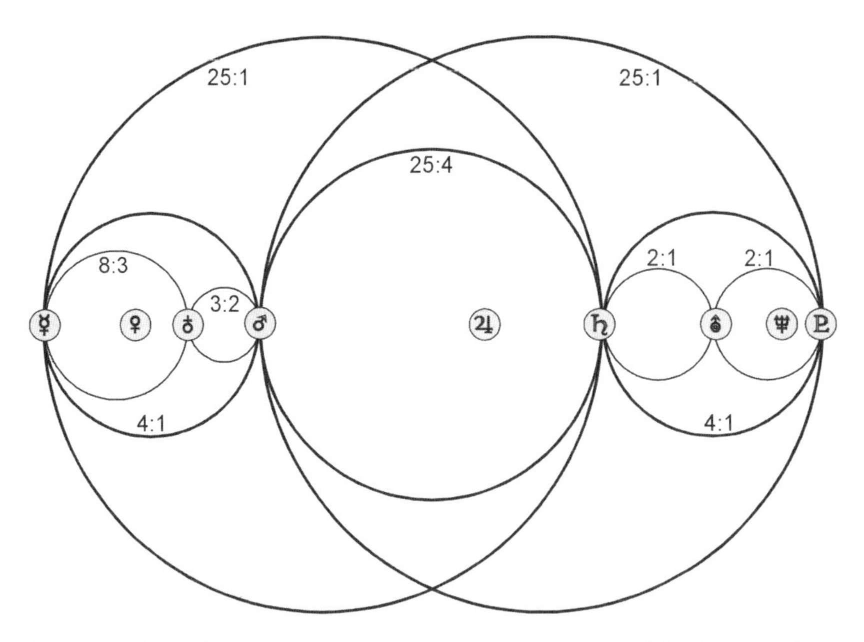


**Fig. 1.5** *Approximation of the intervals of the semi-minor axes of the planets. Left to right: Mercury, Venus, Earth, Mars, Jupiter, Saturn, Uranus, Neptune, Pluto*

*'. . . and so in the end the patterns were opposite on all sides.*
*For thus has He who is before ages and to all ages embellished the mighty works of His wisdom: nothing is redundant, nothing is deficient, and there is no place for any criticism. How desirable are His works, and so forth, all balanced one against another, and none lacks its opposite; of every one He has established (He has confirmed with the best arguments) the goodness (their furnishing and comeliness) and who shall be sated with seeing their glory?'*

Johannes Kepler[9]

The clarity of the order is captivating, with the same intervals evolving outwards from within and inwards from without. One could also describe it as a mirror image around the central region where Jupiter sits on his throne even though he is not in the exact centre (as seen also in Fig. 1.3 which depicts the planetary elements). In detail the distances closer to or further away from the Sun are differentiated, just as the sounding together of two octaves is differentiated from that of a fourth raised by an octave and a fifth. Nevertheless, whether the ratios are musical or not, the four mirrored intersecting main circles manifest a well-ordered structure so clearly that there cannot but be a deeper reason for it. Even those who are only prepared to credit purely physical causes with the creation and formation of our planetary system (of which more in Chapter 5) will henceforth be obliged to include the obvious order in the ratios of the semi-minor axes in their considerations.

There now remain the two planets second from the outer edges (Venus and Neptune) and Jupiter, which are not included in the ordering as shown in the diagram. So let us now look more closely at the planetary orbital periods (see Table 1.1 and Appendix 6.1).

When we observe the planets only, without taking account of the Sun, the Jupiter/Saturn constellation forms the crucial point of our planetary system. According to the law of gravity discovered by Isaac Newton, bodies always attract one another. The masses of Jupiter and Saturn combined make up *c.* 92% of the total mass of all nine planets, whereby Jupiter has the lion's share, being 3.34 times more massive than Saturn. Both together, on the other hand, amount to barely one eight-hundredth of the mass of the Sun. Nevertheless the planets exercise their pull on the Sun, too, causing it to move, however minimally, around the common centre of gravity also known as the barycentre (to be discussed in more detail in Chapter 9). In addition, it is obvious that the planets also attract one another, depending on the measure of their mass and distance from each other. Whereas the masses are constant, the distances vary continually as do therefore their mutual gravitational effects. When two planets are closest to one another their mutual attraction is at its highest. Now if two planets were to meet repeatedly at the same point or at a few always recurring points, as would be the case if the resonances of the orbital periods* were exactly the same, then this could lead over very long periods to a summation of the forces working between them and even to irreversible changes in their orbits. I shall be dealing in detail with all this in the coming chapters. For the present it is sufficient to confirm that for the reasons stated the ratios of the orbital periods play an extremely important part also in terms of physics, especially those between Jupiter and Saturn, but also those between Jupiter and the other planets.

So let us look more closely at them. The approximate ratio formed by the orbital periods of Saturn and Jupiter is 5:2, actually 2.4833:1. This very slight deviation from resonance is highly significant for the wonderfully ordered interplay of the whole planetary system. This statement must be taken at face value for now; the proof will be dealt with later when we shall also discover criteria concerning the estimation of commensur-

---

* Resonance (or commensurability) are astronomical terms for ratios lying close to the fractions of small whole numbers.

abilities between the planets, since it seems that one always finds certain deviations from one hundred per cent resonances (except in the case of the moons). Initially we can begin by accepting the interval of 5:2 which corresponds to that of a major third raised by one octave since the difference from the actual ratio amounts to only $\frac{1}{149}$th. It is also noticeable that the ratios of the orbital periods of the outer planets beyond Saturn, and with relation to Jupiter, amount to *c.* 7:1, 14:1 and 21:1, whereby the deviation decreases with increasing distance. (The exact figures are 7.0832, 13.8906 and 20.8801, the differences from seven and its multiples being 1.188, 0.788 and 0.574%.) Mercury, too, relates to Jupiter in an approximate ratio involving the number seven (49.251 instead of $7^2$, difference 0.513%).

Jupiter's ratio to Earth is just under 12:1 (11.862; since the relation is with Earth in this case, this is the same as Jupiter's orbital period in years), but here with the best will in the world one cannot speak of a resonance, since the deviation amounts to more than one per cent. However, the various periods are such that this deviation, and the also rather high deviation of the Jupiter/Uranus interval almost balance one another in that one Uranus orbit takes 84.019 years,* which amounts to less than 0.03% compared with 7*12 = 84. Let us now look at these ratios in a depiction which includes both the orbital periods already mentioned and the intervals of the semi-minor axes (see Fig. 1.6).

In addition to the Jupiter ratios already mentioned we have the 3:2 interval of Pluto and Neptune and especially the 13:8 resonance between Earth and Venus. (The exact ratio of their orbital periods is 1.6255:1; the difference from 13:8 is merely 0.032%.) This means that Venus makes 13 journeys round the Sun in 8 Earth years—one speaks of 13 Venus years—and that the two planets meet five times during this period.

And now, in order to explain what moved me to add the cross between our home planet and its cosmic sister in this diagram, I must begin somewhat further back.

The ratio between Earth and Venus is relatively close to 1.618.., the value of the golden section. The golden section arises when the division of a length (or a number or an angle) is such that the smaller part is to the greater as the greater is to the whole. In the past this was also known as the 'divine proportion' to express the idea that through this special ratio the

---

* The figure results from the previously mentioned intervals of the orbital periods:

$$\frac{Jupiter}{Earth} * \frac{Uranus}{Jupiter} = \frac{Uranus}{Earth} = 11.862 * 7.083 = 84.019$$

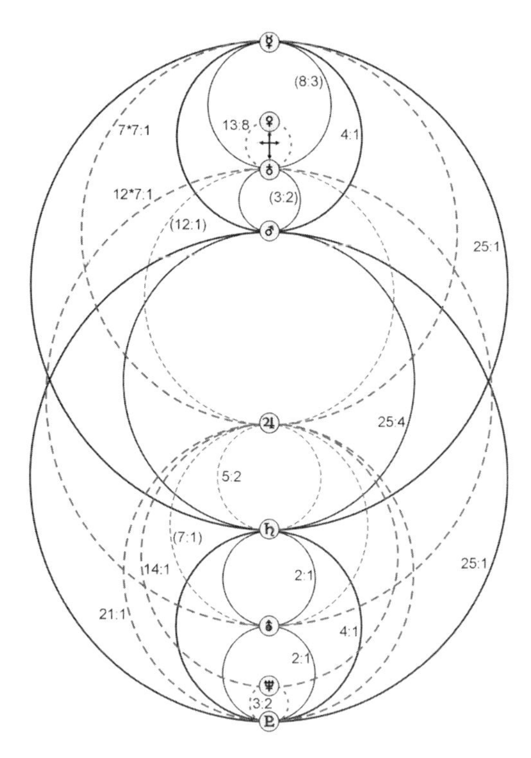


**Fig. 1.6** *Intervals of the orbital periods (dotted) and the semi-minor axes of the planets. From top: Mercury to Pluto. Figures in brackets: deviations of more than 1% between the actual and the depicted ratios*[10]

two parts are also linked with their archetypal origin. In considering the stability of planetary or planetoid orbits, modern mathematicians have meanwhile also come to take the golden section into account. It is relevant because two celestial bodies whose periods of orbit around a common centre have this ratio experience the least possible mutual influence through the gravitation of the other body since theoretically they never meet at the same point. I say theoretically, because via the so-called Fibonacci series it is possible for an ever closer approach to the irrational number expressing the golden section to occur. The interval 13:8 of Earth and Venus orbital periods represents a pair of numbers from that series.

The ratio of the golden section of 1.618034.. possesses what might be termed a natural supplement, namely the number 1.381966.. . Let us call this number the 'silver section' for fun (see Glossary). It arises naturally from the remainder when we divide up the number one by the golden section:

$$\frac{1}{1.618034} = 0.618034; \qquad 1 - 0.68034 = 0.381966$$

We thus arrive at the value of the silver section by adding one or, for example, also from the calculation 3 – 1.618034 = 1.381966; it also has a part to play in the geometrical construction of the golden section (see Appendix 1.2).

Let us now look back to Table 1.1. There for the semi-minor axes of Earth and Venus we find the ratio 1.382332 (this is more accurate here than in the table). Working to only three places after the decimal point, there would be 100% correspondence, but in fact there is a deviation of 0.027%. According to Kepler's Third Law the relation between orbital period and distance from the Sun is defined as:

$$T^2 = a^3 \Rightarrow a = \sqrt[3]{T^2}$$

According to this, with T set to 13:8 the value for a is 1.382194. And so, with the ratio of the orbital periods as it is in the case of Earth and Venus, the approximation of the value of the silver section in the ratio of the semi-major axes is exceedingly accurate; and in relation to the semi-minor axes it is even more so. In other words, the two ratios described cross over in the ratios of the two planets. This is in no way a matter of course, for the mathematical derivation of the golden (and the silver) section and the Keplerian relation are independent of one another. The extent to which the number 1.381966 plays a part in the geometry and in the architecture of the inner planetary system will be investigated further as we proceed.

I shall draw our initial approach to the subject of this book to a close with an illustration depicting the ordering of the universe which originated at almost the same time as Kepler's *The Harmony of the World*. Although not an original intellectual creation, it provides a summary of the image of a harmonious universe that retained its validity up to the early years of the seventeenth century. Robert Fludd who designed it was a physician and not an astronomer or mathematician like Johannes Kepler. Neither did he have at his disposal Tycho Brahe's notes about the movements of the planets. Those were unique in their time and they enabled Kepler to battle through to his laws of the planets and to check and develop further his own ideas and those of others concerning the ordering of the heavens. Nevertheless, images that have arisen purely intuitively also have their value especially when their spiritual essence can be brought into at least some degree of agreement with scientific findings discovered later. Moreover, to this day we do not really appear to have clarified the matter of whose hand it is that tuned or indeed is still tuning the musical instrument of the universe.

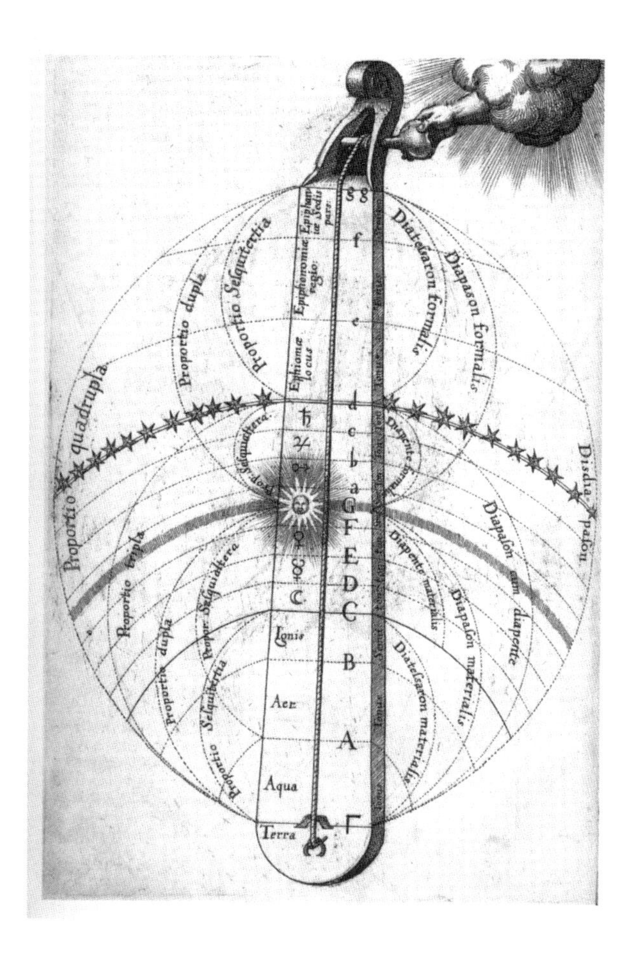


**Fig. 1.7** *Robert Fludd,* Utriusque Cosmi, *Volume 1, Oppenheim 1617*

# 2. Geometry and Harmonics

One of the simplest forms on which a piece of music is based, often quite unconsciously, is that of the lied. A theme is heard, then it is changed or another is added, and finally the original melody returns. This A-B-A structure appears very frequently in lied compositions, and actually it is also the basic form of almost all composed music, though of course the patterns are varied to a greater or lesser degree. But it is not only in the realm of music that this form-principle is found:

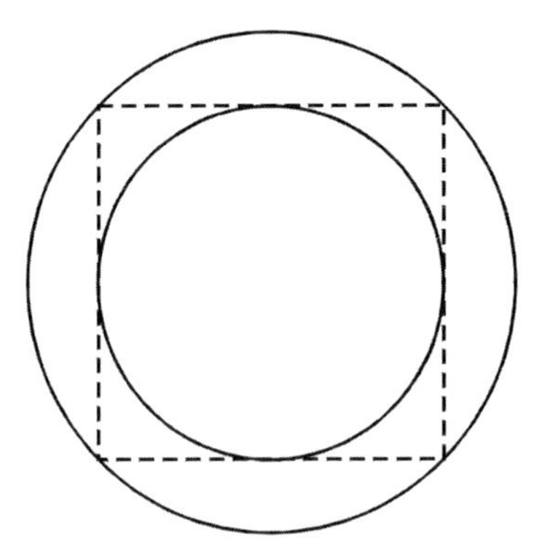

**Fig. 2.1** *The octave of the square*

Those who first noticed the ratio of 2:1 hidden in this simple geometrical construction will no doubt have experienced the delight of genuine discovery. The ratio between the area of the outer circle which contains the four corners of the square and that of the inner circle fitting inside it is exactly 2:1.* Many people, however, are robbed of their astonishment at this wonderful fit between circle and square when they discover the simple mathematical proof for this, which states merely that the square root of two multiplied by the square root of two equals two.

Pythagoras or the early Pythagoreans or whoever it was who discovered that the most consonant, i.e the most perfectly harmonious interval, the octave, arises when one divides the string of an instrument in the ratio of 2:1, gave to humanity an impulse the effects of which continue to be felt to this day. It encourages us all to ask why it is that

* A more detailed algebraic treatment of this and the following constructions is contained in Appendix 1.3.

numbers—in the widest sense mathematics and the laws comprehended by physics—and the musical sensitivity of the human being felt throughout body and soul are linked one with the other. It encourages us to ask where the realms in which this universal harmony, in the sense of the correspondences between seemingly entirely disparate things, can still be found today.

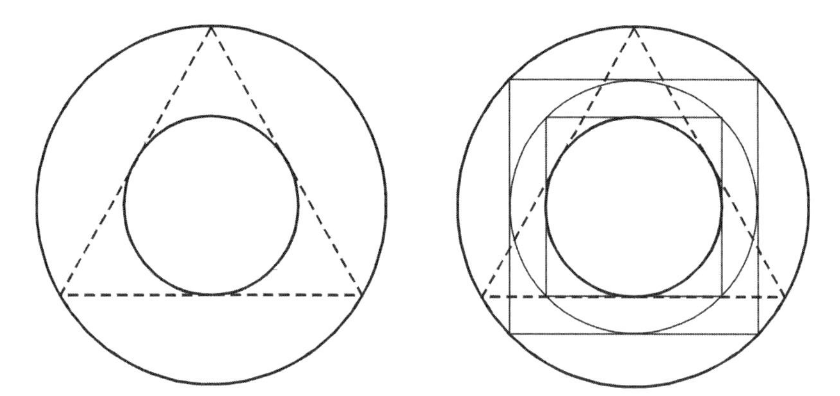

**Fig. 2.2** *The double octave of the triangle*

In an equilateral triangle the ratio of the areas of the outer and inner circles is 4:1. By means of the squares one sees that there are two octaves or one double octave. Here the result is the same; triangle and square are in complete harmony in this ratio. But there is a great difference between arriving at the outer circle via a triangle or via two squares. Furthermore the right-hand drawing shows that two squares, one around and one within a circle, also have as the ratio of their areas the interval 2:1 (for the mathematics see Appendix 1.3). The same goes of course for the triangle and the ratio 4:1.

The next intervals (see Fig. 2.3) arise via the hexagon, to which we owe three intervals in all. In the left-hand figure our point of departure is the outer circle into which we inscribe a hexagon. This is a basic construction we may remember from our schooldays. The radius stepped out with compasses onto the circle gives the corner points. The outer circle and the circle inside the hexagon—here again, and in what follows, I refer to the areas—have the ratio 4:3, i.e. the interval of the fourth. The circle inside the hexagon and the inner circle form the fifth, 3:2. The outer and the inner circle have the ratio 2:1, as you can see in the dotted square in the figure on the right. Here you also see a second hexagon, with its inner circle emphasized. This and the inmost circle make the major second which has the ratio 9:8.

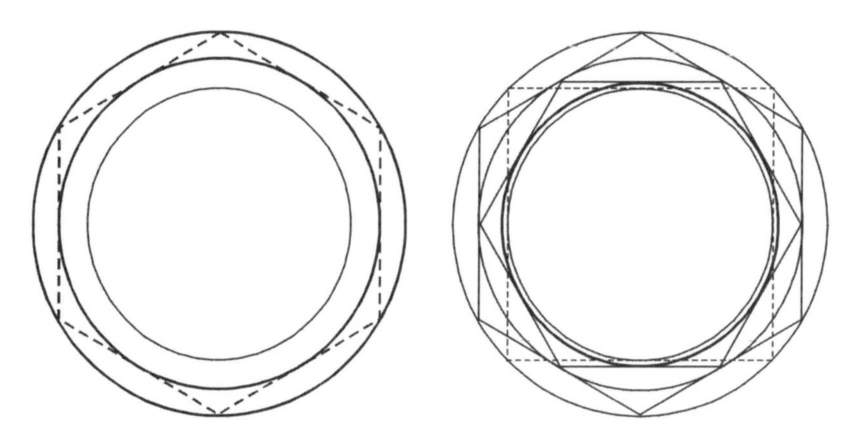

**Fig. 2.3** *Fourth, fifth and major second of the hexagon*

We have thus opened the way to those intervals which had the largest part to play in Grecian music and its musical theory, and also still in the Middle Ages. It is a universally human phenomenon for octave, fifth and fourth to be experienced as being harmonious and for singing and music-making to involve the use of musical intervals (exceptions being only the most basic productions of sound and some forms of twentieth-century music). Research in various cultural settings and among primitive peoples has shown that this tonal basis of music carries significantly greater weight than do the differences which of course also exist.[1]

Even physiologically, the way human beings hear is an utter marvel*—including of course the way sound is carried from the ear to the brain, and the awareness that goes with it. It seems that small whole number ratios of notes are clearly preferred.[2] If this were not the case there would probably be no such thing as music. That human beings relate to music at all, and are capable of making music and listening to it, is in my opinion one of the greatest wonders and also riddles of what is generally regarded as evolution today. I fail to see any purely biological reason for the need to steer the evolution of ear and brain in this direction. However, this is a broad subject, and it will always be possible to find arguments that go in one direction or the other.

* I mean the structure of the acoustic duct from the eardrum via the three ossicles—the hammer, anvil and stirrup—to the internal ear, and the actual inner ear hidden inside the cochlea. This construction is of course also present in more or less similar ways in mammals. The evolutionary transition from reptiles to mammals is especially revealing in the way former parts of the jaw evolved into two of the ossicles by mysteriously travelling from the jaw to the inner ear while also metamorphosing their shape. What an extra-ordinary process! But it is another story.

Let us now turn to the actual subject of this book, the sky itself, before later looking to see where there may be connections between the various realms. First to be considered are the two largest members of our planetary system. Jupiter and Saturn meet each other on average every 19.86 years, i.e. they are then both in line with the Sun. In the case of all planets, the general term for this interval of time is 'synodic period' (for the calculation see Appendix 3.4). The encounter between two planets is known as a 'conjunction', but in this book I shall confine the use of this term to one specific circumstance as shown below:

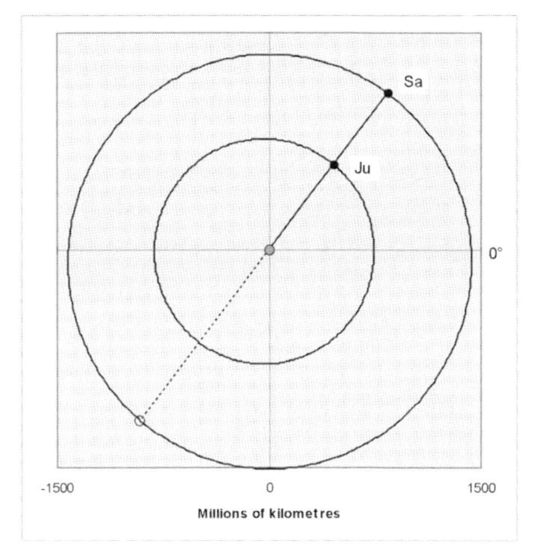


**Conjunction:** Two planets in line with, and both on the same side of, the Sun

**Opposition:** Two planets in line with, but on opposite sides of, the Sun

(Reasons for use of these terms exclusively in this form will be stated in the following.)

0° signifies the point of reference applied in astronomy, the vernal equinox.

**Fig. 2.4** *Jupiter/Saturn conjunction on 23 June 2000*

The concepts of conjunction and opposition stem from the time when people still saw the Earth as the centre of the universe. If we imagine it in place of Jupiter in the above drawing, then from that position Saturn and Sun are in opposition in the usual meaning of the term. Since this appears to me to be confusing when applied to the heliocentric view of the universe (for this would mean speaking of an Earth/Saturn conjunction in the above case) I use the two terms as defined to the right of the diagram above. Another advantage is that in referring to the inner planets—as seen from the Earth—one does not have to use the otherwise usual further differentiation into 'superior' and 'inferior' conjunctions. (According to my definition 'inferior' means conjunction and 'superior' means opposition.) But the main advantage of using only two definitions lies in the fact that each term is valid from the viewpoint of either planet.

With reference to Jupiter the joint synodic period is 1.674 times its orbital period, i.e. a little more than 5/3; relative to Saturn it is 0.674, i.e. about 2/3. At the next conjunction both planets have, according to this, moved forward by an average of *c.* 242.7° and are once again in line with the Sun, except that Jupiter has meanwhile completed an additional orbit around the Sun. So the sequence of conjunctions can be depicted in a circular diagram, whereby it is only necessary to mark one of the planets, since two planets in a conjunction always occupy an identical position on the ecliptic:

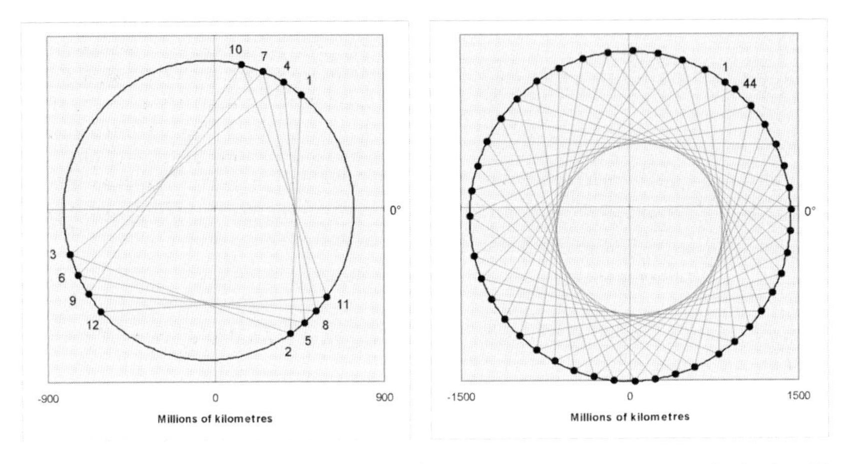


**Fig. 2.5** *Diagram showing the positions of the Jupiter/Saturn conjunctions beginning 23 June 2000; on the right the Jupiter orbit, on the left the Saturn orbit*

In both these depictions two sequential conjunctions are combined. After three conjunctions an approximate triangle arises, for which in this context the term trigon is also used. However, it is not very accurate but remains open to a marked degree or, in other words, it turns by an average of 8.09° every time. In this way one rotation of the trigon in the ecliptic occurs after every 43 conjunctions (point 44 in the right-hand diagram). This takes an average of 853.92 years during which Jupiter has circled the Sun 72 times and Saturn 29 times. However, as you can see, points 1 and 44 do not quite coincide, which leads to the supposition that there must also be some other factor connected with this most important constellation of our planetary system. The sequence shown was of course discovered long ago, for the Jupiter/Saturn triangle is similarly visible in the sky also from the geocentric point of view. A depiction resembling the right-hand diagram in Fig. 2.5 with the Sun at the centre even appears in Kepler's early work *The Secret of the Universe*[3] brought out in 1597. Among other things it was this trigon that persuaded him to look for geometrical connections in the planetary system.

Let us now consider another conjunction that leads with greater accuracy to a further geometrical figure among those discussed above.

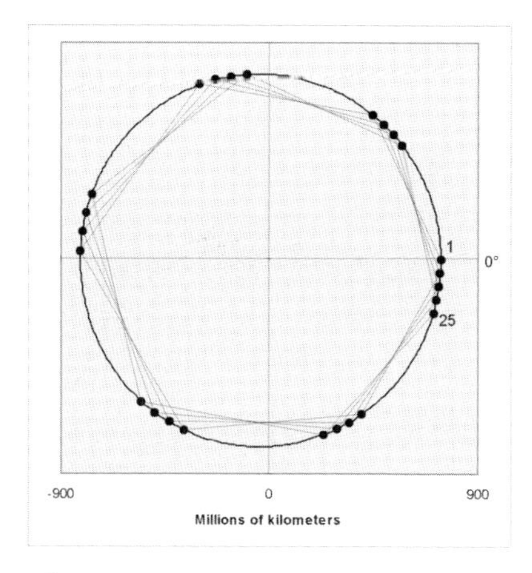


Six planets, including the Earth, have been known since antiquity. Uranus was the first to be discovered in modern times, actually in 1781 by Friedrich Wilhelm Herschel, 151 years after Kepler's death. The unique characteristic of Uranus is that its axis of rotation is inclined at 98°, so it is close to the planetary orbital plane. It appears to roll and twist its way through the universe.

**Fig. 2.6** *Jupiter/Uranus conjunctions in the Jupiter orbit beginning 24 September 2010 (first conjunction after 1 January 2000)*

The 7:1 ratio of the orbital periods of Jupiter and Uranus is metamorphosed into the figure needed for three musical intervals. They do not complete the hexagon exactly in that, after 6 conjunctions, there is a loss of an average 4.92°. Despite this the hexagon is a good deal more resonant than the trigon. For its construction it does not quite require one Uranus orbit, but exactly 82.87 years. The relationship between the two planets can be very accurately shown by the following ratio: in 12 Uranus years Jupiter circles the Sun 85 (84.998) times during which there are 73 (72.998) conjunctions.

We have thus now found in the sky of our planetary system two of the geometrical equivalents to the musical intervals mentioned at the beginning of this chapter. But to find the square we shall have to take a short detour. I therefore now refer to the great discovery of the Pythagoreans who found out the relationship between harmonious intervals and simple number ratios with the help of the monochord.* This is an

* Whether it was Pythagoras himself or one of his successors who first made use of the monochord can no longer be ascertained historically and is anyway irrelevant here. The (now disputed) story goes that he first noticed the connection between the pitch of a note and the resonating body emitting it when passing a smithy. The different sized hammers pounding the anvil produced notes of varying pitches.

instrument with a single vibrating string divisible at any point by a variable stop. (The principle is shown in Fig. 1.7.) Halving the string results in the octave 2:1; one third to two thirds gives the fifth 3:2 etc. The latter means that the whole string gives the lower or fundamental note and two thirds of it the higher note with a frequency 150% above the fundamental. (I point this out expressly because the ratio 3:2 could be ambiguous; one might think that the division of the string into 3 and 2 sections is meant, i.e. 60% of the total length. For example 5:4 means division of the string at 80% of its total length and a vibration of around 125% above the fundamental. This interval, the major third, can be depicted in geometry as follows:

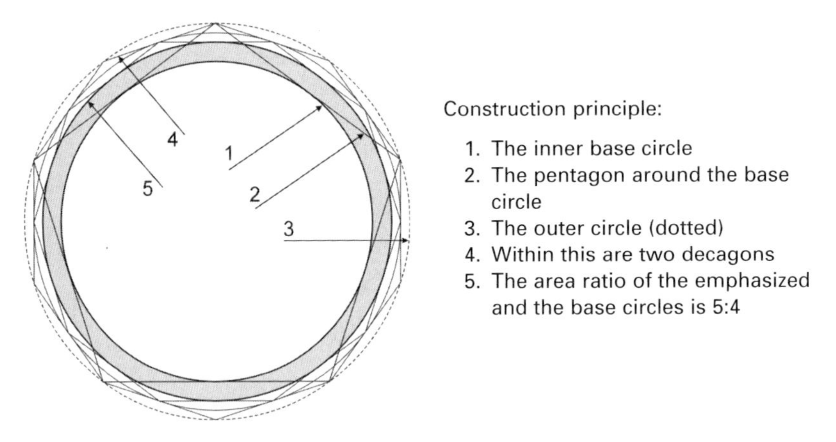


**Fig. 2.7** *Major third of pentagon and decagon*

That the intervals of the third are connected with the pentagon and the decagon is not surprising, since their numerical ratios are formed by the number five. For me the way this happens is, I admit, one of the small miracles of geometry. The construction, though more complicated than the previous ones, is perfectly clear. What I find so fascinating is that the ratios of the pentagon and of the decagon are irrational numbers, as is also the relationship between the two. The terms in the calculation are such that they resemble the golden section, i.e. expressed in terms of the square root of 5. It is indeed surprising that a combination of several of these irrational numbers results in the rational ratio of the major third 5:4 (mathematical derivation, see Appendix 1.6).

All the other musical intervals* can be attained by a combination of the

* Meant here are the 12 basic intervals of harmonically pure tuning. More in Appendix 1.5.

processes discussed thus far (they are shown in Appendix 1.6). Let us now turn to the significance of the thirds and their connection with the number five in order then to see what further correspondences there are with the movements of the planets. The fact is that the thirds (and sixths) only established themselves as consonant intervals over many centuries during which the major and minor keys gradually emerged. This process had more or less reached its conclusion around the lifetime of Johannes Kepler, though there was still a good deal of resistance to the changes taking place. Kepler used the terms 'major' and 'minor' for the most part in their modern sense. The tonal ratios based on the number five (which 'refuses to have its rights of citizenship in the origin of the musical intervals torn from it', as Kepler put it[4]) were still looked upon as dissonant intervals in Greek music and the medieval music theory founded on it. The various interpretations are connected with the development of music from one to several voices; one can say roughly that the former is more appropriate for the Grecian system and the latter for the new view.[5] Thus the tendency followed historically by the development of music is closely bound up with that of humanity as a whole.

It is, then, not surprising that someone like Johannes Kepler also addressed himself to these questions. He was the first to endeavour to find a firm footing for the modern system at least in the sense that he offered something to counter the old view, which was that the ratios of the numbers were the deeper reason for the consonance or dissonance of an interval. For him the real reason for consonance lay in the fact that it must be possible to construct an interval geometrically, i.e. with ruler and compasses.[*] In his view, this meant that it corresponded to an archetype within the human soul which is recognized anew—whether consciously or unconsciously.

Within limits—for Copernicus, Galileo, Newton and very many others were also involved—one can say that Kepler was the founder of modern astronomy. In other words, he gave us a new way of looking at the stars. At the same time he was, perhaps, the last observer of the stars who still thought consistently in terms both of geometry and of music in the way that had until then been taken for granted. For example the medieval

---

[*] Kepler selects those geometrical figures arising from division of the circle which according to his system are suitable to bring out intervals corresponding with musical consonances: diameter, triangle, square, pentagon, hexagon and, with limitations, octagon. Put simply, he derives the seven consonant intervals octave, fifth, fourth, major third, minor third, major sixth and minor sixth from the ratios of the circular arcs to the undivided whole. Imagine a vibrating string bent round to form a circle. Thus Kepler's way is set according to lengths.

university course known as the quadrivium combined the subjects of music, arithmetic, geometry and astronomy.

When a child, or even an adult, draws a picture of a star the result is star-shaped. Why is this? Surely one knows that stars, or their light, as we see them, are small points, or little discs if you look at a planet through a telescope. Is it merely the refraction of the light on our retina, or a twinkle caused by the atmosphere?

In the world of music the struggle for the number five's 'rights of citizenship' had begun. It stood for something new that was making itself felt and wanted to find its way in the world so as to be able after two or three hundred years to pave the way here for such unheard-of works as Mozart's piano concertos, Schubert's sonatas and the symphonies of Anton Bruckner—music, you might say, that can reach up and touch the stars.

What, then, are the characteristics of this number—which appears to be so intimately connected with the evolution of humanity—characteristics not possessed by the numbers that precede it? Is it special merely because it is one or two sizes larger than the four or the three?

In order to throw some light on these questions let us now look more closely at the two planets which stood out in Fig. 1.6 as being so close to the golden section.

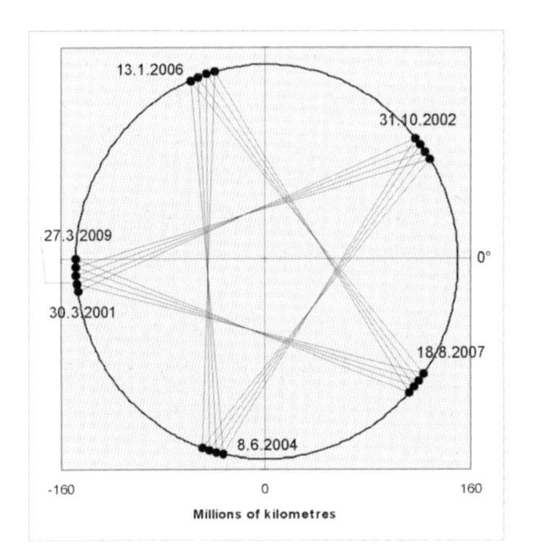


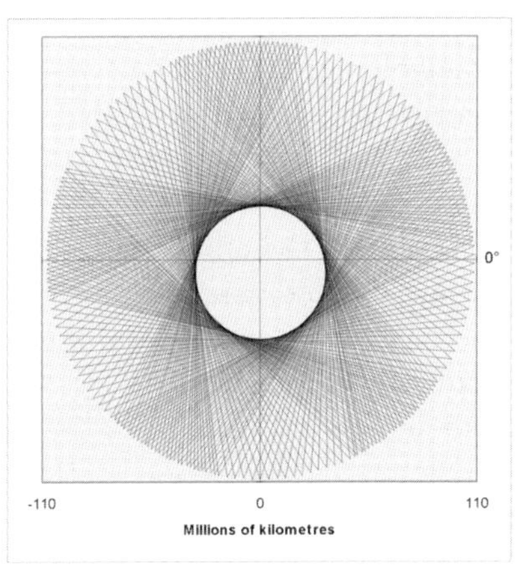


**Fig. 2.8** *Venus/Earth conjunction, shown left in the Earth's orbit (21 times), and right in the orbit of Venus (220 times), beginning 30 March 2001*

I find it a shame, or indeed sad, that hardly anyone knows about this. Specialist literature is almost completely silent concerning the exceptional nature of our relationship with our neighbouring planet. I have only managed to locate the *Pentagramma Veneris*—as it was already termed in the Middle Ages—in a few publications more or less closely connected with anthroposophy.[6]

The orbital periods of the Earth and Venus with their ratio of 13:8 thus mean that one of the few archetypal images—if not the only genuine one—derived by humanity from the stars is inscribed in the firmament, not in luminous paint perhaps, yet recognizable by the human intellect. Even if no special importance is attached to the pentagram and its appearance in the sky, surely the fact that this happens is at least worth a mention. So why this refusal to take note of it? Presumably it is simply ignorance in its original sense of 'not knowing'.

Music and geometry disappeared quite rapidly from astronomy after Kepler's time, with geometry serving at most as an aid to arithmetic which now came to be accorded the high degree of importance which is unquestionably its due. One of the greatest achievements of the human intellect to which generations of mathematicians have contributed is for me the fact that we are able today to calculate with incredible precision the movements of the planets either forwards or backwards for thousands of years. Because of this and also because of all the other advances in the art of observing the heavens, one half of the quadrivium has fallen into oblivion. One-sidedness, it seems, is a condition of outstanding achievements, but this always brings with it also a degree of impoverishment. Yet just as in the heavens everything is on the move and will one day return, so, let us hope, will it also be on Earth.

**Fig. 2.9** *A Huichol (Mexico) painting depicting the Creation with the first beings. The five serpents are the Mothers of Water representing terrestrial waters. At the right the first plant appears, bearing both male and female flowers. At left the Sun Father is flanked by the Morning Star*[7]

Let us now have another look at Fig. 2.8. The synodic period of Venus and Earth is 583.92 days. So it takes five conjunctions, i.e. 8 years except for a deficit of 2.44 days, for the pentagram to be formed. This time the deviation between two neighbouring points is 2.41°, so the resonance is improving all the time. Thus during the course of *c.* 150★5 conjunctions the pentagram turns once about itself or about its centre, where the Sun is located. (The figure of 150 arises when we divide the 360 degrees of the circle by the deviation.) However, the 150th pentagram has overshot the 360 degrees by a small margin. The average for one turn is 149.447★5 conjunctions, so that the total time taken is 1194.6 years. The right-hand image shows how the pentagrams are progressively overlaid during a period of *c.* 350 years.

Of course the pentagram is closely linked to our musical point of departure, the major third arising from the pentagon and the decagon. In addition, when one point is skipped in joining the corner points, the pentagon permits the first division of the circle leading to a star-figure. In this way the number five does indeed bring a new dimension into the world of geometry, just as the interval of the third made possible something new in music. In addition, the pentagon shape has very surprising links with the double octave of the equilateral triangle, as is shown in more detail in Appendix 2.1. And it also has a harmonious relationship with the square, the final figure we shall consider in connection with the movement of the planets. The square that is derived from the height of one point of the pentagram and the square that can be drawn around the inner circle of the pentagram also show a ratio of 5:4 (see also Appendix 2.1). In this way it fits perfectly that also in the planetary system the square appears in connection with the pentagram.

So now it is the turn of Earth's outer neighbour: Mars. All we have to do is enter its position whenever Venus and Earth have completed one pentagram.

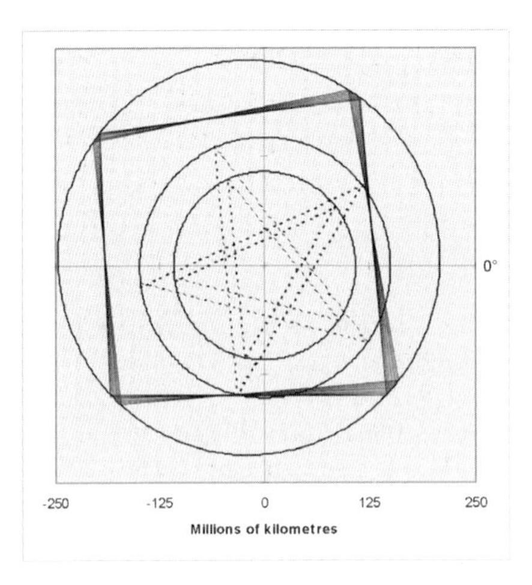


Superimposed and linked are (outside) the sequential positions of Mars after every 5 Venus/Earth conjunctions, corresponding to a pentagram. The time taken for 4 pentagrams is almost identical with that of 17 Mars orbits. This means that Mars completes 17/4 revolutions during the period of time required for one pentagram. So after 4 sequential pentagrams Mars has completed 4 1/4; 8 1/2; 12 3/4 and 17 orbits. It is situated each time at the corner points of the developing square.

**Fig. 2.10** *Mars at Venus/Earth pentagrams (150 times); from the centre: Venus, Earth and Mars orbits*

Mars as it were gives the pentagram its framework, and it does this with a resonance that is almost perfect. We see that the period of time portrayed here is 1195 years, i.e. the period required for the 150 sequential pentagrams (for a complete turn of the circle). This is why the *Pentagramma Veneris* is here shown only with dotted lines since Venus and Earth are now executing circles consisting of the closely positioned points appearing in the left-hand drawing of Fig. 2.8. The square drawn by Mars—let us call it a square even though the eccentricity of Mars brings about a deviation from the ideal shape—is almost static in the sky, for it would take 100,000 years for a single revolution to be completed. This arises from the ratio of the intervals of time of 5 Venus/Earth conjunctions to one Mars orbit of 686.98 days. As already mentioned, this amounts to 17:4. This minimal deviation from exact numerical proportion results in the period of time mentioned (for the calculation see Appendix 3.5). It is especially felicitous that the figure of the octave, the most perfect musical interval, also finds expression in this most precise way in the movements of the planets.

We have shown in this chapter that all 12 musical intervals can be depicted in the ratios of areas of four simple geometric figures. For the geometrical construction of the 12 musical intervals it is even possible to omit the triangle corresponding to the double octave (which in its conformity with the Jupiter/Saturn conjunction is anyway rather approximate) and thus to manage with the circle, square, pentagon (in combination with the decagon) and hexagon.

In the figures that arise directly* from the planetary conjunctions, the triangle, the hexagon and the pentagram arising from the pentagon—which in its construction includes the decagon, see Appendix 1.2—occur with increasing accuracy. The square also comes into play indirectly yet logically with phenomenal accuracy. Other simple geometrical figures (right up to the dodecagon, or their corresponding star-figures) do not occur in the direct ratios of the conjunctions.†

---

* 'Directly' in this context refers to the relationship of two planets and their corresponding synodic periods. 'Indirectly' refers to the square arising in connection with Earth/Venus conjunctions.

† The next direct figures are the 13-sided polygon of Jupiter and Neptune and the 14-pointed star-figure of Mercury and Venus which I shall introduce in Chapter 6. One might also mention the approximate 11-sided polygon of Earth and Jupiter. Which figure the planets actually form is of course always a question of how accurately one works, as is shown in the example of Jupiter/Saturn. However, I have found a good and more or less compelling measure for this in the cycle-resonance, by which I mean the deviation in degrees which arises between, for example, one pentagram and the next (see Appendix 3.5 and Glossary).

Another step we have taken in this chapter is to reconfirm that the intimate connection between arithmetic, astronomy, geometry and music theory is more justified than is generally assumed nowadays. The attainments of modern times are enabling us to raise to a new level the ideas of former ages which were often purely intuitive and of course also frequently erroneous. Perhaps we shall indeed gradually succeed in drawing closer to the timeless essence

*'. . . of those secret powers*
*Which hold together this world of ours'*

as Goethe put it in *Faust.*[8] And it would perhaps not be wrong to regard the principles at work in 'this world of ours', and in its elementary building blocks, as being all of a kind with those that bring order into the cosmos as a whole.

I shall bring this chapter to a close with a scene from *Faust* as an example of what the pentagram, the 5-pointed star, must have meant to people in former times. But first let us glance at something which might seem to be an irrelevance in connection with the *Pentagramma Veneris* shown in Fig. 2.8. One completed figure with five points always lags behind the previous one by about 2.4°. Looking more closely we notice that actually a small portion of the 360° of a full circle is missing rather than being in excess since the planets move counter-clockwise. This, however, causes the pentagram itself to be properly completed.

Near the beginning of *Faust*, Part One, Mephistopheles approaches Faust in the hope of striking a bargain with him. This spirit 'of perpetual negation' has crept into the room in the shape of a poodle. But as he makes to leave again he is brought up short by an obstacle:

*MEPHISTOPHELES*
*I must confess that on the floor,*
*Across your threshold, you have put*
*A certain obstacle—a witch's foot—*

*FAUST*
*You mean, that pentagram I drew*
*Hinders a gentleman from hell?*
*Then how did you get in? Well, well!*
*How did I fool a sprite like you?*

*MEPHISTOPHELES*
*It's not well drawn; look closely, sir!*
*One of the outside angles—there,*
*You see? the lines do not quite meet.*

*FAUST*

*How curious! how very neat!*
*And so you are my prisoner.*
*A lucky chance, I do declare*

# 3. A Backward Glance and a Glance up to the Stars

Was it indeed merely chance—both in the one instance as well as in the other?

Undoubtedly this is what our modern scientific view of the world supposes regarding the origins of the universe, the planetary system, life and the human being; or rather, more comprehensively, that it was chance combined with necessity determined by the laws of nature. The orbital periods of Jupiter's four larger moons, for example, are so regular that this cannot be seen scientifically or mathematically to be a matter of chance in the usual sense* of the word but must be the result of some natural law that has simply not as yet been discovered. It is claimed that the universe as we know it today began with what people call the big bang. All the knowledge and theories of physics and astronomy—it is hard, sometimes, to distinguish between the two—have furthermore shown that the existence of our universe is exceedingly improbable, to a degree, in fact that its existence could be said to be virtually inconceivable. (See Chapter 11 for more.) In order to wriggle out of this dilemma—seeing that our universe does indeed exist—some modern theories now propound the possibility of unimaginably many and varied universes, so that somewhere and at some time one like ours would have had to arise, or at any rate could have arisen.

Whatever the case may be, one thing appears to be certain, and that is the fact that the universe as we know it did at some point have a beginning. Regarding our solar system this can even be regarded as proven. So thus far the modern scientific view concurs, so I believe, with almost all the creation myths that have come down to us from the past. The only difference is that people in the past assumed the cause for the creation of the world to have been the will and governance of supersensible powers or God. Music and numbers as principles of order also have a special part to play in those myths. By way of illustration, here are two short examples from the cultures of China and India. (This theme will, however, not be further explored in the present work.)

---

* They behave very precisely as 2:1, 2:1 and 7:3.

*'That wherefrom all beings arise and have their origin is the great Oneness; and that through which they are formed and perfected is the Twofoldness of dark and light. As soon as the seeds begin to stir they congeal into a shape. The bodily form exists within the world of space, and all that is spatial possesses a sound. The resounding note arises out of harmony. Harmony arises out of conformity.'*

*'In the beginning this All was Brahman. His seed prevailed and became the Brahman. In his spirit he contemplated in silence . . . His head became the sky, his breast the realm of air, the mid-part of his body the sea, his feet the earth. He created the whole world and gave to it the Saman as food.' (The text then tells further about this Saman, a musical medium, consisting of seven notes which one after the other serve gods, men, animals etc. as their nourishment.)*[1]

We shall begin our backward glance in more detail with Plato who was the first to give exact measurements and numbers with reference to the order of the universe. I shall confine myself to those passages in the *Timaeus* which deal directly with the ratios in the planetary system. Then I shall introduce the Titius-Bode law and—in Chapter 4 and Appendix 5—also some twentieth-century endeavours to comprehend the harmony of the spheres in numbers. Johannes Kepler will of course be the subject of our main considerations since in these matters his life's work occupies a unique position in history. For this reason I have to some extent modified the historical sequence.

## *Plato*

For Plato the universe is a Living Being. He describes in the *Timaeus* how God worked to create the universal soul, constructing it out of a previously prepared primordial mixture in accordance with specific ratios.[2] Thereafter he shaped that which he had created in such a way that two circles were formed, an outer and an inner one. The inner one he divided into six parts, so that seven circles existed, into which—'to define and preserve the measures of time'—he later set the Moon, the Sun and the five planets. The circles of the planetary orbits were unequal, we read, and they were arranged 'in intervals that were double or triple'.

Plato had already mentioned these intervals. They become relevant in the following way during the creation of the universal soul. First one portion of the primordial mixture is taken, followed by specific multiples of that first portion, namely the

2-fold, 4-fold, 8-fold and the
3-fold, 9-fold and 27-fold, amounts, i.e. the squares and cubes of 2 and 3.

Then 'both the twofold and the threefold intervals' are filled in so that new intervals arise, namely in the ratios of 3:2, 4:3 and 9:8. What is interesting here is the structural principle contained in these figures since we have already met with the number ratios of the musical fifth, fourth and major second in connection with the hexagon in Fig. 2.3. The configuration shown there on the left can now be redrawn as follows:

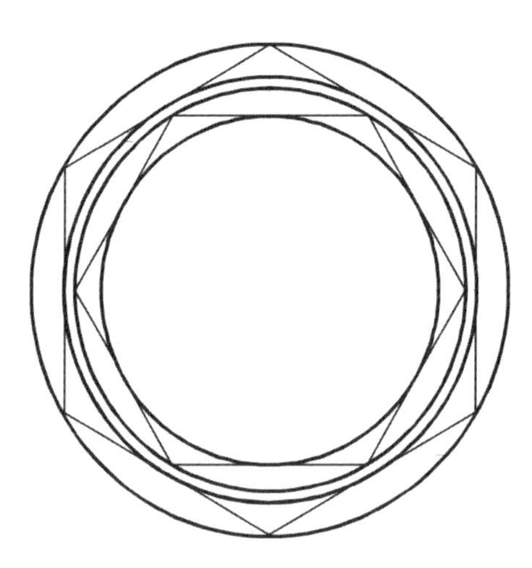

*'In this way . . . there came into being night and day . . . ; the month, complete when the moon has been round her orbit and caught up the sun again; the year, complete when the sun has been round his orbit. Only a very few men are aware of the periods of the others; they have no name for them and do not calculate their mathematical relationships. They are indeed virtually unaware that their wandering movements are time at all, so bewildering are they in number and so amazing in intricacy.'*[3]

**Fig. 3.1** *4:3 — 9:8 — 4:3 as area ratios given by the hexagon*

The formative principle expressed here can be made more obvious by distinguishing between the four circles which are stepped out by the two hexagons in an 'octave circle' (where the inner and outer circles have a ratio of 2:1), as shown in Fig. 3.2.

Looking at this purely arithmetically, one checks out the two simplest calculations on the right, understands the ratios and then, alas, often regards the matter as closed. The geometrical view reveals a good deal more, namely the interpenetration of similar ratios working from the inside outwards and from the outside inwards which we have already encountered in connection with the semi-minor axes of the planets (Fig. 1.5), only with different numbers. The musical view, if we may call it that, demonstrates exactly the constructive principle of the Greek musical scale which consists of two tetrachords.[4] A tetrachord is nothing other than a sequence of four notes in which the two outer

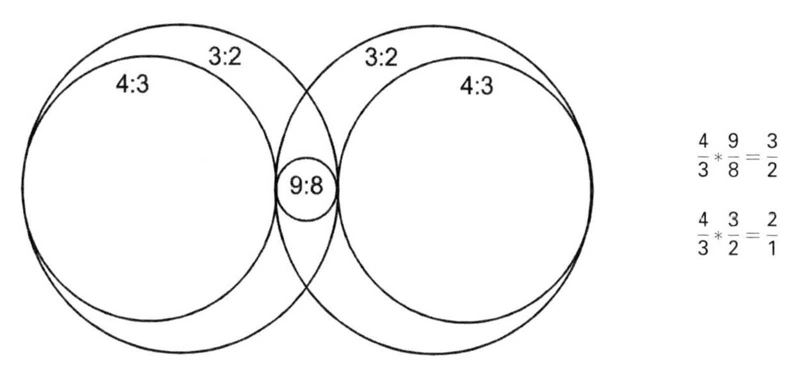


**Fig. 3.2** *Tetrachord structure*

ones form the interval of the fourth, 4:3, and the two inner ones can have varying positions depending on the musical mode. (Major and minor, for example, are musical modes or species; the ancient Greeks had others.)

It is quite obvious that Plato modelled his account of creation on an existing musical system. But it also seems certain that a similar principle permeates the music made on the one hand by human beings and on the other by the planets in the solar system in which those same human beings live. However, since one should be circumspect in claiming something to be truly certain, let me now draw this consideration of Plato to a close with a final quotation from the *Timaeus*:

> *Don't therefore be surprised, Socrates, if on many matters concerning the gods and the whole world of change we are unable in every respect and on every occasion to render a consistent and accurate account. You must be satisfied if our account is as likely as any, remembering that both I and you who are sitting in judgement on it are merely human . . .*[5]

## The Titius-Bode law

This law, named after Daniel Titius and Johann E. Bode, is a formula (published in 1772) for the mathematical calculation of the relation between the mean distances of the planets from the Sun. At first glance it has two small disadvantages: firstly, in some instances it depicts the actual facts relatively inaccurately; and secondly, it takes no account of Neptune and calls for a trick to make it apply to Mercury. On the positive side, however, it also has a great advantage in that, especially in its updated version, it is sufficiently accurate to express in a language recognized by science the fact that some hidden order is indeed woven into the solar system. Another way of describing this advantage is to say that theories expounded regarding the creation of the planetary system do have to take into account, among other things, the fact that 'the orbital radii of the planets and the asteroidal belt very roughly follow an exponential distance rule'.[6] To date, no one has succeeded in discovering a direct theoretical background to this law,[7] so the establishment of those theories is a tricky matter (though of course not only for this reason, as we shall see in Chapter 5).

To gain an understanding of how perplexing the phenomenon of an exponential increase in distances actually is, one might imagine an avenue of trees in which the first two are 100 metres apart, the third is 200 metres from the second, the fourth 400 metres from the third and so on, so that the distance from the eighth to the ninth tree already amounts to 12.8 km. In the case of the planetary system the mean growth factor is not 2 but about 1.66 (if one counts the distance between Mars and Jupiter double, or includes the asteroids in the calculation, otherwise it would be *c.* 1.86). In addition, the increase is of course not constant but varies around the stated mean—with a tendency to increase even more in most cases. The increase in the distances signifies that the ratios of neighbouring planets remain (on average) constant. So it appears that we must look to the ratios, or one could say the intervals, to find what constitutes the order in the structure of the planetary system.

The Titius-Bode law, which approximately replicates this arrangement, has the charm of being based on a series of small whole numbers. Beginning with Venus, they are the numbers $n = 0$ to 7 linked by a simple formula.* In the case of Mercury, n must be made to equal a large negative number, e.g. minus infinity (which, in the strict mathematical sense, is a trick, as I mentioned above), while the number three corresponds to Ceres, the largest inhabitant of the asteroid belt; and Neptune is ignored. The resulting correspondences range from very good to relatively inaccurate. The extremes are Jupiter with a difference of *c.* 0.02% and Mars with one of *c.* 5%.

In our century this formula has been improved in a number of ways by using numbers other than 0.4 or 0.3 in the equation or by furnishing it with an additional periodic function in order to accommodate mathematically a degree of oscillation of the numbers.[8] So if one depicts the series geometrically in a revised form by entering the dis-

* $r_n = 0.4 + 0.3 \star 2^n$.

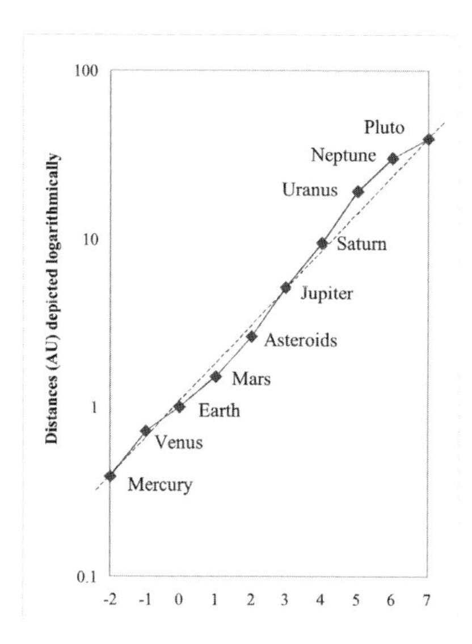


**Fig. 3.3** *Titius-Bode law in a new form after Nieto (1972)*[9]

tances (in any scale) logarithmically on the y axis against small whole numbers (here Venus = −1) on the x axis, a mysterious oscillation arises. (And, by the way, Neptune fits very well into this modern version of the series.)

It ought, we are told (no doubt rightly), to be possible to find the reason for this oscillation in the early times of our solar system. Perhaps asteroid impacts were what led to the situation we see today. Whatever the case may be, it shows an oscillation with Jupiter at its centre which as far as we can tell was effective even in the earliest times of our cosmic home.

## *Johannes Kepler*

We now come to Johannes Kepler who was born on 27 December 1571 at Weil der Stadt in Württemberg, Germany. At the latest by the time he entered university he had reached the conclusion that the heliocentric picture of the world as outlined by Nicolaus Copernicus and published in 1543, the year of his death, must be true. It is difficult to imagine what this new view of the sky must have meant to people when what had been believed for thousands of years was now said to be incorrect. It called into question all the other ideas that had been held with regard to the heavens in the religious sense as well, which makes it clear, although not necessarily convincingly, why the Catholic Church rejected the Copernican revolution and persecuted those who championed it with varying degrees of severity—though Kepler, for example, was spared such treatment.

Kepler faced the question of how to find an up-to-date way of presenting the harmonious order of the celestial bodies, established for the most part by Plato and Aristotle and adopted by the Christian world, while also taking into account his own firm belief in the divine creation of such order. In addition to the new astronomical view of the universe, his other main concerns were the developments taking place in the world of music as described here in Chapter 2. One could say that in their different ways both these realms represent an all-round outlook on human evolution. In all his searching Kepler remained firmly convinced that it must be possible to discern God's hand in the script of the visible heavens.* What does this mean for us, especially in the way we look at things today?

Imagine standing in a deserted region and hearing a single note in the distance or several different notes, perhaps uttered by the wind or by a bird we cannot identify clearly. When a melody then begins to take shape, or even a whole symphony, we might justifiably assume that we are listening to something created by the human spirit. Or suppose we find a few lines drawn in the sand on the seashore. Perhaps some process of nature made them. But if they represent a complicated geometrical pattern it is exceedingly unlikely to have been made by a twig blown by the wind across the surface of the sand.

The first model Kepler developed to explain the order among the planets circling round the Sun was an interlocking construction of the five Platonic solids.† These are the only spatial bodies all of whose faces are in each case the same regular polygon.

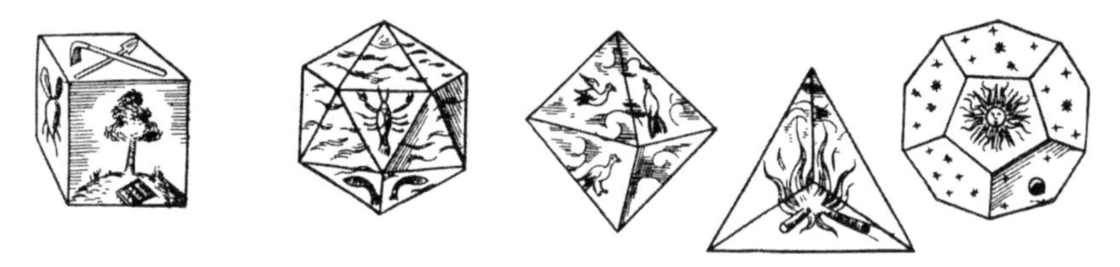

**Fig. 3.4** *The Platonic solids from Kepler's* The Harmony of the World*: cube or hexahedron (6), icosahedron (20), octahedron (8), tetrahedron (4), dodecahedron (12). The figures in brackets show the number of faces from which they derive their names*

Each one of these solids can be enclosed in a sphere that touches each of the corners, and an inner sphere can be added that touches the centre of each face. The outer and inner spheres have specific ratios as regards radius, area and volume. By utilizing the radii in the correct sequence and taking the outer sphere of the one as the inner sphere of the next Kepler

---

* Outstanding insight into Kepler's life and the way he thought may be found in the very sensitive and thoroughly researched biography published in 1948 by Max Caspar, who also translated into German the works Kepler had written in Latin.

† Thus named because they were described by Plato, in the *Timaeus*. It is thought that they were also known to the Pythagoreans. They were assigned to the elements of fire, earth, air, water and the heavenly-etheric substance.

managed to construct a model showing the distances of the planets from the Sun. The ancient idea of crystal spheres to which the planets were attached also fitted well here.

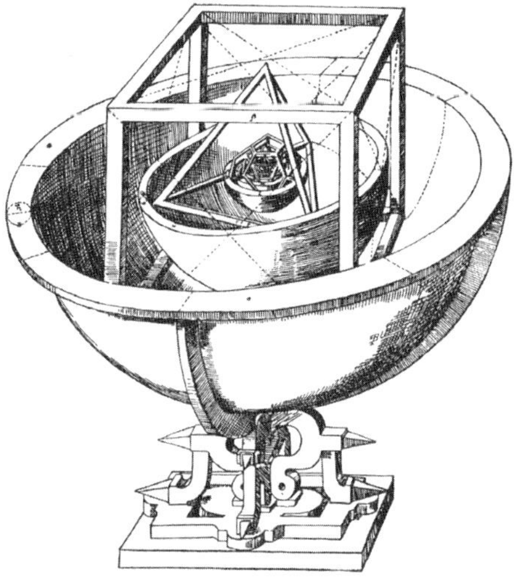

**Fig. 3.5** *Illustration from Kepler's* Secret of the Universe

*'From the modern standpoint, it may seem hard to believe that two intellectual giants of the caliber of Plato and Kepler should have proposed such crackpot theories. What drove them to seek connections between the regular solids and the structure of the universe?'*[10]

(What drove them, thinks the author of this quotation, was their conviction that the patterns and the orderings of the universe can to some extent be recognized and described by mathematics.)

This picture, which occurred to Kepler in 1595 but was not published until 1597, appears to be imbued with something that has caused it to remain unforgotten for 400 years even though it gives only an inaccurate account of the actual situation.

**Table 3.1** *Comparative ratios in Kepler's* Secret of the Universe *with orbital data as seen today*

| | Cube | Tetrahedron | Dodecahedron | Icosahedron | Octahedron |
|---|---|---|---|---|---|
| Circum- to insphere radius | 1.7321 | 3 | 1.2584 | 1.2584 | 1.7321 |
| Associated with the distance between | Saturn-Jupiter | Jupiter-Mars | Mars-Earth | Earth-Venus | Venus-Mercury |
| Ratio of semi-major axes | 1.8342 | 3.4146 | 1.5237 | 1.3825 | 1.8686 |
| Deviation (line 3/line1) | 1.0590 | 1.1382 | 1.2108 | 1.0986 | 1.0788 |

With the data available to him at the time, Kepler of course arrived at different numbers. Realizing that his initial material was not very accurate, he had hopes of working with the Danish astronomer Tycho Brahe to correct the deviations from the ideal shown in his model. People knew that Brahe had for decades been gathering data which were uniquely accurate for the time but that he also guarded his findings as closely as a mother hen her chicks. Fate seemed to be making sure that the very man to be given access to those figures was the one who would really know how they should be handled. Brahe and Kepler met for the first time in 1600. Brahe died a year later, and Kepler was invited to succeed him as court mathematician to Emperor Rudolf II in Prague.

He then discovered that planetary orbits were ellipses, as already mentioned in Chapter 1. This represented a revolution as far-reaching as that of Copernicus, for now the circle as the symbol of perfection was banished from the sky where it had resided for millennia, just as earlier Earth had been banished from its centre. Having asked himself during the sixteenth century why there were exactly six planets at specific distances from one another, now, in the new century, Kepler had to wonder what reason the Creator might have had for making the ellipses so very different from one another. It was to take him more than ten years of intensive work before *The Harmony of the World* could slowly begin to take shape. The greatest jewel of all, his Third Law $T^2 = a^3$, was almost a kind of by-product in all this. It 'popped up in my head', as he described his inspiration, only once the construction of the main building was all but complete.

*The Harmony of the World* consists of five books (in one volume). The first deals with the geometry of regular figures in a plane which are the basis for the harmonic ratios; the second with the regular spatial bodies (among other matters); the third with the musical ratios; the fourth with the influence of the stars, for example on the weather. Regarding the fifth I shall quote its original subtitle: 'On the most perfect harmony of the heavenly motions, and on the origin from the same of the Eccentricities, Semidiameters, and Periodic Times'. We are of course here concerned solely with this fifth book and shall be getting to know the main features of musical correspondences which Kepler believed he had discovered in the planetary system. Whether modern methods and means can show these correspondences to be correct will be discussed in Chapter 4.

Kepler's method was to exclude one after another those planetary ratios which showed no or only slight harmonious relationships: the orbital periods, the distances and the velocities.* In the case of the latter two he also investigated the extreme ratios, i.e. the points closest to and farthest from the Sun (perihelion and aphelion). The extreme distances brought,

---

* Some of Kepler's terms are different. For example instead of velocities he spoke of 'daily distances travelled'.

as he put it, a first ray of light. He found a far higher degree of harmony in the extreme values of the angles calculated from the point of view of the Sun which the different planets cover in the same unit of time (called by physicists the angular velocity).

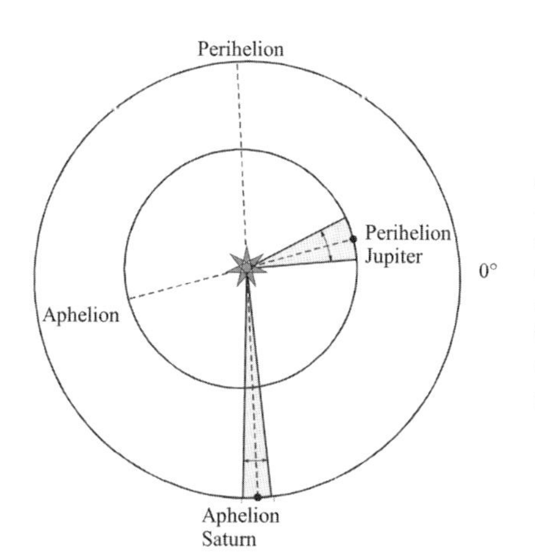


Jupiter is at perihelion, Saturn at aphelion (or, more accurately, they cover part of their journey before and after passing their extreme point). The angles covered in the same period of time are shaded. Their ratio is (approximately) 3:1, the interval of a twelfth (octave + fifth).

**Fig. 3.6** *Kepler's angle, sketch of schematic diagram*

The reverse ratio of Jupiter at aphelion and Saturn at perihelion works out exactly (according to Kepler's and almost exactly according to today's data) as the interval of the octave, 2:1. To make comparisons at the extreme points is appropriate because the values arising from movements change continuously, i.e. the relevant planetary ratios, in this case the angles, cover all the intermediate values between 3:1 and 2:1 during the course of the orbits round the Sun.

As a first step, Kepler compared neighbouring planets at points where one was at perihelion and the other at aphelion, and vice versa. (He spoke of converging and diverging movements.) These ratios formed the main structure shown in the middle of Fig. 3.7. He then noticed that at most of the points where correspondence was not quite so good (with the musical ratios shown at the side in each case), he could gain better results by comparing perihelion with perihelion or aphelion with aphelion. Here, too, he characterized the resulting ratios in two steps: all the intervals of the main structure were so close to a harmonious sound that 'if the strings were tuned in that way, the ears would not easily detect the imperfection'.[11] The only exception is the Jupiter-Mars interval which I have shown in brackets. Having described the ratios of perihelion/perihelion and aphelion/aphelion he then noted the 'perfect harmonies' of both comparisons (shown by the solid lines in the diagram).

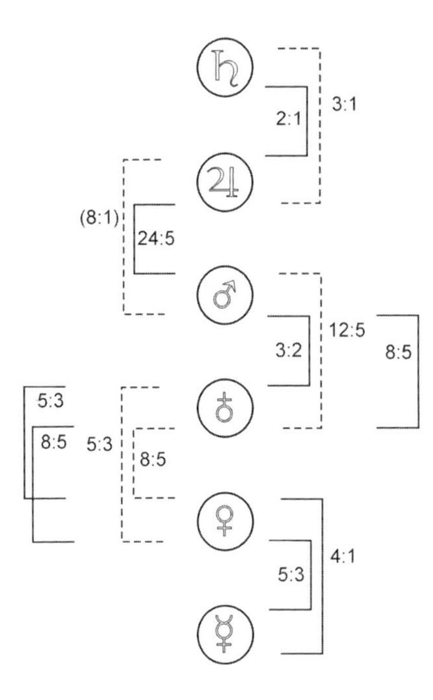


**Fig. 3.7** *Planetary harmonies after Johannes Kepler, from top downwards: Saturn, Jupiter, Mars, Earth, Venus, Mercury; upper point aphelion and lower point perihelion in each case*

'*[Kepler] wrote a long book,* Harmonies of the World, *to try to show that each planet possesses a distinctive range of voice—bass, tenor, contralto and so on—and that the music it emits is determined by its distance from the common centre of revolution, the sun, as the length of a string on an instrument determines its sound. The main theme of the book is pure fantasy . . .*'[12]

'*The fundamental difference between Kepler and later science lies in the structure of the laws for which he is looking. The Keplerian type of law has to do with the planetary orbit as a whole. Motion and orbit are considered in combination, and these orbits, as Kepler found, are arranged on harmonic principles. They are such as they are [made by the Creator] in order that the harmony of the spheres may exist and resound (for the adornment of the world, he says).*'[13]

The following table shows the exact ratios arising from the calculations.[14]

**Table 3.2** *Differences between Kepler's harmonies and the actual values*

| | Kepler's value | Today's value | Interval | Musical term |
|---|---|---|---|---|
| Ju pe/Sa ap | 3.113 | 3.057 | 3 | Fifth (* 2) |
| Ju ap/Sa pe | 2.000 | 2.018 | 2 | Octave |
| Ma pe/Ju ap | 8.448 | 8.404 | 8 | Octave (* 4) |
| Ma ap/Ju pe | 4.770 | 4.763 | 4.8 | Min. third (* 4) |
| Ea pe/Ma ap | 2.337 | 2.335 | 2.4 | Min. third (* 2) |
| Ea ap/Ma pe | 1.501 | 1.502 | 1.5 | Fifth |
| Ve pe/Ea ap | 1.711 | 1.703 | 1.667 | Major sixth |
| Ve ap/Ea pe | 1.547 | 1.551 | 1.6 | Minor sixth |
| Me pe/Ve ap | 4.049 | 4.015 | 4 | Octave (* 2) |
| Me ap/Ve pe | 1.680 | 1.696 | 1.667 | Major sixth |

'*. . . thus did the mature scientist later become filled with enthusiasm for Pythagoras' harmony of the spheres, adapting it to his own calculations of planetary movements while, however, getting lost along the way in what today has to be seen as an aesthetically matchless fantasy cosmos. The most astonishing aspect of this is the fact that the notes of the individual planets are graded in a way that makes them accord with the intervals of the musical scale.*'[15]

So we see that the conformities do indeed appear to be fairly accurate. An assessment as to whether these results might be purely accidental follows in Chapter 4. Opinions evidently differ considerably. One very obvious fact, however, is that the values arising from the data available to Johannes Kepler at the time in most cases differ only minimally from those arrived at today, which shows, furthermore, how excellent and valuable were the observations made by Tycho Brahe.

Having ferreted out these ratios, Kepler then asked himself whether he had merely arrived at a motley collection of notes or whether a further order might be detectable. In order to find out he reduced the intervals to a directly comparable measure (namely the octave space 1 to 2; the simple procedure for this is given in Appendix 1.5). The result he arrived at showed that, apart from one instance, all the intervals of the major scale made their appearance when he set the movement of Saturn at aphelion as the keynote, i.e. when he related all the ratios to the angle which arose as shown in Fig. 3.6. It is perfectly reasonable to base one's calculations on the value of Saturn at aphelion since this marks the slowest of all planetary movements. Keplar took bottom G as his starting point, as was usual in his day.[16] But one might just as well take any other note since the ratios, our sole concern here, remain the same both in the major and the minor scale regardless of which keynote one might choose. (More in Appendix 1.5.)

He arrived at all the notes in the minor scale, again with the exception of one, by taking the value of the angle as the keynote G which arises from the movement of Saturn at perihelion. This is best shown by his original illustrations:

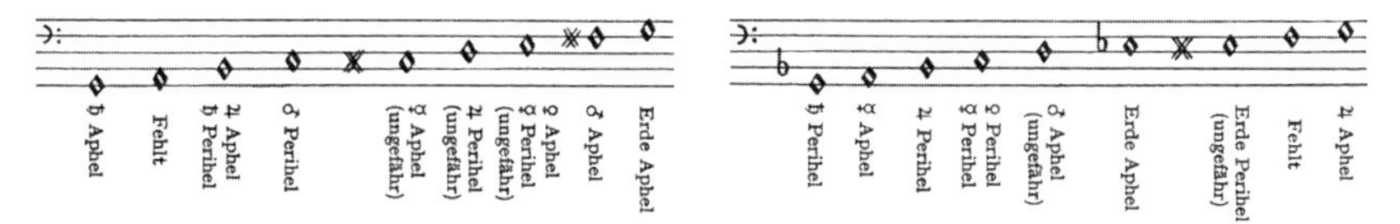


**Fig. 3.8** *Planetary scales from Johannes Kepler's* The Harmony of the World*; left major, right minor*

We have now covered the main elements of the musical correspondences in the planetary system and have thus laid the foundation for the analysis we shall embark on later. One new aspect which Kepler brought to humanity's ancient ideas about a music of the spheres is the way he showed that, whereas previously each planet had been accorded one constant note (see an example of this in Fig. 1.7), now varying songs with several voices could resound (in the imagination of the observer).[17] In view of the forward strides made in astronomy it is now quite possible to conceive of such a thing. Parallels between developments in music and astronomy are also thinkable from this point of view. And I shall throw some light on a similar connection with geometry at the end of this chapter.

Here we shall touch only briefly on the other analogies brought forward by Kepler. The various eccentricities of the planets mean that each planet also passes through its own musical scale (which Kepler compared with the old music for one voice). For example by comparing the movement of planet Earth at perihelion and aphelion one arrives almost at the value of a semitone of 16:15. The more eccentric planets naturally show more comprehensive scales. Earth's song thus consists of no more than two notes, but compared with the earlier single note this is a clear gain. Based on the keynote of G, see above, the Earth thus constantly sings Mi-Fa-Mi (in the old Latin depiction of the notes, corresponding to our E and F). Kepler explained in a footnote, perhaps with his tongue in his cheek, how '*the Earth sings MI FA MI, so that even from the syllables you may guess that in this home of ours Misery and Famine hold sway.*'[18]

There then follow explanations of chords that arise when several planets occupy specific positions. The greater the number of planets that are included in a chord, the rarer will that chord be, since of course the condition is that each one must be exactly at its appointed place at perihelion or aphelion. The instances when all six planets resound together are separated by aeons of time, says Kepler, speculating that such a harmony perhaps determined the beginning of all time, with this constellation having been present at the creation of the world. We have already seen that a special oscillation must indeed have taken place.

Kepler further endeavoured to explain the deviations of the actual ratios from the ideal by making a connection between his earlier image of the Platonic solids one within the other and the harmonies depicted. He also saw in this synthesis the reason why the various eccentricities of the planetary orbits are as they are—or as they ought to be, as he put it. He did this in an exceedingly detailed and complex manner. But since, with regard to the exactitude of the intervals, it is merely a matter of doing away with the 'inconsistencies in small details', as he put it, I shall here mention only one further aspect which will then constitute my starting point for a different picture that will show in a most surprising manner the structure of our planetary system on the grand scale, rather than in small details.

In addition to coming upon his three Planetary Laws and giving an abundance of indications regarding a mysterious and harmonious structure of our solar system, Kepler also discovered two new stars. These were, however, bodies previously unknown in the firmament not of astronomy but of geometry. Taking one's departure from the two Platonic solids in which the number five has a part to play one arrives very easily at the following figures:

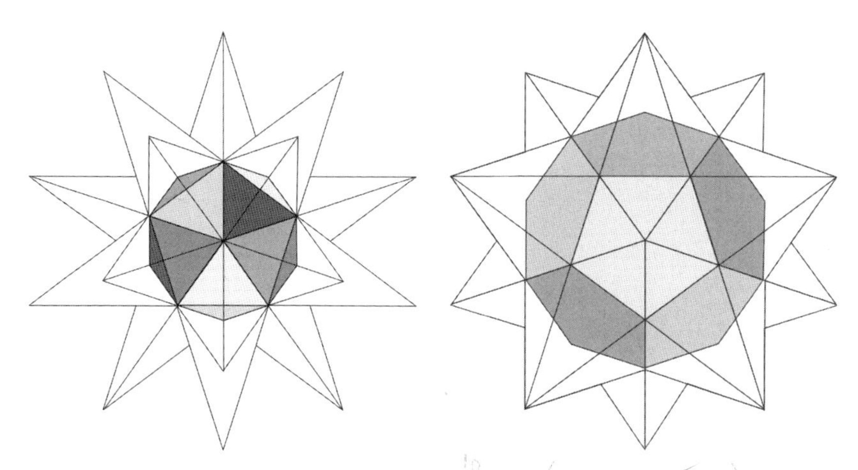

**Fig. 3.9** *Stellated figures derived from the icosahedron and the dodecahedron: Kepler's stellar solids*

These wonderful figures arise quite simply when one extends the edges of the inner bodies. It is strange, very strange, to find that no one did this before Kepler even though the Platonic solids had been known for about 2000 years. And another question we should ask is where these stars were before Kepler turned his gaze up to the firmament of geometry in the year 1599. As we know today, the four moons of Jupiter discovered by Galileo 11 years later had been moving along their orbits in the heavens for millions of years. So where were Kepler's stellar solids? One cannot claim that he invented or constructed them. He uncovered their existence by drawing a few simple lines in his imagination or on a sheet of paper. Did he thus perceive an already existing purely spiritual archetypal idea? If so, what was the space into which he looked? Or were these stars already present as possibilities in the initial figures, like blossoms in the seed of a plant? We may tend to give preference quite quickly to the second answer. But if we do, we shall have to ask ourselves where the initial figure was located before it was seen for the first time ever by a human being.

Perhaps there is no generally convincing answer to these questions. But they show us nevertheless that one cannot so easily and without more ado deny that a purely spiritual world must exist in addition to that of the human mind. Let us leave the question open as we pay closer attention to these starry figures. Assuming that comparison with the plant realm proves to be correct, we shall now look to see in what way the blossom is able to metamorphose and what fruits it is capable of bearing. Otherwise we shall find ourselves observing a solely spiritual space. The icosahedron-star has as many points as the dodecahedron has corners, namely 20. And the dodecahedron-star has the same number of points as the 12 corners of the icosahedron. (This seems to have been Kepler's favourite, for otherwise why would he have given it a nickname—'Hedgehog'?) Moreover, as is right and proper for every descendant of the Platonic solids, the arrangements of all the points are perfectly regular. This means that these figures are capable of the following metamorphoses:

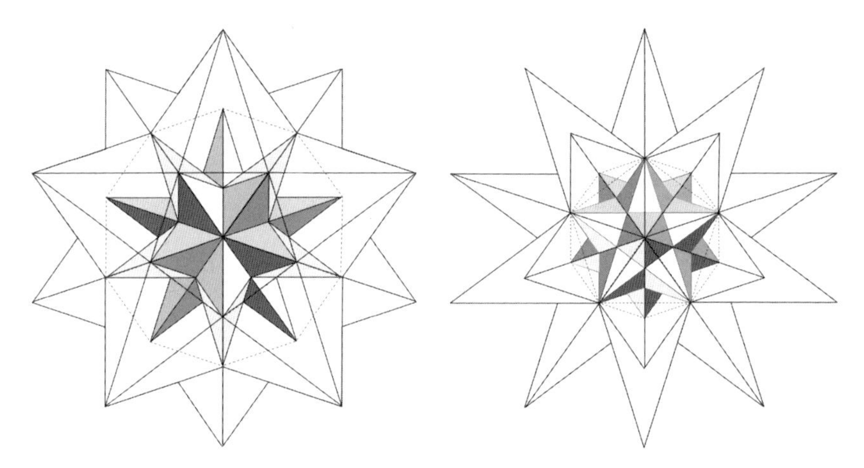

**Fig. 3.10** *Metamorphosis of Kepler's stellar solids one into the other*

Looking more closely into this mysterious world from the depths of which Johannes Kepler drew up his stellar solids we realize that the space within it is utterly filled with these mutually interchanging figures. The metamorphoses can be infinitely reproduced both inwards and outwards. This is therefore the same capacity in space as that which the pentagram, the 5-pointed star-figure, manifests in the plane. Marvellous though this may be, it is no miracle—as the mathematician will tell us, since he can calculate all the ratios. And the geometrical eye will recognize that each of the two starry figures is constructed from 12 pentagrams that are merely arranged differently in each case. There are 24 pentagrams, or perhaps 25 if one counts the archetype, and these are sufficient for the construction of a whole universe.

I do not know whether Kepler was fully aware of this astonishing capacity of his stellar solids in all its perfection although of course he did realize that the points of the icosahedron-star form a dodecahedron, and vice versa. But he did have an inkling that there must be something special about them in connection with the planets, since he used the 'Hedgehog' in the combination already mentioned of his geometric and harmonic models.*

However, it was necessary for all nine planets of our solar system to be discovered before a more comprehensive correspondence could be found. For this we do not need an infinite number of Kepler's stellar solids but only three, an icosahedron-star, a dodecahedron-star and another icosahedron-star which metamorphose into one another. So here we have once again a variation on the musical form of the lied, A-B-A. The radii of the spheres which alternately separate the stellar solids (i.e. the outer sphere of the solid in question and the sphere surrounding the Platonic solid within it) have the ratios shown in the following table. For the sake of completeness I have also included those of the corresponding areas and volumes. Oh yes, and then there are also the three numbers which we encountered much earlier on when we were seeking to gain a first overview of our solar system and found that the Pythagoreans attached such great importance to the number ten.

**Table 3.3** *Kepler's stellar solids—ratios of the circumspheres and the underlying Platonic solid, and the overall ratios in the planetary system*

| | Radius | Area | Volume |
|---|---|---|---|
| Icosahedron-star | 2.383963 | 5.683282 | 13.548735 |
| Dodecahedron-star | 1.776901 | 3.157379 | 5.610351 |
| | | | |
| Ico * Hedgehog * Ico | 10.098631 | 101.98235 | 1029.88213 |
| (sequence of Fig. 3.11) | | | |
| | Mean velocities v | Mean distances a | Orbital periods T |
| Pluto/Mercury (at v: Me/Plu) | 10.09960 | 102.00185 | 1029.27080 |

We shall close this chapter with a glance up to the stars, those stars which owing to a mysterious law mirror the overall ratios of all three

* The ratio between the Sun's distance from Mars at perihelion and the mean distance of Venus (1.9098) equals, apart from a deviation of 0.4%, that between the radius of the inner and the outer sphere of the 'Hedgehog' (1.9021). By inner sphere I mean the sphere that touches the centres of the sides of the initial pentagons, i.e. of the edges of the dodecahedron from which the solid is derived. Regarding the calculation of Kepler's stellar solids, see Appendix 2.4.

parameters of the two planets at the boundaries of our system with exceedingly astonishing accuracy, i.e. all those parameters not influenced by the eccentricities (or of course the masses, etc.). The law in question, expressed mathematically as $T^2 = a^3$, revealed itself just under 400 years ago to an earthly star-seeker, perhaps the greatest there has ever been. He would, I believe, have been especially delighted to hear that the splendid conformity of his combined 'stellar children' with the overall ratios of the planetary system has been discovered exactly 400 years after he beheld them.

**Fig. 3.11**

# 4. Analysis of the Harmonic Relationships

The music of the spheres can be approached from two directions. On the one hand we have a few reports—some more, some less credible—from people who claim to have heard the celestial music themselves or who have set out to describe the experiences of others said to have been granted this gift by the gods. In the final analysis such accounts must remain unverifiable, though they may contain ideas that could throw some light on the subject. On the other hand one can investigate the question with which this chapter is chiefly concerned: Is it possible to find among the ratios that hold sway in the heavens any correspondences that accord with the harmonic intervals present in the music made by human beings?

Let us formulate this second approach more specifically. In a number of intervals such as can always be found between measurable aggregates in a coherent totality (e.g. the orbital periods of the planets), there will always be more or less close approaches to ratios like 2:1, 3:2, 4:3, etc. Such interval series can be obtained by means of a random number generator, and in accordance with the laws of statistics one will always find several such 'harmonies' gathered around a specific mean. So we shall have to determine what can be counted as a correspondence in the case of musical ratios and above all whether the ascertained percentage of these correspondences deviates from a random distribution, and if so, by how much.

As far as I am aware, no such systematic investigation has hitherto ever been published or indeed undertaken. But without it we shall forever be feeling our way in the dark when trying to form a judgement about the planetary music discovered by Kepler or the efforts made by others in this field. Rejection or acceptance is then nothing more than mere belief. Rudolf Haase, one of the most unwavering champions of the validity of the harmonic correspondences discovered by Kepler, considers that '*Kepler's method is correct . . . as is the validity of the results obtained by it even today. So Kepler's proof of a musical world harmony is entirely successful, and the planetary harmonies discovered by him are realities.*'[1] Haase refers to a number of authors who have checked Kepler's calculations—including modern calculations of planetary movements—and who have in essence also reached the same results.[2] This is quite right, and Table 3.2 shows a distillation of those and also my own calculations. But absolutely nothing is said about any proper assessment, let alone an elucidation by a mathematically scientific method, of the degree to which the ratios calculated differ from pure chance and therefore in very truth represent something special. However deserving the efforts of Haase and others to validate Kepler's lifelong work may be, it would be better not to base talk of a 'proof' on such a weak foundation.

Similar criticism must also be directed towards those who see the harmony of the spheres as the product of a vivid imagination or suchlike without first having investigated the situation in ways akin to those suggested above. For example the author of a book (well worth reading in other respects) on recent calculations of long-term developments of the planetary system maintains that Kepler's 'vision of a harmonious, orderly solar system eventually foundered on the shoals of chaos'.[3] He means mathematical chaos which in the first instance merely states that at some point the behaviour of a system can no longer be predicted in advance. I shall look in more detail in Chapter 5 at the results of the supposed or perhaps actual chaos in the solar system. As to whether the ideas of Kepler and others concerning the good order existing among the planets have indeed foundered, surely if such a statement is not to remain in the realms of fantasy this can best be tested by means of calculations which it is now possible to carry out.

But before embarking on an analysis of this second, verifiable aspect of the music of the spheres, let us listen to a witness of the first kind, beginning with a question: What do we actually hear when we listen to music?

This may sound like a simple question, but it is likely to take more than five minutes to find an answer to it. Perhaps we shall only ever arrive at approximations when seeking for the real core of music or of what we experience when we listen to it, especially when endeavouring to put our experience into words. In the company of others we can listen, attentively and without allowing ourselves to be distracted by anything else, to a short piece of music, perhaps a movement from a symphony or a sonata by a great composer. But when each of us then tries to describe his or her impressions we discover how very different they usually are, though perhaps not always. Depending on who the listener is, the same music can lead to entirely different reactions. Whatever comes from the notes created by the vibrations of the airwaves, which can be objectively calculated, always receives an individual colouring. The poet Wilhelm Heinrich Wackenroder (1773–98) put this fact into words when he said: '*The human heart learns to recognize itself in the mirror of the musical sounds.*'[4]

Why, then, should something that holds good when listening to earthly music not also apply when perceiving celestial sounds? Pythagoras is said to have possessed the unique ability to hear the music of the spheres while wide awake and fully conscious (although this was only chronicled centuries later); but on the whole people speak only of dreamed or dreamlike experiences. The oldest authentic such communication about a listening experience that has come down to us stems, once again, from Plato. In his *Republic* we read about a fallen warrior who was, however, only apparently dead. In that state he underwent an experience of the next world where, among other things, he heard the harmony of the spheres resounding. The celestial spheres are described moving in circles '*on the knees of Necessity*' with the planets, Moon, Sun and fixed stars attached to them:

> *On the top of each circle sits a siren, who is carried round with the circle, uttering one note at one pitch, and the eight together make up a single harmony. Round about at equal distances others are seated, three in number, each on a throne. These are the Fates, daughters of Necessity, Lachesis and Clotho and Atropos. They are clothed in white and have garlands on their heads, and they sing to the accompaniment of the sirens' harmony, Lachesis of what has been, Clotho of what is, and Atropos of what shall be.*[5]

Let us assume that someone really did hear something, whether in a dream or a near-death experience or some other state of consciousness, and whether it was Plato himself or someone who told him about it. Two things are clearly expressed in the way this perception is described: the idea Plato himself had of the structure of the cosmos, and his opinion of what truly noble music should sound like. We know that he attached little or no importance to purely instrumental music because '*it is almost impossible to understand what is intended by this wordless rhythm and harmony, or what noteworthy original it represents*'.[6]

So in Plato's description of the harmony of the resounding cosmos the song is all about time and fate; themes of true noteworthiness are presented to the ear and the sense of the listener. The planets as it were provide the accompaniment, whereby each one brings forth only a single note. As we saw in Chapter 3, what now appears to us to be a rather limited idea only came to be broadened two thousand years later with the parallel developments in music and astronomy. We shall find this path of humanity mirrored also in later reports from individuals who claim to have heard celestial music.

But for now let us turn once more to the instrument from which the music of the spheres resounds, the planetary system, regardless of whether its task is merely to accompany the songs of goddesses or angels or whether like an orchestra it produces instrumental music that is sufficient unto itself. Our task now is to test how well the celestial instrument is tuned. To do this we need to know what measure to apply to each degree of correspondence. In other words we must decide which musical system to use. As a beginning we shall choose the one that has emerged from a two-thousand year evolution, the one that has harmonically pure tuning with seven clearly-defined intervals within one octave and 12 possible semitones. (For a description of this see Appendix 1.5.) However, we need only take this as a working hypothesis because, as we shall see, the following investigations will show whether a different system would be more appropriate.

We shall enumerate the criteria by means of which we can reach a reliable estimation of whether the celestial harmonies—or any other possible conspicuous order of the planets—determined by a specific model can be taken seriously:

- There must be a system on which the model is based. This could be a musical system or a mathematical law such as the Titius-Bode law or a geometrical model like that of the Platonic solids in Kepler's *Secret of the Universe.*
- The assertions must be uniform and provable statistically. This means that any statement relating to this matter must contain at least a certain number of coherent and exactly known values such as, for example, the distances of the planets from the Sun.
- The model must be clear and as simple as possible. Kepler's early ordering of the planets would surely soon have been forgotten if, instead of the Platonic solids, he had used other solids constructed in any possible way even if the conformity with the actual situation had been better.
- Conformity with the basic system of order must be as accurate as possible. Statements claiming that the actual conditions always differ slightly from the ideal, as one sometimes reads in the literature, are of little help here.
- And now comes the be all and end all of our whole consideration: there must be an adequately clear deviation from a random distribution. And we must also be clear that when such deviations are found we may still be a long way from knowing what their cause may be.

Before turning to our present main concern, the analysis of the harmonic ratios discovered by Kepler, we shall first use some rather simpler statements to show the need for the approach sketched above. Hans Cousto, for example, has provided us with a schedule of so-called planetary notes.[7] These arise when the reciprocal of the orbital periods in seconds is taken and the resulting number is equated to a frequency and finally transposed into a reference note by means of octavation. Here is an example:

> Orbital period Mercury: T = 87.969 days = 7,600,521.6 seconds
> $1/T = 1.315.. \star 10^{-7}$ [Hz]; $1/T \star 2^{30} = 141.27$ Hz
> (Hertz = 1/sec = cycles per second)

When related, as Cousto does, to a keynote A′ of 432 Hz, this yields the note D (or, to be more accurate, a note somewhere in the region of D). From the point of view of physics there is nothing wrong with this (see Appendix 1.5 on 'Octavation') but let us see what it means musically. To do this we have displayed the results for all nine planetary notes and related these to the note C = 129.6 Hz (whether A′, C or any other note is used is irrelevant for the following analysis).

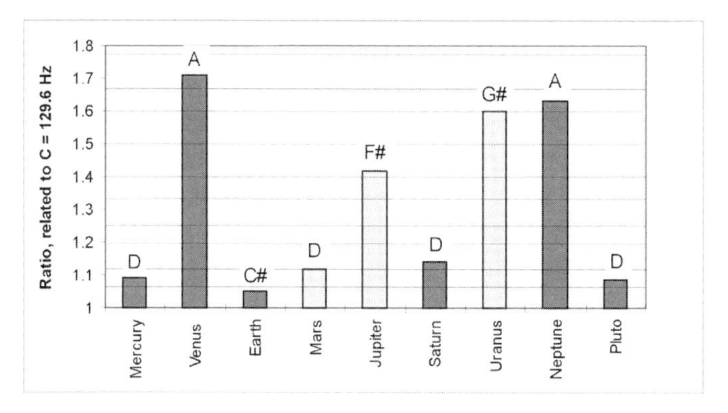


**Fig. 4.1** *Deviations of musical ratios of the frequencies from the orbital periods, related to C = 129.6 Hz or A′ = 432 Hz. The values of the musical intervals are shown as horizontal lines*

We can see at first glance that there are three or four good to very good correspondences and five to six intervals that can be located fairly accurately between musical notes. In other words, one does not need any further statistical calculations to show that this instrument is not optimal but actually rather out of tune. For it was not without reason that the orbital periods were the first to be rejected by Kepler in his search for harmonic ratios in the solar system.[8] In view of this it is all the more worrying to read in Cousto that '*Kepler's allocation depends on the first random step in allocating the notes while here, through the law of the octave, every note is given by nature itself*'.[9] Put plainly, Cousto's model lacks both accuracy and systematic treatment. He establishes the relationship to musical notes by taking as his point of departure a well-defined keynote—indeed the natural* keynote A′—but then attaches no importance to the tuning in his handling of the frequencies that arise from the procedure he uses. The 'octave space' is divided into 12 sections which extend roughly from the middle of one semitone interval to the middle of the next, and the allocation of planetary to musical notes is set according to the boundaries of those sections. In this way every resulting frequency inevitably becomes a 'note'. By this method, the orbital periods of the members of any fictitious planetary system constructed on the basis of random numbers could not be expected to generate anything other than planetary notes. Quite apart from the fact that the attribute 'random' thus becomes deservingly applicable, this also means that one would have to abandon, perhaps unconsciously, the notion which is basically inseparable from the idea of a music of the spheres, namely that such a thing—if it indeed exists—can hardly be anything other than the work and expression of a creative intelligence in the cosmos.

In the context of Pythagoras' report (see opposite) I find the question of what, if anything, he heard to be less important than the manner in which his way of listening involved a conscious act of will. There are some evident parallels with statements made by well-known composers about how their own works arise. Something mysterious always comes into play—out of a realm that can probably never be fully plumbed: an internal listening, an inspiration or whatever else one might like to call it. If something truly great is to be born, untiring and lucidly alert effort is essential. The obvious conclusion to be drawn from this is that what the modern listener receives from music that has come into being in this manner—when he 'exercises and works towards becoming irrigated and well arranged by its most musical ratios'—is in some way a gift from the gods such as is normally confined to the realm of legend.†

But let us continue with our analytical project. We have seen how important it is to be clear about the degree to which an interval arising out of planetary ratios and given a musical assignation 'hits the note'. What is actually meant when Kepler calculates an interval of 1.68 between the values of Mercury at aphelion and Venus at perihelion (see Table 3.2, last line), so that the corresponding value for the major sixth assigned to it is 1.666 . . ? Is this an exact correspondence or not?

---

* We shall not here discuss further the justification for calling the keynote A′ tuned to 432 Hertz a 'natural keynote'. Other tunings are also used, but in the present context this can be ignored since there would be hardly any difference in the results.

† The art of listening hinted at here is demonstrated at the Musicosophia School for Conscious Listening to Music at St Peter in the Black Forest, Germany, where I have learned an incomparably more profound way of accessing the treasures that lie hidden in classical music.

**Fig. 4.2** *Athanasius Kircher,* Musurgia Universalis, *1650*
*Bottom left: Pythagoras pointing to the smithy said to have inspired him. Bottom right: earthly music. The sphere in the centre symbolizes the music of the spheres with the divine music of the nine angelic choirs above it*[10]

*'Pythagoras [conceived, however] that the first attention which should be paid to men is that which takes place through the senses; as when some one perceives beautiful figures and forms, or hears beautiful rhythms and melodies, he established that to be the first erudition which subsists through music, and also through certain melodies and rhythms from which the remedies of human manners and passions are obtained together with those harmonies of the powers of the soul which it possessed from the first. [He] did not, however, procure for himself a thing of this kind through instruments or the voice; but employing a certain ineffable divinity, which it is difficult to apprehend, he extended his ears, and fixed his intellect in the sublime symphonies of the world, he alone hearing and understanding, as it appears, the universal harmony and consonance of the spheres, and the stars that are moved through them, and which produced a fuller and more intense melody than any thing by mortal sounds. This melody also was the result of dissimilar and variously different sounds, celerities, magnitudes, and intervals, arranged with reference to each other in a certain most musical ratio, and thus producing a most gentle, and at the same time variously beautiful motion and convolution. Being therefore irrigated as it were with this melody, having the reason of his intellect well arranged through it, and as I may say, exercised, he determined to exhibit certain images of these things to his disciples as much as possible, especially producing an imitation of them through instruments, and through the mere voice alone.'*[11]

The question as to whether a correspondence is exact or not indicates one of the core points of any thorough consideration concerning the harmony of the spheres, so I shall now introduce at least the basic ideas one meets along the path leading to a solution of the question. Details needed for the necessary calculations will be found in Appendix 4.1.

Having first considered the individual 'note', we shall then establish the connection for every interval determined according to a specific procedure, such as for example Kepler's arrangement.

A planetary interval will always lie somewhere between two notes of whichever musical system is used for purposes of comparison. So there will always be a note that is closest to this interval and another, neighbouring one that is the second smallest distance away. There are only two theoretical exceptions to this: either the planetary ratio coincides exactly with the ideal value, or else it lies exactly halfway between two notes and is thus equidistant from both. We can imagine that these two situations are equally unlikely to occur. So the mean probability for the members of an accidentally occurring series of values to be examined would also be the mean between the extremes. Just a clearly—or, as statisticians say, 'significantly'—deviating distribution could indicate that the ratios in our solar system genuinely accord with a musical order.

So the following principle (only slightly simplified) can be applied in working out the probability for one particular value of a deviation from a random distribution:

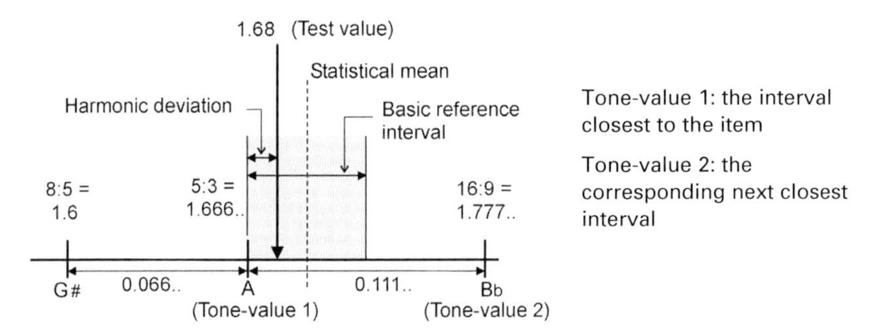


**Fig. 4.3** *Simplified diagram for calculating the harmonic probability for a particular value*

The test value falls between the intervals 5:3 and 16:9, or in the semitone step between A and B♭ (related to the C major scale). Thus it lies in a space of which the sum can be stated as the difference between the two intervals (= 0.111 ..). Unless one wants to delude oneself with doubly favourable results, one must of course apply only half of the value, i.e. the shaded portion. A value to the right of this would apply to the next tone-value; and for a test value lying to the left of it the calculation needed would be with G♯, or 1.6 as tone-value 2.

If we now imagine the theoretical possibilities in whole numbers, there can for example be 1111 values between A and B, of which half, 555, must be taken into account when working out the probability. Between note A and our test value, 1.68 − 1.666 . . = 0.0133 . . (i.e. 133 values) are thus possible. The probability arises by dividing 133/555 = 0.24 (or 24%). So for a single value the harmonic probability, as I shall call it for the sake of simplicity, can be calculated by a very simple procedure. To assess the number arrived at in this way one must take into account that the above extreme cases would deliver values of 0 (exact correspondence) and 1 (exactly in the middle between two notes). The mean value is 0.5 or 50%.

So with this Mercury/Venus interval we find ourselves pretty accurately in the middle between absolute average and complete harmonic correspondence. However, a listener would quite happily accept the corresponding note played on an instrument as being A. This is because the value of A in the well-tempered tuning system is 1.6818, which equals a harmonic probability of 0.277 (the maximum value of all well-tempered notes is 0.283 for a minor third). It is clear, however, that nothing can be derived statistically from a single case: one swallow does not make a summer. So let us now look at all the values arrived at by Johannes Kepler according to the procedure described in Chapter 3. It is essential that an assessment always uses all the possible values arrived at by the method chosen.* So following Kepler's example we shall take all the perihelion and aphelion values for the angles as seen from the Sun and relate them to the smallest value, Saturn at aphelion, i.e. allocating to Saturn the lowest note as the keynote (see Table 4.1 on p. 64).[12]

There was no theory of probability in Kepler's day, thank goodness, one might say! The reason for what is now a very obvious difference compared with today's values does not arise from there having been any noticeable change in celestial circumstances over time. Changes in the angular velocities are so gradual that they would have virtually no effect on the probabilities calculated. We can therefore confidently base our considerations on the fact that the mean value of *c.* 0.4 was already valid long before the beginning of the seventeenth century and will continue to be so for a long time to come.

Being at the forefront of mathematics as it stood in his day, Kepler would himself surely have rejected the results and conclusions he reached regarding the existence of a harmony among the movements of the planets. On the basis of a mean (derived from the data observable at the

---

* This remark is directed not to Kepler but to all those who think that one can deduce improbabilities from certain significant cases (whether of the orbital periods or of the distances between the planets, or anything else) without relating these special cases to the totality of all possible combinations.

**Table 4.1** *Harmonic probability of the angles according to Kepler (passed through on one day, as seen from the Sun). (In calculating the mean values, the values of Saturn at aphelion are of course not taken into account.)*

| | Angles according to Kepler (°) | Current value | Related to Saturn at aphelion | Current value | Trans-posed into an octave | Current value | Note, current value, rel. to C or G | Harmonic prob-ability | Current value |
|---|---|---|---|---|---|---|---|---|---|
| Me. per. | 6.4000 | 6.3465 | 217.3585 | 211.7743 | 1.6981 | 1.6545 | A–E | 0.5753 | 0.3617 |
| Aphel | 2.7333 | 2.7553 | 92.8302 | 91.9395 | 1.4505 | 1.4366 | F♯–C♯ | 0.9589 | 0.6571 |
| Ve. per. | 1.6269 | 1.6241 | 55.2547 | 54.1941 | 1.7267 | 1.6936 | A–E | 0.9046 | 0.4921 |
| Aphel | 1.5806 | 1.5805 | 53.6792 | 52.7396 | 1.6775 | 1.6481 | A–E | 0.1978 | 0.5510 |
| Ea. per. | 1.0217 | 1.0192 | 34.6981 | 34.0081 | 1.0843 | 1.0628 | C♯–A♭ | 0.6133 | 0.1156 |
| Aphel | 0.9508 | 0.9534 | 32.2925 | 31.8141 | 1.0091 | 1.9884 | C–G | 0.2787 | 0.1830 |
| Ma. per. | 0.6336 | 0.6347 | 21.5189 | 21.1806 | 1.3449 | 1.3238 | F–C | 0.3223 | 0.2255 |
| Aphel | 0.4372 | 0.4364 | 14.8491 | 14.5632 | 1.8561 | 1.8204 | B♭–F | 0.3830 | 0.8886 |
| Ju. per. | 0.0917 | 0.0916 | 3.1132 | 3.0573 | 1.5566 | 1.5286 | G–D | 0.8541 | 0.5823 |
| Aphel | 0.0750 | 0.0755 | 2.5472 | 2.5203 | 1.2736 | 1.2602 | E–B | 0.5753 | 0.2482 |
| Sa. per. | 0.0375 | 0.0374 | 1.2736 | 1.2489 | 1.2736 | 1.2489 | E–B | 0.5753 | 0.0426 |
| Aphel | 0.0294 | 0.0300 | 1 | 1 | 1 | 1 | C–G | 0 | 0 |
| | | | | | | | Mean | **0.5672** | **0.3953** |

time) that shows an even lower degree of correspondence with musical ratios than does a purely random distribution, it would not have been possible to substantiate the existence of celestial music; and the idea of a music of the spheres would then have been even less likely to survive until our scientific age. In this respect it is perhaps a good thing, too, that those who later took up cudgels on behalf of Kepler's ideas likewise did not get involved in those particular considerations. It is sometimes quite remarkable how things work out.

Two more questions now present themselves. How do the three planets discovered after Kepler's time fit into this picture? And, above all, what is the significance of the actual mean of *c.* 0.4? We may remember that the closer a value comes to zero (from 0.5 downwards), the better is the correspondence with musical ratios.

By means of the t-test, a statistical method that investigates a relatively small number of random samples (explained in more detail in Appendix 4.2), one can determine that in the case of the harmonic correspondences derived from today's values there is a 10.6% likelihood that these are values arising from randomly distributed data. In other words, one can be about 90% sure that the values investigated are not from a random distribution. However, to be realistic one has to say that a 90% statistical certainty is not all that good. Truly acceptable statistical significance begins at 95%, but 99% or higher is even better. With the degree of significance arrived at one might expect to find one among ten planetary systems similar to our own that had this degree of harmonic structure. To assume anything other than mere chance as the cause seems to be rather too bold, though it cannot be discounted entirely. Since it is likely to be some time before any other planetary systems can be investigated along these lines there will (unless better correspondences turn up elsewhere) be nothing for it but to describe the idea of finding a musical order in our solar system backed up by numbers in Kepler's sense* as no more than a tentative hope for the future.

And our disillusionment continues. We shall now investigate how the correspondences with musical proportions show up when we apply the values not to Saturn at aphelion but to other planetary positions. Although Saturn at aphelion or Mercury at perihelion offer themselves respectively as the lowest or highest 'note' from the musical point of view, any other possibility would do just as well from the purely mathematical point of view.

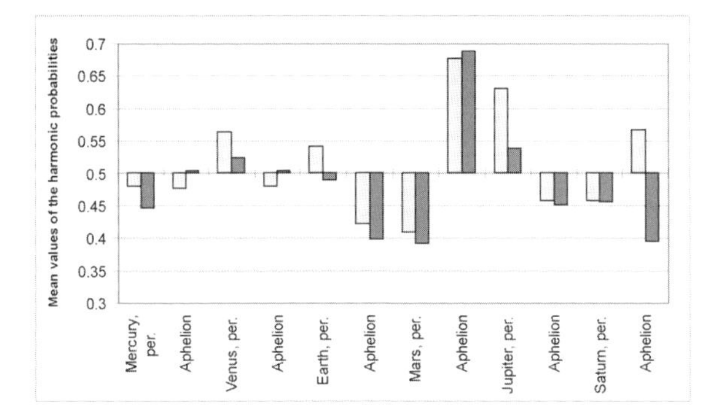


Pale shading: calculations based on the values available to Kepler

Dark shading: calculations based on current values.

**Fig. 4.4** *Harmonic probabilities (mean values) of the angles according to Johannes Kepler, related to all possible planetary positions out to Saturn*

Kepler's choice of keynote was a good one, as is made obvious by the present-day values. Saturn at aphelion (together with two other values) stands out as the best correspondence with harmonic intervals. With the data available in Kepler's day Mars at perihelion would have delivered a more favourable result. However, to justify the choice of this particular value as the keynote still requires a certain amount of mental gymnastics. The first or last value of any series here appears to be much more plausible. On the other hand, without any need for further calculations, one also sees that the deviations upwards and down-

* The correspondences between the musically harmonic ratios of the areas in basic geometrical figures and the appearance of these same figures in the conjunction positions of the planets must be seen as being more qualitative in kind (see Chapter 2).

wards (from the mean value of 0.5) more or less balance one another. Let us now take a look at the same depiction as that in Fig. 4.4 for all nine planets:

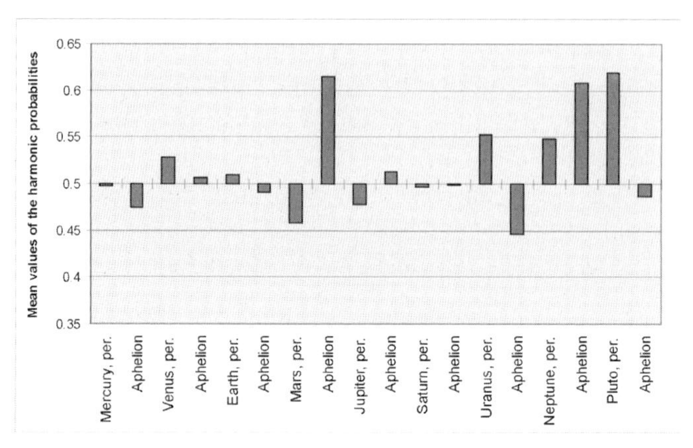


**Fig. 4.5** *Harmonic probabilities (mean values) of the angles according to Johannes Kepler, related to all possible planetary positions out to Pluto*

The three planets discovered since 1781 bring about a further disharmonization. Uranus at aphelion provides the most favourable value. With the help of the t-test, however, one arrives only at a probability of almost 80% for a deviation from a random distribution. Statistically this is as good as nothing. Neither does a different combination of the values provide any materially different results. For example an investigation of the 'main structure' shown in the middle of Fig. 3.7 yields a mean value of 0.46 for the data used by Kepler and 0.477 for today's astronomical data. Please be assured that I have also tried out a number of other possible combinations without success. One would have expected this, since there is a further procedure (to be discussed later) that can be applied to working out the overall harmony of a model of the music of the spheres, regardless of the keynote selected. According to this, using present-day astronomical data, the result arrived at for the Kepler angles right out to Saturn is a mean of 0.48 or, when the planets out to Pluto are included, a little over 0.5.

To sum up we can say that Kepler's allocation of values for the planets visible from Earth with reference to Saturn at aphelion is the only one that shows some deviation from a random distribution. But one must then add that there is a specific reason for this restriction.

So where do we go from here? I can well imagine that all this amounts to a shattering result for many a champion of Johannes Kepler's ideas. Yet nothing can be done to change it. Anyone rechecking the results shown here will arrive at the same results (perhaps with a few variations following the decimal point). Two other models of planetary harmony are also discussed in the literature. Although in some instances they show remarkable correspondences they remain unconvincing when subjected to the criteria put forward at the beginning of this chapter. These are a musical scale suggested by Hans Kayser arrived at via logarithms of the mean distances from the Sun, and an allocation by Thomas Michael Schmidt based on the differences between the synodic periods (more in Appendix 5).

I repeat, where do we go from here? What would Kepler say if he were to get wind of these investigations? Well, surprisingly, he may already have referred to them four centuries ago. Towards the end of *The Harmony of the World* there is a passage stating that he has noticed a certain, though small, discrepancy that shows up in the combination of his model of the Platonic solids (extended to include the 'Hedgehog') and the harmonies of the angular velocities, about which he does not wish to '*remain silent*':

> *Nevertheless, by this example I challenge as many of you as will chance to read this book and are imbued with the disciplines of mathematics and knowledge of the highest philosophy: come, be vigorous and either tear up one of the harmonies which have been everywhere related to one another, change it for another one, and test whether you will come as close to the astronomy laid down in Chapter IV; or else argue rationally whether you can build something better and more appropriate on to the heavenly motions, and overthrow either partly or wholly the arrangement which I have applied. Whatever contributes to the glory of our Founder and Lord is equally to be permitted to you throughout this my book; and I have assumed that I myself am permitted up to this hour freely to change anything which I could discover which was incorrectly conceived in the preceding days if my attention nodded or my enthusiasm was hasty.*[13]

When I first began investigating Kepler's music of the spheres systematically—having failed to find any such investigation in the relevant literature—I did so on the assumption that he would probably turn out to be right. On the whole, utterly fanciful ideas do not continue to live on in human culture for four hundred years or more. And Kepler's ideas, certainly in essence, have not died. One sign of this is the way in which many other treatises or dissertations about him, even those merely recognizing him as the discoverer of his three Planetary Laws, are embellished with his picture of the Platonic solids neatly stowed one inside the other. The idea of a music of the spheres is indeed indissolubly linked for all time with only two names: Pythagoras and Kepler. Even that of Plato lacks an equivalent resonance since his ideas in this regard derive more or less from those of the Pythagoreans. Kepler has truly saved the idea and brought its flame right over into our own age, which is so unenlightened in this respect. The fact, as the present work shows, that the angles and also the angular velocities employed by him do not deliver what he believed to be true can be regarded as almost immaterial, as can also the fact that no one today thinks of asking for exact correspondences in his geometrical model. This would soon lose its importance if one were to succeed in showing that he was right after all, right not only in the realm of ideas that remain unrelated by proof to astronomical reality, something that might well be debated at length and in detail, but right also in the meaning he himself attached to the celestial harmonies as he stood on the threshold to our modern age of science. As the quotation above demonstrates, he knew that the last word on the subject had not yet been spoken and perhaps never would be spoken.

Well then: let's us take him at his word!

With this in mind we shall create the powerful instrument needed to throw light on the overall harmony of the intervals derived from the angular velocities, orbital periods, and all the other parameters. It is obvious that firstly we must include all possible relationships in our considerations and that secondly we must make ourselves independent of a keynote the choice of which has such a decisive influence on the end result. For every direct—i.e. without first finding a common denominator for the values—possible combination of planetary data we shall define the harmonic probability by using the procedure shown in Fig. 4.3. In the examples of the 12 values used by Kepler, 66 possibilities (1 + 2 + 3 + ... + 11 = 66) will then arise out of the angles at perihelion and aphelion. Taking only the 11 direct adjacent combinations might result in an inaccurate assessment, as demonstrated by the small fictitious example in the figure on p. 69.

This method shall serve as a filter. In making our calculation of statistical significance in accordance with it with the help of the t-test we shall take into account that the t-test—as well as other statistical procedures—assumes independent values if a correct result is to be reached. However, we have no more than 11 (in Kepler's example: the 'direct' intervals). But for an initial assessment—for all possible ratios—the t-test can be used, and so long as the ratios are in the region of mean probability the differences anyway play no part. But if our instrument should signal

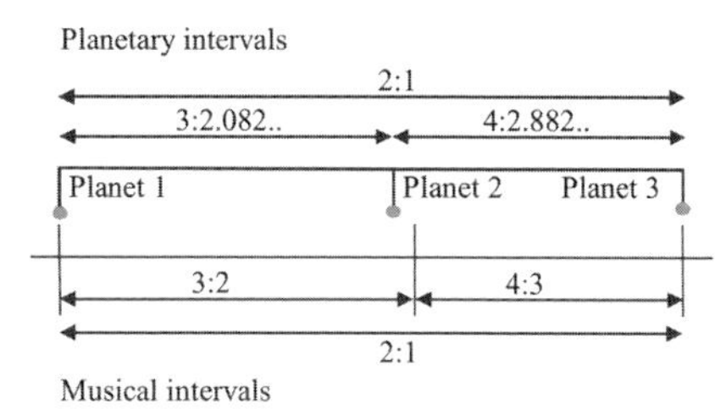


**Fig. 4.6**

When exclusive account is taken of the two intervals of neighbouring planets, a mean harmonic probability of 0.62 results. (The values of 3/2.082 = 1.441 and 4/2.882 = 1.388 deviate considerably from musical intervals.) An objective overall view ought to show that the interval of Planet 3 to Planet 1 corresponds exactly to the octave. This results in a mean value of 0.42 formed from all three possible probabilities.

some improbable events of a harmonic kind in the planetary system—and it is the optimal instrument for this purpose—then we shall take a closer look at whatever triggers the alarm.

So now we are ready to embark on a little fishing expedition upon the lake of cosmic harmonies. We shall cast our net not in the hope of catching some theoretically conceivable exotic specimens—such as, for example, the intervals from the third root of the logarithm to the base 18 of the synodic periods relating to Neptune, or something similar—but in the hope of landing some perfectly normal, everyday planetary fish just like those for which Kepler was also looking. Here is the result of our first trip:

**Table 4.2** *Preliminary probabilities for a correspondence conforming to a random distribution of the 36 possible ratios of various orbital parameters with harmonic intervals. (v: velocity; value T: calculation step for determining the probability from the t-test; minus signs show mean values <0.5. When T is positive the probability also approaches zero and thus as it were shows some degree of poor tuning.)*

| | Number of intervals | Mean of harmonic probability | Standard deviation | Random number T | Probability |
|---|---|---|---|---|---|
| Perihelion distance | 36 | 0.47714 | 0.33006 | −0.41559 | 0.340124 |
| Aphelion distance | 36 | 0.48735 | 0.28400 | −0.26718 | 0.395451 |
| Semi-major axis a | 36 | 0.54772 | 0.27092 | 1.05693 | 0.148892 |
| Semi-minor axis b | 36 | 0.46227 | 0.25022 | −0.90466 | 0.185916 |
| v, mean | 36 | 0.52475 | 0.29821 | 0.49803 | 0.310790 |
| v at perihelion | 36 | 0.50593 | 0.29760 | 0.11963 | 0.452730 |
| v at aphelion | 36 | 0.34189 | 0.23352 | −4.06255 | 0.000130 |
| Orbital period | 36 | 0.51178 | 0.28882 | 0.24468 | 0.404068 |
| Linear eccentricity a*e | 36 | 0.46159 | 0.32241 | −0.71477 | 0.239744 |

For percentage data the values in the final column should be multiplied by 100.

There you are! At our very first attempt we appear to have caught quite a nice specimen in our net. The values for the velocities at the point most distant from the Sun (v at aphelion) show with a probability considerably less than 0.1%, related to a random distribution. Even though this resulting probability will still need to be corrected, we are certainly on the right track. However, before embarking on the precision work, let us cast our net once more. But where, in what now appears to be a fairly productive lake, should we do our fishing? Whereas the velocities at aphelion point to a greater degree of harmony, the values at perihelion and also the ones in between show absolutely no significance. So how about investigating a combination of the various velocities?

| | Number* | Mean | Standard deviation | Value T | Probability |
|---|---|---|---|---|---|
| v pe/v ap | 81 | 0.45911 | 0.29988 | −1.22730 | 0.11165 |
| v pe/v mean | 81 | 0.51438 | 0.27921 | 0.46337 | 0.32218 |
| v ap/v mean | 81 | 0.50139 | 0.27598 | 0.04532 | 0.48198 |

So our second fishing expedition has been less successful. But why should the velocities at aphelion, and only these, be expected to exhibit a harmonic order? On its own this would be meaningless. Despite the high value of the t-test it would tend to hint at chance since if you take the corresponding probabilities for all velocity intervals at every possible angle which the planets can attain in their orbits,† a minimum—and also at another point a maximum—must occur somewhere. A calculation, which I shall not demonstrate here, would show that the smallest probability, i.e. the greatest correspondence with musical proportions, would indeed come about at aphelion, whereas the most inharmonious ratios would show at about 20° from perihelion. Taking a sum of all the angles of course roughly balances out the values since the velocities change continuously so that in consequence when the ratios are formed from two values all the intermediate values must occur between harmonic intervals. Although it is worth noting that the aphelion shows the most musical point, this is not a circumstance which on its own can lead us to deduce that the music of the spheres arising from this is more than our choice of an inevitable probability minimum.

'*Argue rationally whether you can build something better and more appropriate on to the heavenly motions*,' advises Kepler. An ellipse is determined by two parameters, and if his idea is correct that the form of the planetary ellipse results from the intention to create harmony between the movements,

---

* Here and in the following we have investigated the intervals for example between all nine velocities at aphelion and all perihelion values. One could omit the ratios formed by each planet from its own two values, since this would be insignificant for the overall result.

† The angles are counted from the perihelion (0°) and run continuously up to the aphelion (180°) and then back again. This means that the distance from the Sun and the velocity are the same at, let's say, 210° as they are at 150°.

then there must also be a second value which can form musical ratios with the velocities at aphelion. Yet we have already investigated all the obvious possibilities, and I have said that we don't intend to cast our nets for exotic specimens. A planet at aphelion or perihelion marks the extreme points of its orbit. When it is travelling at its mean velocity (see Appendix 3.3) it is at its mean distance from the Sun, the semi-major axis. Yet already in Chapter 1 we stated that this has no significance for a logically structured planetary order where the important thing is the semi-minor axis b. So what is the velocity of a planet when its distance from the Sun equals its semi-minor axis? As a precaution, the mean value of perihelion and aphelion velocities has been added in Table 4.3, since errors can be caused not only by supposed probabilities but also by various misconceptions.

**Table 4.3** *Velocities (km/sec)*

| | Mean | At perihelion | At aphelion | At distance b | Mean of ② and ③ | Deviation ⑤:④ |
|---|---|---|---|---|---|---|
| | ① | ② | ③ | ④ | ⑤ | ⑥ |
| Mercury | 47.8721 | 58.9759 | 38.8590 | 48.9062 | 48.9175 | 1.00023027 |
| Venus | 35.0207 | 35.2597 | 34.7833 | 35.0215 | 35.0215 | 1.00000002 |
| Earth | 29.7847 | 30.2836 | 29.2943 | 29.7889 | 29.7890 | 1.00000292 |
| Mars | 24.1293 | 26.4980 | 21.9722 | 24.2350 | 24.2351 | 1.00000462 |
| Jupiter | 13.0642 | 13.7103 | 12.4482 | 13.0794 | 13.0792 | 1.00001412 |
| Saturn | 9.6430 | 10.2005 | 9.1284 | 9.6579 | 9.6644 | 1.00067279 |
| Uranus | 6.7990 | 7.1238 | 6.4976 | 6.8062 | 6.8107 | 1.00065379 |
| Neptune | 5.4319 | 5.4738 | 5.3977 | 5.4321 | 5.4358 | 1.00067738 |
| Pluto | 4.7400 | 6.1131 | 3.6775 | 4.8917 | 4.8953 | 1.00073626 |

So that's how it is, then. The velocity at the distance of the semi-minor axis b from the Sun (in brief: 'v at b') is the actual mean value or at least very close to it.* So if the velocity at aphelion and 'v at b' or, respectively, the actual mean show harmonic ratios, this, together with the role played by the semi-minor axis in the spatial ordering, would also be temptingly logical . This would then also give us in a certain sense an expansion of the multi-voiced music of the spheres introduced by Kepler, inspired by the points at perihelion and aphelion. This is because every planet reaches the distance b from the Sun twice during each orbit, which in combination with the aphelion amounts to three harmonic points. So now we can land our fish:

---

* What constitutes the minute difference is explained in Appendix 3.3. The mean velocity is naturally just that, though it shows the geometrical mean ($= \sqrt{x * y}$), whereas mean value or respectively the average value usually refers to the arithmetical mean ($= (x + y)/2$).

| | Number | Average | Standard dev. | Value T | Probability |
|---|---|---|---|---|---|
| v ap/'v at b' | 81 | 0.35724 | 0.25008 | −5.13794 | 0.00000096 |
| v ap/mean value v | 81 | 0.35417 | 0.24811 | −5.29013 | 0.00000052 |

And, sure enough, I can now not only make use of Kepler's favourite utterance but also join him in exclaiming that the dawn '*of that most wonderful study*' is beginning to show on the horizon. For these are midway to being 'astronomical' probabilities, or perhaps one should say improbabilities since it is exceedingly improbable that we are here still looking at a random ordering in the harmonies of the movements of the planets. We also see that it now makes little difference whether we take our departure from the velocity 'at b' or from the mean value, thus fully complying with the requirement that the music of the spheres is not to be sought in exceptional circumstances. Furthermore, it is basically quite logical from the point of view of physics that the velocities have turned out to be the bearers of the harmonies of the spheres. Like a musical instrument, the structure of the planetary system must be capable of generating vibrations that relate to one another in resonant ratios. One can really only think of this in connection with the orbital periods or the varying velocities. The orbital periods, however, as we have already discovered, are, for reasons of the stability of the planetary system, highly unsuitable, for here the resonances of small whole numbers could have disastrous consequences (as we shall see in Chapter 5). In addition, we know today that, far from being a total vacuum, the spaces between the planets are in fact filled with the solar wind. Although perhaps somewhat speculative, it is quite possible to picture to oneself the spread of the vibrations thus generated. Although I would not be so presumptuous as to speak of a proof, we do have here for the first time a confirmation not only of humanity's ancient idea of a cosmos filled with music but also of Kepler's endeavour to demonstrate somewhat more scientifically, with the help of the planetary laws he discovered, the possibility of such a music-filled cosmos, even if audible only in spirit.

The precision work still to be done concerning the actual probabilities will—I might as well point out in advance—reduce the above values somewhat, but we shall nevertheless be left with a goodly series of noughts while also being able to clothe the naked numbers in, as it were, unmistakable musically resonant garments. Since, however, it will be as impossible for us to hear these as it has been for us to hear all previously assumed harmonies, it is now time once again to listen to the words of aural witnesses to the music of the spheres. And the two individuals quoted here assuredly heard something.

*'... because I am aware of what I want to do, the fundamental idea never leaves me; it ascends, it rises up, I see and hear the image in all its breadth like an outpouring before my inner eye ... Whence do I receive my ideas? I cannot answer this with authority; they come uncalled for, indirectly, directly, I could seize them with my bare hands, in the open air, in the forest, when I go walking, in the stillness of the night ...'*

*'... When I gaze astonished in the evening upon the sky and the host of shining bodies resonating within its bounds, what we call suns and earths, my spirit soars across all those stars, millions of miles distant, to the archetypal fountain of them all from which flow all created things and from which new creations will continue to flow in all eternity.'*

Ludwig van Beethoven

*'... Can you not conceive of how this takes hold of one's whole being and of how one can be so deeply immersed in it that one appears dead to the outside world. And now imagine to yourself a work so great that in it the whole world is mirrored—one is, as it were, merely an instrument on which the universe is playing. At such moments I no longer belong to myself ...'*

*'I have just finished my eighth* [symphony] ... *Imagine the whole universe beginning to ring and resound. These are no longer human voices but planets and suns revolving.'*

Gustav Mahler[14]

Of course neither Beethoven nor Mahler nor any other great composers have claimed—as a way of describing what resounded within them—that they had heard the music of the spheres. Perhaps they felt as we do when we now listen to their works outwardly performed by musicians. There are some passages which, when we hear them, give us a sense of listening to the cosmos filled with sound. In the final movement of Beethoven's Ninth Symphony, for example, as they accompany the words '... *surely a loving father dwells above the starry canopy* ...', the violins, soaring ever higher and becoming ever more ethereal, seem (to me, at any rate) to conjure up an 'archetype' of celestial music. Another way of putting this might be to say that at such a moment the music reaches its highest spirituality before descending again to the everyday realm.

Let us now investigate the harmonic realities that have emerged from the statistical numbers. In the following my starting point will be the ratios of the velocities at aphelion and 'at b'. (As I have said, the minimal difference from the arithmetical mean is negligible.) For musical reasons it is practical to relate these intervals to a common keynote. The following picture arises for all 18 possibilities:

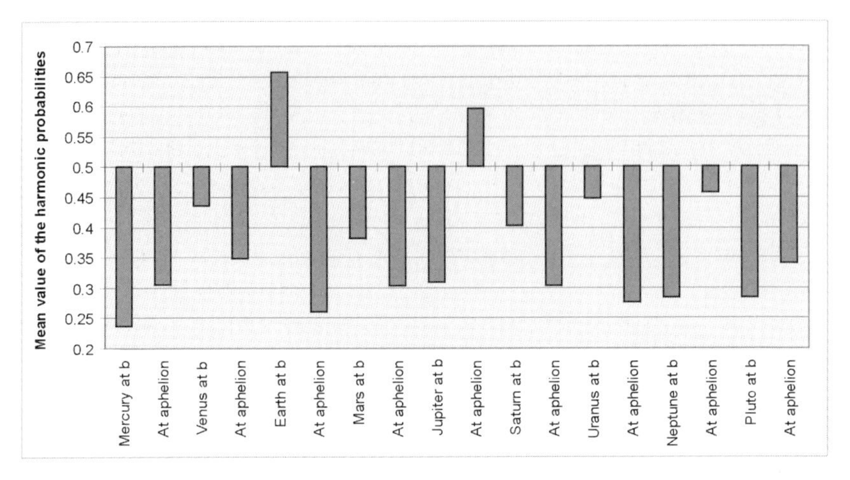


**Fig. 4.7** *Mean values of the harmonic probabilities for the intervals at aphelion and the 'at b' velocities, related to all possible planetary positions out to Pluto. All the ratios and harmonic deviations leading to this depiction are shown in Appendix 6.4*

So the movements are at their most harmonious when we relate them to the greatest velocity according to this ordering—corresponding to the highest note—i.e. the velocity of Mercury, the messenger of the gods, when its distance from the Sun is that of its semi-minor axis. We note that there would have been no better choice as regards the tuning of the musical instrument. Let us once again imagine the planetary system to be an actual musical instrument, in this case one that always emits a note when two planets are located at one of their three possible 'musical points'. This condition is, of course, relatively rare, and such a constellation will be all the rarer the greater the orbital period of the two planets involved. Accordingly, the positions in which Mercury is involved will be the most frequent, and Mercury 'at b' even occurs twice, on average every 44 days. So to tune the instrument relative to this note will surely be seen as resulting from most careful consideration, since this alone guarantees the maximum harmony in the present arrangement. It is perfectly permissible to speak of such a harmony here, for the mean value of harmonic probability with Mercury 'at b' is 0.23738, which is even lower than some of the relevant values in the well-tempered system of tuning mentioned above. The harmonic deviation in accordance with the same procedure can be worked out, and the resulting numbers lie between 0.036 and 0.283 (with an average value of 0.172).

The probabilities shown in Fig. 4.7 represent for all the planets the mean value of the harmonic deviations of, in each case, 17 possible intervals (e.g. Mercury 'at b' to 8 other planets, and 9 aphelion values). Taken separately, each row consists of values that are independent of one another, so that it is now possible to work out the actual probability with the help of the t-test. As we see in Appendix 4.2, the probability is *c.* 0.062% or 1:1600 that the values investigated reveal the harmonies described purely by chance. Or, in other words, we can state definitively that with 99.9% probability an influence was at work in our solar system which brought about an ordering involving the musically harmonious ratios shown to have come into being. Despite the reductions applied, this can still be taken as a statement of very high statistical significance.

Let us now ask ourselves how the harmonic correspondences look when shown as musical ratios. To do this we shall transpose them into an octave related to Mercury 'at b':

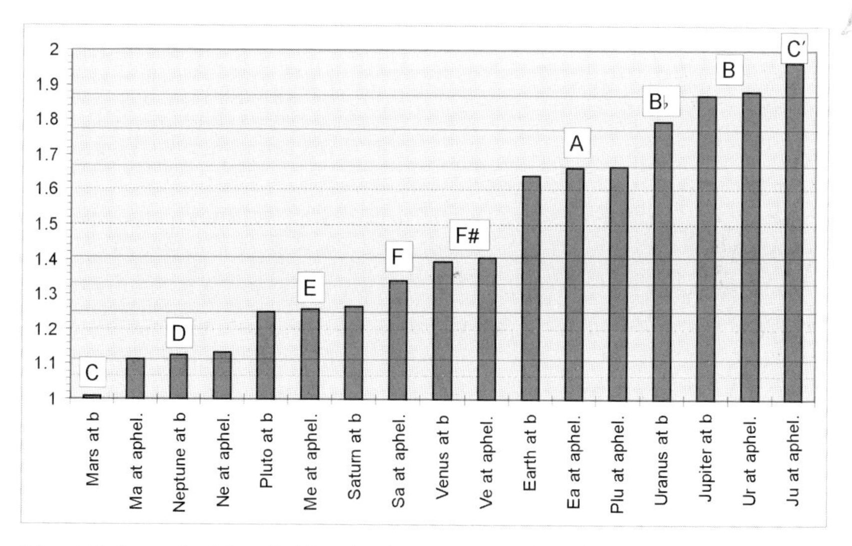


**Fig. 4.8** *Intervals of the velocities related to Mercury 'at b' as the keynote (= C, exactly at 1). The musical intervals are shown by the horizontal lines*

Regardless of all probability analyses we can now observe clearly that, almost throughout, the correspondence between planetary and musical intervals is good to very good. Furthermore, with two exceptions, we have in this ordering the structure of a major scale (related to C, the F♯ tritone appears in place of the G fifth; in addition, the diminished seventh

B♭ also comes into play). This mode of depiction would, by the way, also show us whether a different musical system than the one used would give a more suitable description of the ratios. This would be recognizable in that the columns of the intervals would show a structure that looked conspicuous in a different way. However, having scrutinized a good many similar diagrams, I have not found any more suitable system arising from the ratios of other parameters (though I have to say that I was not expressly searching for such a system). The one I did finally find comes very close to the musical ordering that has taken shape during the course of human evolution both in the masterpieces of classical music and in the melodies of much folk music.

However, this system only provides us with the stock of notes available to us which would still have to be transformed by the actual movements of the planets into something musical or, speaking more neutrally, into a sequence of notes. Whether such a sequence, if we were capable of hearing it, would be anything like genuine music is unclear and would have to be tried out. Such a trial has indeed taken place using the ratios of Kepler's model, albeit with horrendous results.[15] This was caused not by Kepler's ideas but by the fact that in this kind of planetary music all the intermediate values that are possible between two adjacent intervals also resound. In consequence what one hears is not music like that made by humans but rather a cacophony of howling sirens shrieking away at various pitches and graded velocities. In order to avoid such a performance, Kepler based his ideas primarily on the fixed values of a planet at perihelion and aphelion. Without proper tuning of the notes, 'pure melodies' and 'noblest song', such as those perhaps heard by the aural witness quoted here in a dreamed excursion into the heavens, are impossible at least for human ears.

*'Never have I sensed the magic of music so powerfully as I did yesterday evening; indisposition and gloomy melancholy had taken hold of me . . . And then, all of a sudden, I came upon Pergolesi's* Salve Regina, *as though sent by an angel; I sang it, and the heavenly* O dulcis opia *filled my soul . . . I felt relieved, the feverish nervousness left me; I sank into a refreshing stillness, and had a feeling not of cheerfulness but of well-being . . . Then came a dream; a memory of it floats towards me as though through a mist. The material veil fell from my eyes, I departed the earth and floated at once in the measureless spaces of the universe. Suns, planets, stars all around me, innumerable, in indescribable beauty; such enchantment filled my ears! In oft surmised pure melodies the spheres wheeled in noblest song—the greatest unity in most manifold variety, audible only to a spiritual ear . . .'*[16]

There exist in mathematics axioms, established principles, that can never be proven but which are so self-evident that one need not feel called upon to prove them. In the case of a music of the spheres—if such a thing were to exist or if someone were to try to emulate it, as is said of Pythagoras—the axiom one could apply would be that it must resemble or indeed surpass the most enchanting and dignified, the most noble and joyful music made by human beings. However, when people's sense of the grandeur of music is lost, or has been buried under the rubble of modern life, then any music of the spheres would also sound accordingly. The witness quoted above was still able to hear some of the grandeur, perhaps because a piece by the Italian composer Pergolesi (1710–36) had set those strings vibrating in him during the evening leading up to his dream journey. Pythagoras, too, we are told, possessed the unique ability to listen in to the music of the spheres; and he endeavoured to give his pupils an idea of what it was like. Meanwhile, people of today might perhaps be put in mind of music that could help them imagine unearthly sounds if they could learn to understand the greatest works of a musical evolution reaching back two thousand years.

This chapter's analysis or respectively its consideration of two sides of the music of the spheres may have been somewhat lengthy and perhaps also rather wearisome. We had to do away with Johannes Kepler's ideas about planetary harmonies derived from the angular velocities before resurrecting them once more in the intervals of perfectly normal velocities. Copernicus related his idea of a music of the spheres to the Earth; Kepler then allowed imaginary dwellers on the Sun to enjoy the celestial harmonies. And now in a certain sense we can claim that through their movements the harmonies of the planets permeate the whole of the solar system. That there is also a correspondence between the structures of human music and the velocity relationships among the celestial bodies of our cosmic home which can be depicted through physics has now also been proven with a high degree of significance—in so far as this is possible within the framework of statistical probability. As regards that other aspect of celestial music—claimed to have been heard by some very few individuals—we have determined that there is always a very close connection between earthly music that has been found meaningful by those individuals and the way in which they describe the heavenly harmonies. So now, to continue in this vein, let us close this chapter with the testimony of another very credible witness, the writer Heinrich von Kleist:

> *From time to time, as I stroll into the breeze of the west wind, and especially when I close my eyes, I hear whole orchestras, from the sweetest sounds of the*

*flute to the rumbling of the double bass. I remember especially as a boy of nine years walking upstream beside the Rhine and into the face of the evening wind, so that waves of air and water reverberated round me, hearing a melting Adagio with all the magic of music, with all possible melodic turns and a whole accompanying harmony. It was like an orchestra . . . Indeed, I believe that everything the wise men of Greece told of concerning the harmony of the spheres was nothing more delicate, nothing more beautiful or more heavenly than what I heard in my strange daydream.*

*And whenever I so wish I can repeat this concert for myself, without musicians; but the minute I think about it, everything disappears, as if made to vanish by magic . . . the melody, the harmony, the way it sounds; in short that whole music of the heavenly spheres.*

*So now, from time to time, as twilight falls on the street, I open my window and my lungs to the inflowing evening breeze, closing my eyes as it ruffles my hair; thinking of nothing, I listen intently . . .*[17]

# 5. Current Scientific Thinking

What is a planetary system? Or, more specifically: What is this configuration of nine gigantic spherical bodies that have been journeying around the centre of our cosmic home for millions or billions of years?

It is a Sun which for the same length of time, or perhaps somewhat longer, has been sending out a huge and more or less constant stream of energy that appears to be perfectly adjusted to the needs of Earth, 150 million kilometres distant, as well as to the life upon it; neither too much nor too little, yet not entirely regular, so that certain fluctuations of sunshine and heat—in combination with the slight inclination of the Earth's axis—could not have been better chosen for the purpose of achieving a dynamic evolution and multitude of life forms on this third planet of the system. We are only just beginning to understand the processes inside the Sun, which these days is regarded as a kind of cosmic nuclear fusion reactor. These processes, and also those which bring about a constant transport of the resulting energy to the surface, are so complex that they remain utterly astounding.

The nine wandering stars, slightly flattened spheres of varying composition, have masses so monstrous—with Earth, for example, amounting to 6,000,000,000,000,000,000,000, i.e. six thousand billion billion tonnes—that it continues to be entirely beyond our comprehension. These masses are held firm in their orbits by a law called the law of gravity of which the physical background—or one might say its very nature—although eminently amenable to calculation, has also to this day almost totally escaped our understanding. Isaac Newton stated that although he could describe this force he was not able to construct any hypotheses as to what causes it. What remained a riddle for Newton was similarly not solved by Albert Einstein's theory of relativity, although in this context it did make possible an improvement of calculability in extreme conditions, e.g. in the vicinity of gigantic masses or at very, very high velocities.

The interplay of the planets, which are subject to the mutual attraction exercised by their gravitation, is—we can say without exaggeration—optimal, since it can by no means be taken for granted that a gravitational system of this kind, with such manifold reciprocal actions, is likely to remain functional over aeons of time. One is almost tempted to talk of collaboration between the various celestial bodies. The ruler of our solar system occupies exactly the location which enables it to be a guardian for the inner planets. With its gigantic mass, Jupiter intercepts most of the comets or hurls them away from the planetary system. Without Jupiter, says an astronomical journal, the Earth would have received so many catastrophic hits that no worthwhile evolution could ever have become established on it.[1] The Moon stabilizes Earth's axis; without this, over long periods of time, it could have wavered out of control, which would in turn also not have been conducive to further evolution of life. And how we acquired this satellite in the first place is a mystery, so that its origin is at best only the subject of theories.

Today science claims that the celestial bodies solidified in some way out of an amorphous cosmic cloud of dust and gas. On Earth, over aeons, continents, mountains, rivers, valleys and fertile plains arose, while in a parallel development life appeared, having emerged from lifeless matter. How this could have happened is, once again, entirely unclear. In the 1950s and 60s, and even later, people felt optimistic about finding an answer to this question quite soon. Their expectation arose from the discovery that in a reconstruction of what was thought to have been the 'primordial soup', which would have constituted Earth's earlier atmosphere, certain organic chemicals came into being of their own accord. However, the journey from those substances to even the most primitive form of life is, one must suppose, still longer and more difficult to negotiate than that from the cosmic cloud of gas to the Sun and the nine planets. Recent realistic assessments of research in this field of so-called prebiotic evolution admit our complete ignorance in this respect.[2] So all consideration of how life might also have developed on any imaginable number of planets circling billions of stars is nothing more than mere speculation. It is impossible to compute what the probability might be.

And how did music find its way into the world? Or perhaps it would be better to ask: How did something that was already manifest in a different form in the movements of the planets find its way to the Earth? In a footnote near the beginning of Chapter 2 (p. 26) I mentioned that for this to happen it had been necessary, among many other aspects, for a metamorphosis to take place in the acoustic duct during the transition from reptiles to mammals. Today we know a good deal about the extraordinarily complicated process that is involved in the genetically regulated growth of an embryo. All the changes which take place must be adjusted to the process in a way that does not impair the functioning of the organism. We should make an effort to imagine how two ossicles moving away from the jaw can have found their way into what had formerly been the single unit of outer and inner ear. Of course this cannot have taken place all at once; many stages must have been necessary, over the course of umpteen generations. And at no stage in the process can this remarkable sequence of mutations have been allowed to impair the ability to hear, otherwise the creature in question would more or less certainly not have survived. Can it really be the case that this evolutionary step—and of course many others—took place in an uncontrolled succession of accidental mutations, as so many people believe today?

At any rate, human beings with all their musical and other capacities, including not least the ability to think scientifically, did eventually come into being out of the primordial cloud of dust. Although today we may assume that we have evolved from the animal kingdom, this in no way explains the origin of what we call the human spirit. It was possible for human beings to be born because a multitude of exceptional circumstances permitted this to happen, because our Sun has sufficient energy to last for some billions of years, and because the planetary system has been organized for long aeons of time in the way now known to astronomy. The most marvellous thing about the various disciplines of science today is perhaps their capacity to open up an overall view—however tentative and provisional this may be. In his novel *Confessions of Felix Krull* Thomas Mann applied his inimitable literary style to describing the scientific view of the early decades of the twentieth century by giving an account of the three archetypal events of creation: the leap from nothingness into existence, the appearance of the animate out of the inanimate, and the coming into being of human consciousness. But now, in a new millennium, the only conclusion to be drawn despite several centuries of scientific thought is that in the final analysis we still know virtually nothing about those three original events of creation. However many theories may argue to the contrary, the actual discoveries we have made to date speak in more modest terms. This applies even to the supposed big bang, or rather to the various endeavours which have been undertaken to find some way of making that 'leap from nothingness' plausible, if only from the point of view of physics.

Things that cannot be explained in terms of natural laws often come to be described as miracles. So the conclusion I am obliged to draw is that, overall, everything we now know about our world amounts to one stupendously immense miracle. As their name implies, the natural sciences seek explanations for things on the basis of natural laws. Yet the quintessence of all they have discovered urges us to acknowledge the existence of a creative and purposeful force in the universe which *cannot* be comprehended by scientific methods. Surely there must be *a meaning* in all that has happened!

## *From the past to the present day*

The professor permitted by Thomas Mann to talk about the three primordial events of creation did not mention the appearance of a sun as such. But he was no doubt right in the now generally accepted sense that stars come into being and then fade again, albeit over aeons of time. This chapter will now discuss what science has to date found out about how our planetary system came into being as well as what it has to say about the system's future. In the latter case I do not mean what is likely to happen to the planets when the Sun ceases to exist or first of all swells up to become a red giant, but how the planetary orbits are developing and what is likely to happen in the long run to the stability of what are at present such well-ordered conditions. Of course I shall only be able to provide some sort of overview on these two subjects, although what I have to say is based on detailed publications by various authors in the field. So let us look first at the past.*

According to present scientific knowledge stars arise under specific conditions in gigantic cosmic dust clouds, so one may assume that our Sun, too, came into being in this way. The material of those cosmic clouds, at least in part, is derived from matter returned to the cosmos by stars, especially towards the end of their life, in nova and supernova explosions. In this respect a cycle of coming into being and dying away exists in the heavens just like that with which we are familiar on Earth. So what are suns? Are they living beings? And the same question can be asked in respect of planets. In many ancient cultures the celestial bodies of our solar system were worshipped as divinities. Do we really have the right to deride the essence of this idea that these are beings greater than ourselves? Is the Sun truly nothing more than a machine resembling an excessively huge nuclear reactor as people think today even though they may not put it quite so baldly? Are we closer to the truth merely because thanks to the instruments and equipment we have created we are now able to describe the physical side of reality better than we could in the past?

Let us look once again at how the stars come into existence. Depending on temperature and mass, conditions must be such that parts of a previously condensed cloud are able to separate off (one speaks of fragmentation) and continue to contract under the influence of their own

---

* *Die Entstehung von Sonnensystemen—Eine Einführung in das Problem der Planetenentstehung* by Hans J. Fahr and Eugen A. Willerding (1998) is a very welcome book because it does not disguise existing gaps in current knowledge. References here to this work are shown thus: (F/W + page number).

gravity without causing the resulting opposing pressures to become too great. On account of the rotation that accompanies the condensation, in many cases the fragment of cloud flattens more and more to form a disc. At the core of this disc which grows ever more dense, and thus ever hotter, the new star is created if, that is, everything goes well and the critical temperature for a nuclear reaction is attained. Of course in nature this is a very complicated and astonishing process which, furthermore, takes place inside dust clouds that are hidden from our sight. The reader will have to consult the literature[3] for further details of this intriguing subject. Here we are mainly concerned with what is said to take place outside the core of the disc: the creation of planets.

It has been possible to prove the existence of a protoplanetary disc in the case of 40–50% of the young stars investigated with this in mind. They can now even be shown optically, although initially they were only detectable in infrared:

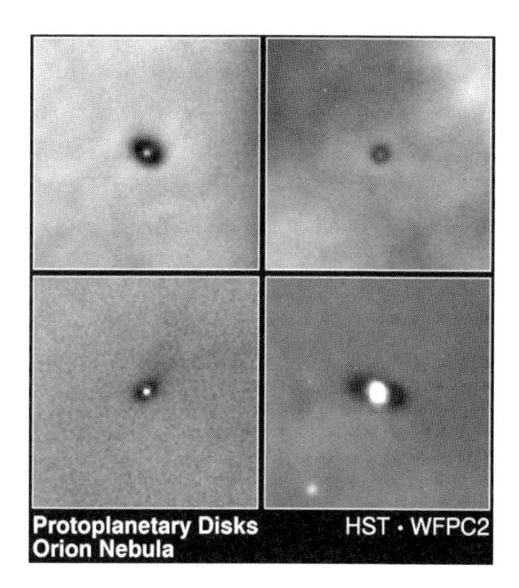


**Fig. 5.1** *Protoplanetary discs. Photo by the Hubble Space Telescope*

So how was it possible once upon a time for a small number of gigantic planets moving along specific orbits to arise out of an unstructured conglomeration of gas and minute dust particles such as must at some time have surrounded our Sun? A theory designed to provide a satisfactory answer to this question will have to take account of the conditions that exist now in our planetary system. Among these are, for example, the Titius-Bode law already mentioned, showing the distances between the planets, and the differentiation between the inner and outer region with regard to their size and make-up. Since we now also know more about the protoplanetary discs we must also take account of other facts meanwhile discovered. One in particular states that the discs have relatively low temperatures and that they only exist for about 10 million years, since older stars do not appear to possess them.[4] This indicates that on the cosmic scale the planets must have come into being relatively rapidly, or at least that the dust must have been transformed into the initial material needed for further development in that relatively short span of time. One speaks in this connection of planetesimals, meaning clumps of rock about one kilometre in diameter. That these did indeed once exist is proved above all by the craters on the moons which are the result of numerous hits by large blocks early on in the evolution of the planetary system.

*'. . . This interdependent whirling and circling, this convolution of gases into heavenly bodies, this burning, flaming, freezing, exploding, pulverizing, this plunging and speeding, bred out of Nothingness and awakening Nothingness—which would perhaps have preferred to remain asleep and was waiting to fall asleep again—all this was Being, known also as Nature, and everywhere in everything it was one . . . Our human brain, our flesh and bones, these were mosaics made up of the same elementary particles as stars and star dust and the dark clouds hanging in the frigid wastes of interstellar space . . . He would not, however . . . conceal what it was that distinguished Homo sapiens from the rest of Nature, the organic and simple Being both, and which very likely was identical with the thing that had been added when Man emerged from the animal kingdom. It was the knowledge of Beginning and End . . . Being was not Well-Being; it was joy and labour, and all Being in space-time, all matter, partook if only in deepest sleep in this joy and this labour, this perception that disposed Man, possessor of the most awakened consciousness, to universal sympathy. "To universal sympathy," repeated Kuckuck . . .'*

Thomas Mann[5]

All in all the view most widely held is that the planets must have evolved over time through a process of gradual growth. As early on as the eighteenth century the philosopher Immanuel Kant (1724–1804) was of this opinion. At present the growth of the particles in the original disc of dust is thought to have taken place in three stages. First the particles would have had to clump together to form blocks of about one metre in diameter, for they can only begin to detach themselves from the gas travelling round with them once they have reached this size or more. Especially in the case of the inner planets this first phase must have occurred very rapidly, for only as far out as Earth's orbit, for example, the gas had disappeared after merely 10,000 years and drifted off into the Sun. And any dust still attached to that gas would have been taken along by it (F/W, 171). Once they had reached a metre in diameter, the bodies remained in the orbits determined by Kepler's laws and were ready for the next stage. This ended when the planetesimals mentioned above had formed.

From here on we have computer simulations to show us how countless such blocks might have come to form a handful of large planets. Of course the results of such trials always depend to a great extent on the parameters initially fed into the system. But it is claimed that 'all similarly designed numerical studies tend to reach very similar conclusions' (F/W, 180). This signifies that a body roughly the size of Earth would have taken about 80 million years to form. After some initial difficulties and with one exception, it was also possible to model the formation of the gas giants with their solid cores successfully. A core of about 10 Earth masses in the Neptune orbit would, however, take far more time to form than has thus far run its course in the life of our solar system.

It would take too long to enumerate fully all the difficulties involved in reaching a plausible view of the period during which our solar system came into being or, as Fahr and Willerding put it rather neatly, to 'reach an understanding or, more modestly, merely an inkling, of the genesis of a system as complex as this one' (F/W, 318). So let me mention only one more point. As of June 1999, the existence of planets had been discovered in connection with 18 stars mostly 50 to 150 light years distant from the Sun. In all cases these are approximately the size of Jupiter (0.4 to 11 Jupiter masses*). Of course this does not mean that there are no smaller planets elsewhere but only that our means of discovering them are insufficiently sensitive. To discover planets of the size mentioned requires instruments of phenomenal precision, since the presence of such distant bodies is indicated by the minutest shift in the light spectrum of the relevant star which indicates that the star is making small but regular movements brought about by the gravitational pull of an orbiting planet. The degree and the rhythm of the shifts then show the mass and orbital period of that planet.

Most of the giant planets discovered are, surprisingly, closer to their sun than is our own small Mercury. This glaringly contradicts the ideas based on our own solar system about how planets come into existence. According to these, giant planets ought to be found only in the more distant regions. And this cannot be all that wrong, since close to the Sun it would not be possible to accumulate enough material to make a Jupiter.

---

* As of February 2009 planets have been discovered belonging to 290 stars; some of these also have smaller masses right down to only five times that of the Earth (Source: Internet encyclopedia Wikipedia). My further discussions on this subject are not affected by this.

So one assumes that those giant planets were indeed formed further out before travelling inwards on elliptical or spiral orbits. The next question to be asked is: What brought the travelling to a standstill and why was the process different in our case? The answer to the first question might be that the rapid rotation of the young Sun brought the planet into a stable orbit (through gravitational influences similar to our own Earth/Moon system). And, as regards the second question, after enumerating a number of possibilities, an article in a scientific journal reaches the following conclusion: 'We should consider ourselves lucky that Jupiter ended up in a nearly circular orbit. If it had careened into an oval orbit, Jupiter might have scattered Earth, thwacking it out of our solar system. Without stable orbits for Earth and Jupiter, life might never have emerged.'[6]

Quite so. Yet even though it is of course still too soon to draw definitive conclusions from the observations made to date concerning the formation of planets in general, as the authors just quoted emphasize, nevertheless there are indications that actually there might—possibly—have indeed been very special influences at work at the time when our solar system was coming into being.

In drawing our consideration of past research to a close I shall now introduce the ideas which Hans Fahr and Eugen Willerding developed in their endeavour to come a few steps closer to solving the problems. Expressing themselves with considerable caution, they wonder whether one way of reducing the gaps in their explanations would be to assume a process in which circular waves of pressure passed through the gaseous disc from its centre outwards. Under certain circumstances these could become shock-waves like waves undulating in water and forming crests which break. Such shock waves could have promoted the growth of clumps of matter. A sequence of several such waves would also provide an explanation for the development of a progression of distances analogous to the Titius-Bode law. Such waves might be stimulated in the border layer between the about-to-be born or newly born star and the disc of dust and gas surrounding it. Since the rotation of the star would be slowed down by magnetic fields (F/W, 319) while that of the disc remained constant, sudden eruptions of waves could result. Astronomers seem to agree now that shocks caused by waves of pressure, only on a much grander scale, also stimulate the coming into being of stars in the spiral arms of a galaxy.

I cannot judge whether this theory is justified; and Fahr and Willerding themselves also write that only future discoveries will be able to show how far their supposition concords with reality. Intuitively I would say that something like this may indeed have occurred. In another passage the two authors speak not of waves of pressure but of sound waves for the simple reason that the way in which both types of wave propagate can be described in physics in similar terms.

So in the final part of our look at scientific thinking concerning our origins, let us briefly give our imagination free rein. Perhaps what travelled through our solar system all those aeons ago were not shocks but

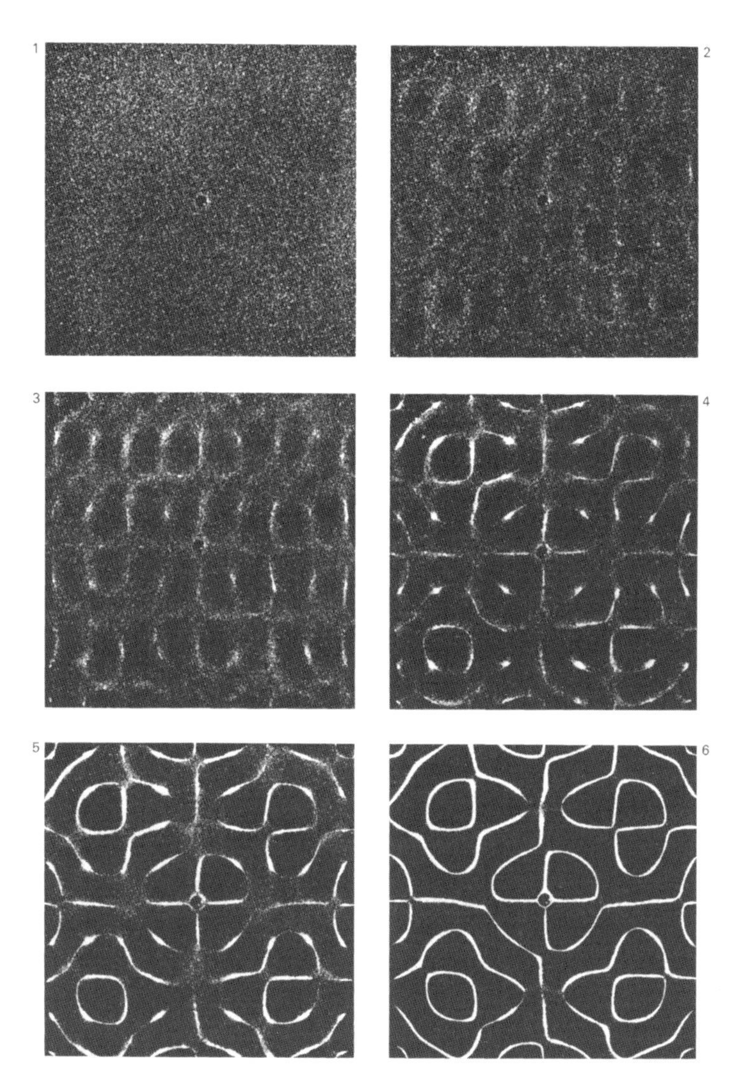


**Fig. 5.2** *Sonorous figure according to Hans Jenny.*[7] *Pictures 2–6 show a simple sonorous figure taking shape on a steel plate strewn with calcined sand (top left-hand picture); frequency 7560 Hertz*

a special kind of sound more appropriately described as the music so mysteriously generated in accordance with the harmony which Johannes Kepler placed at the dawn of all time. However hypothetical this may sound, we have known since Ernst Chladni (1756–1827) performed his experiments that musical notes can have an effect on matter and cause it to take on specific orderly formations. This is demonstrated by the so-called Chladni figures arising in scattered sand on a metal plate that is caused to vibrate by means of a violin bow.

Hans Jenny took up the work of Chladni and continued it in the twentieth century, applying the same principle but producing the vibration to be examined not with a violin bow but by making use of piezoelectricity* generated by means of crystals attached to the sand-strewn plate. Noticing the development of the shapes formed by the sand in the sequence shown in Fig. 5.2, one might well imagine how in the chaos of the initially wholly diffuse protoplanetary dust disc vibrations or waves may have been what caused the particles to begin forming into structures. But since our solar system is not a metal plate strewn with sand, one may imagine further that it could have been a harmonious sound—or a melody, or both—instead of a single note which resounded at its creation. That this suggestion may not be entirely on the wrong track is shown by a second sequence produced by Jenny, where a chord consisting of two notes causes the grains of sand to congregate into a kind of planetesimal (see Fig. 5.3).

Such mysterious processes capable of producing a structure of nine planets which began circling the Sun aeons ago and which, at least in historical times, have in addition done so in a way that is strongly related to the basic principles of music, will—possibly—become more comprehensible once music (in the widest sense) has resumed its original place within the science of astronomy. The knowledge we now have about the origins of our solar system indicates at the very least that the mythological  ideas about the creation of the world having been a sonorous event may amount to more than mere fictions of the imagination. This of course then leads to the question: Where is the creative music supposed to have come from or, expressed more scientifically, what set the structure-creating vibrations going in the first place? A hand that wields the violin bow is needed if sonorous figures are to take shape on a plate strewn with sand. So for there to be a stroke of good fortune such as the cosmic home of humanity is increasingly turning out to be, surely the story of its

---

*Many crystals under pressure generate an electrical field, and vice versa an electrical tension causes distortions in the lattice of a crystal which manifest as vibrations.

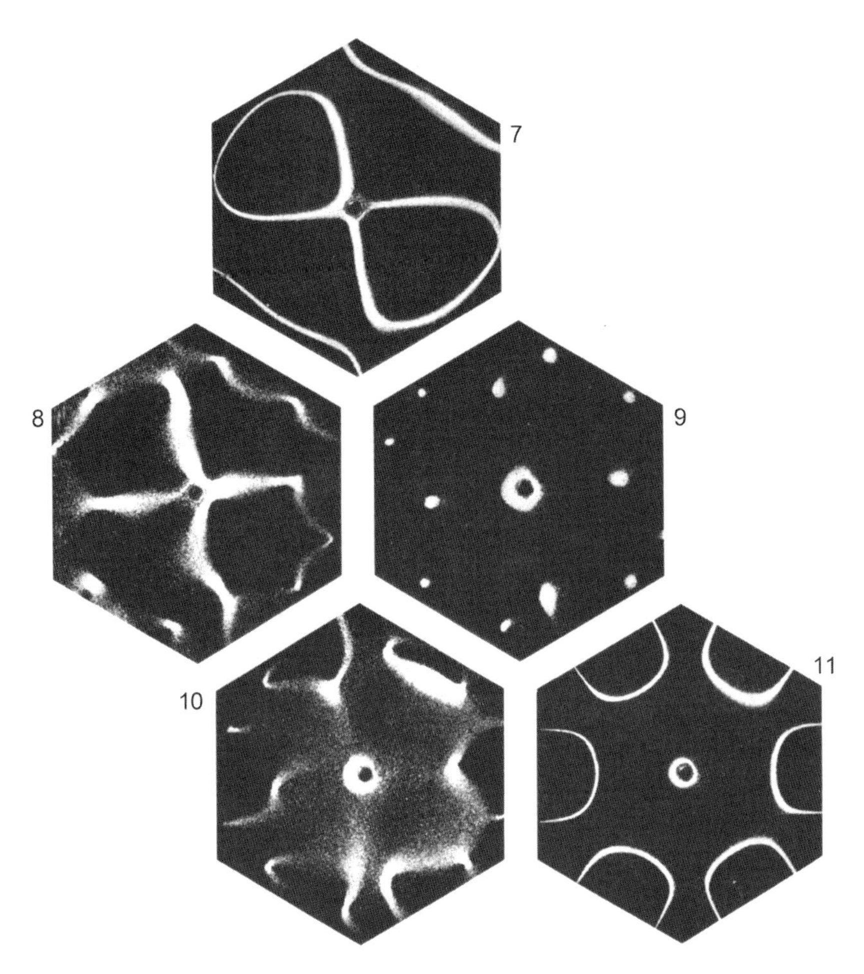


**Fig. 5.3** *Sonorous figure according to Hans Jenny. In pics 7 and 11 a single tone (800 Hertz in 7, 865 Hertz in 11) has produced its own sonorous figure. Pic. 9 shows the result when both tones are sounded* at the same time *and at equal strength; pics 8 and 10 show intermediate stages*

creation must also include more than waves generated in an uncontrolled way at its border layer.

A final glance at Jenny's work and its possible parallels with the genesis of our planetary system will conclude our journey into the past which has now arrived at the present time.

An ancient myth tells us: 'And God *said*: Let there be ...'

Using a different process, Hans Jenny scattered sand or powder onto a membrane and showed how the vibrations generated by a spoken or sung vowel caused the formless material to take on a specific order:

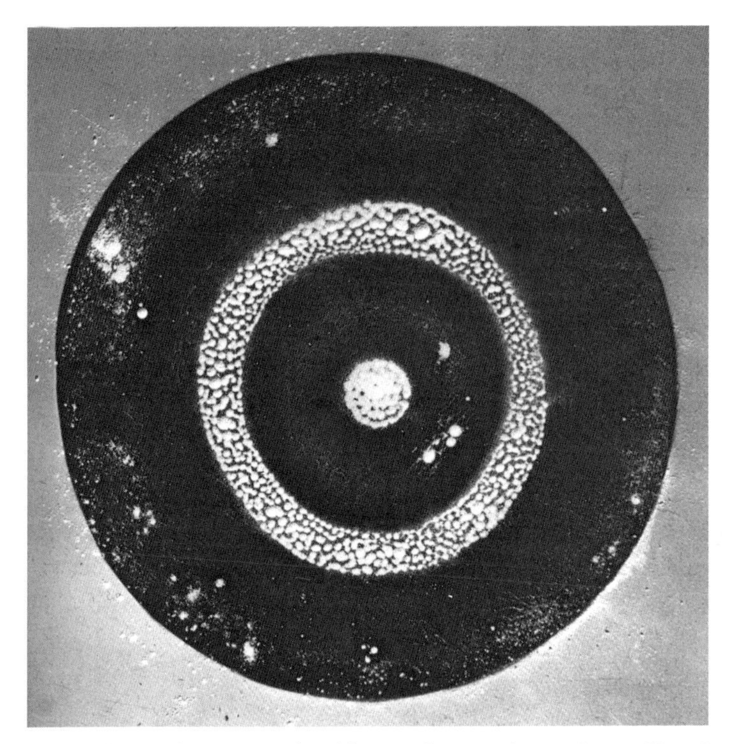

**Fig. 5.4** *Vibrational pattern produced by a spoken vowel, according to Hans Jenny*

## *The future*

There is another feature shared by most creation myths: that the creation of the world was preceded by darkness, the abyss or chaos. The latter accords with our general ideas. Many accounts tell of the powers of chaos doing eternal battle with creation and never being entirely overcome. In recent decades chaos has even entered into the stronghold of abstract order: mathematics. This incursion has proceeded hand in hand with the increase in capability of electronic data processing. In fact it was only possible to discover it properly by means of the speed of calculation made possible by computers, though a mathematical genius, Henri Poincaré (1854–1912), had discovered signs of it on the horizon decades earlier. Nevertheless, in one sense it is no exaggeration to claim that we owe modern chaos to computers.

We all know that as well as being very much an abstract science mathematics is also very well able to describe many concrete processes that take place in the world, sometimes very accurately indeed. What is more, we have heard that as an art mathematics has the advantage of being the last one of astronomy's three former siblings to remain. So, knowing the possessive properties of chaos, we need not be surprised that it did not take long before it began to stretch out its grasping fingers towards that celestial sibling. It was indeed relatively easy for it to do this because great mathematicians, who have always been prepared to employ their abilities also in the service of astronomy, have been somewhat ill at ease ever since the days of Newton. The law of gravity he discovered has enabled us to calculate the orbits of the planets very much more accurately and over much greater intervals of time, which has caused us to ask what is likely to become of those orbits and therefore of the system as a whole.

Newton's epoch-making work of genius made it possible to include in our calculations the manner in which the planets influence one another, causing what we call perturbations by means of their gravitational pull. And our appreciation of his achievement is all the greater when we take into account the fact that the calculations he has enabled us to carry out can only ever be approximations. Whereas it is possible to work out exactly what effect the gravitational forces of two bodies have on one another, when an additional one is included—the three-body problem—more equations become necessary than can be solved. (Only in certain exceptional cases can the three-body problem also be solved.) However, by means of numerical integration very good, and now even phenomenally good, approximations of the true movements of the celestial bodies can be calculated.

The mutual influences exercised by the planets on one another are also subject to periodic fluctuations. The orbits themselves also change over long periods of time, i.e. the eccentricities or the degrees of inclination to the ecliptic, for example, either increase or decrease. In an extreme situation, if these perturbations were to become too great, one planet might cross the orbit of another and at some point collide with it, or a planet might crash into the Sun or else depart from its cosmic home and set off into infinity. Over time, resonances could cause even relatively minor perturbations to mount up, as happens with a child's swing when a push at the right moment can greatly increase the momentum. As mentioned earlier, the ratio of the orbital periods and the resulting conjunctions when the mutual attraction is greatest are therefore especially significant. That is why, soon after the new modes of calculation became available, people began to worry about a ratio of Saturn and Jupiter with a resonance of very nearly 5:2, especially as Jupiter is the most massive begetter of disturbances in the community of the solar system. In addition, alarming orbital changes had been observed in the case of both those planets for many years.* Even an observation of Saturn by the Chaldeans and handed down from the year 228 BC pointed to worrying developments.

The French astronomer and mathematician Pierre Simon Laplace (1749–1827) solved this problem for his time. In 1784 he proved that the perturbations observed are caused by the small deviation of the orbital periods from a true ratio of 5:2 which is evened out during the course of a little over 900 years, so that those perturbations are merely the result of periodic fluctuations (IP, 227). Although this rhythm is close to the equalization period mentioned in Chapter 2 that occurs after just under 854 years (72 Jupiter and respectively 29 Saturn orbits), it should not be confused with it. Using modern astronomical data and electronic possibilities, the results of Laplace's calculations can be depicted as shown in Fig. 5.5.

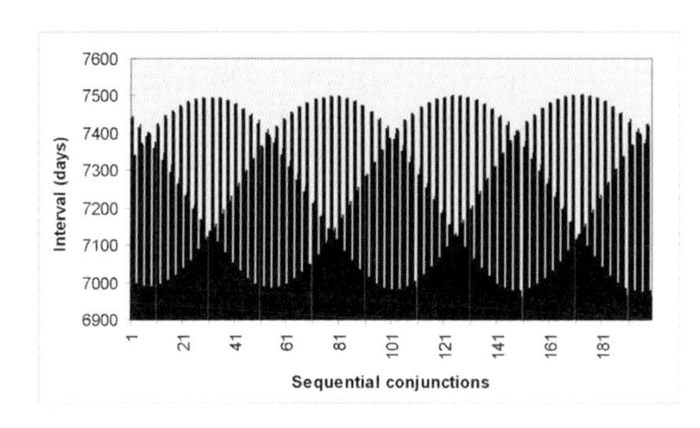


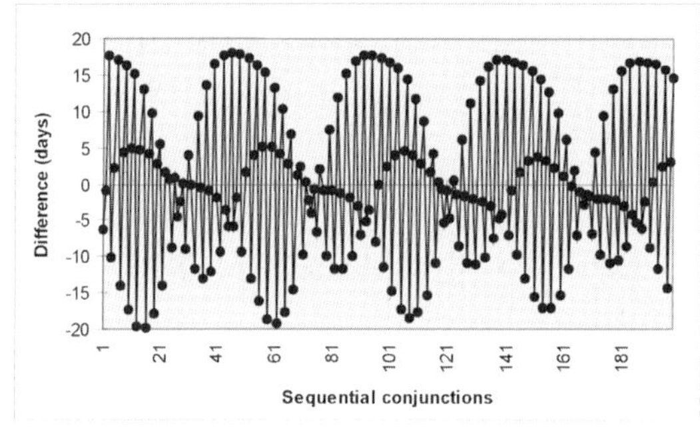


**Fig. 5.5** *Jupiter-Saturn rhythms. Top: interval between two conjunctions. Bottom: difference between perturbed and unperturbed orbit at the moments of conjunction;† 200 times beginning 23 June 2000, time c. 3972 years*

The top figure shows the beautiful threefold vibration of the intervals of the conjunctions where the amplitudes

* The following descriptions are based chiefly on Ivars Peterson, *Newton's Clock; Chaos in the Solar System*. Peterson's depiction is to date perhaps the most comprehensive account for the lay person of the subject suggested by his title. References here are shown thus: (IP + page number).

† Calculated as the difference between the temporal interval between two sequential actual conjunction moments derived from the calculation involving perturbations, and the interval between two conjunctions calculated on the basis of mean orbital data (without perturbations).

result mainly from the eccentricities of the planets. If one looks carefully one notices two very slightly displaced columns for each value. The conjunctions were calculated firstly according to the mean data, i.e. without taking account of the influences of the other planets, and secondly including those influences which accord with the actual values. The difference appears to be relatively small, but the lower graph shows that it is actually ± 20 days, although periodically it fluctuates around zero. So we see that (at least during the span of time shown) the influence of the perturbations does not increase. And perhaps we also gain an inkling of the beauty and complexity of the interplay of the planets which we all too easily describe as perturbations. The various rhythms that are superimposed upon one another are never repeated exactly. We can gain from all this a true impression of what Newton, Laplace, Carl Friedrich Gauss and many others have achieved in bringing all this variety into a mathematical form while also continually optimizing it right up to the degree of accuracy attainable with today's computers.

Laplace was furthermore able to prove that the deviations in the eccentricities and other parameters revealed by astronomical observations are caused by periodic fluctuations and do not exceed specific limits. By applying the simplifications he has elaborated (all calculations concerning the planetary system are and will always be simplifications because it will never be possible to work out, for example, the mass of Jupiter with 100% accuracy), it has been possible to calculate the stability of the whole system for several hundred thousand years, approximately the duration which in his day was thought to be the age of the Earth.[8]

We owe the next stride into a more distant future to Poincaré. Since the three-body problem cannot analytically be solved precisely, mathematicians were concerned to find out whether they might at least discover series solutions that could settle the question of stability once and for all. Mathematically this depends on how such infinite series* behave. In principle there are two possibilities. A series can strive to attain a fixed limit or it can grow eternally, which with complicated series takes place partly discontinuously. So if the movements of the planets could be grasped by means of a series of the first sort, this would be like a periodical fluctuation that will always remain within certain limits such as that calculated by Laplace for a specific span of time. However, the long-term

---

* A series is the sum of infinitely many numbers which follow a specified mathematical law, for example 1 + 1/2 + 1/3 + 1/4 . . . or 1 + 1/2 + 1/4 + 1/8 . . . Whereas the first of these diverges, i.e. it grows ever larger, the second strives to attain the limit of 2, i.e. it converges.

behaviour of series is irrelevant to a practical calculation of positions in a comprehensible span of time because the calculations can be interrupted after a few steps of the series used once the desired accuracy has been reached. The assumption was that the perturbation calculations for approximately similar initial conditions (e.g. the positions of several planets measured with a specific degree of precision at a given moment in time) would deliver approximately similar results.

Poincaré's great discovery was that this is not, or not always, the case. Instead there are equations in which only slightly differing initial values lead, after a certain number of steps in the calculation, to entirely varying results. In some cases they may nevertheless remain between specific limits, but in others they can be utterly irregular. (A detailed description of this complicated subject may be found in IP, 143ff.). It has meanwhile become possible to support Poincaré's theoretical considerations with more or less concrete figures. In a calculation of the orbit, a deviation of Earth's initial position of only 100 metres—which is a fantastically accurate measurement given its 150 million kilometre distance from the Sun—would show, latest after 100 million years, that absolutely no statement of its position would be possible any longer. (The borderline for exact calculability is considerably lower.) And an even more precise statement of position would in principle also not change anything because the differences in the calculation grow exponentially. This also does not automatically indicate that Earth has departed from its orbit; it could still be somewhere on its original orbit but it would not be possible to show at which point. Furthermore an accumulation of perturbations could quite possibly lead to a sudden and abrupt unforeseeable change in, for example, the eccentricity of a planet or asteroid, so that it departs from an orbit that has thus far been stable. In this case, too, the danger of this happening is caused above all by resonances in the orbital periods, although strangely enough not for small whole numbers in all cases. And vice versa, in the case of irrational number ratios, above all that of the golden section and those related to it, a long-term stability of the orbits is more likely.[9]

The mathematical chaos initially suspected by Poincaré (he did not give it this name, and neither of course were all the consequences immediately predictable) is thus something rather different from the picture conjured up by the everyday use of the word. It is very important here to keep this difference at the back of one's mind. One can say that in mathematics chaos includes the possibility of a disorganized, irregular state of affairs. But initially all it denotes is that there are some things which cannot be calculated because the equations necessary are too dependent on the initial situation which can only ever be determined to a certain degree, or because the arithmetical arts have come up against their principle limits.

The significance of Poincaré's work for the calculation of planetary movements sank into oblivion for a while, until in recent decades the continued improvement of electronic data processing began to make it possible to undertake ever longer excursions into the future of our solar system. But what did people hope to find on these difficult journeys that called for tremendous amounts of preliminary work? '*After centuries of searching for regularities, the time had come to seek explicit evidence not of order but of chaos in the heavens*' (IP, 169). And, as a rule, people do find what they are looking for. So let us now have a look at what the researchers have so far brought back from their expeditions.

The promising new beginning found its first justification in a plausible explanation for certain gaps in the asteroidal belt known as Kirkwood gaps after the man who discovered them. The 3:1 gap, i.e. the region in which the orbital period of Jupiter and that of a hypothetical asteroid relate to one another in this way, came about as follows. As predicted by the theory, the eccentricities of the planetoids which should once have inhabited this region change in a desultory fashion. This effect arises at the soonest after 100,000 years of continuous small gravitational impulses from Jupiter; or, expressed differently, one also says that Jupiter thus determines a chaotic zone. But the irregular jerks in eccentricity soon disappear again, i.e. the asteroid in question returns approximately to its former orbit. So Jupiter is only indirectly responsible for the gap; what did away with the former inhabitants was the way the change that had taken place sent them off for a time to cross the orbits of Mars or even Earth. After a long while, and there was surely sufficient time in the early days of the planetary system, all the asteroids from the chaotic zone had either collided with Mars or Earth, or perhaps even with the Moon if it already existed then. However, it is early days yet for research into the causes of the Kirkwood gaps. This model fails to reveal anything with reference to some other gaps, and the cause of the overabundance of asteroids at the 3:2 resonance is also still quite obscure (IP, 185).

But what is the situation with regard to the planets themselves, which are surely more interesting than a few asteroids lost aeons ago? Researchers have not held back from making 845-million-year journeys into the future to tackle this question. One will be tempted to ask how journeys of such length are possible, since we have said that at the latest when one reaches 100 million years no further reliable results can be achieved. However, during such long-term trips it is not a question of laboriously calculating every location of a planet at every moment, but rather of keeping an eye above all on changes in key orbital parameters such as eccentricity and semi-major axis (IP, 237). So the statements that can be made are initially qualitative. But once the researchers find themselves on the trail of chaos they can then follow this up with more specific calculations. (As this shows, the search both for harmonious order and for disintegrative tendencies and chaos calls in principle for the same tricks, although of course this does not suggest that the dimensions are comparable.) A second step then entails making calculations concerning, for example, two different Plutos. This means varying the starting conditions by investigating one Pluto which has its real position on a particular date as closely as this can be given, and a fictional twin which is allocated a minimally different starting point at the same starting time.

So let us begin with Pluto which from the first has looked like a good candidate for chaotic tendencies since its orbital inclination and also its eccentricities are the most extreme of any among all the planets. Although the first method did not reveal any '*obvious bouncing around*' even after 845 million years (IP, 243), the simulations with Pluto's twin did lead to the conclusion that it is not possible to indicate the position in its orbit which the actual Pluto will occupy after 100 million years. In the mathematical sense this means that we have chaos since unpredictability has occurred. In the case of the four large planets—this study only involved the outer solar system—the researchers found '*little that looked chaotic . . . the orbits stubbornly remained regular*'. In the meantime a number of further long-term simulations have been undertaken by various research groups. These have confirmed that in the outer region relatively few indications point to chaos, apart from Pluto as the exception. But the situation in the inner region is different. Calculations have also been undertaken for the solar system as a whole. More or less the same results were obtained, namely that latest after several tens of millions of years exact predictability becomes increasingly difficult and soon disappears altogether. From about 5 million years only qualitative statements can be made for the inner planetary system, i.e. the orbital parameters of the ellipses can be determined but not the positions of the planets.

Strange to say, however, nowhere have concrete indications come to light showing that a planet could change its orbit to an extent that would threaten the existence of the planetary system even if, as is pointed out, one cannot be entirely sure of this. The orbital parameters vary within certain boundaries whereby the variations in the inner planetary system are greater than those in the outer system. Mercury may constitute one minor exception. According to other publications one cannot exclude the possibility that its eccentricity (some day several hundred million years from now) could increase enough to cause it to cross the orbit of Venus.[10] The most far-reaching '*stable chaos*' (IP, 195) that has come to light despite the unpredictability existing in principle is quite a mystery. I shall return to it in a moment, after first investigating another chaotic movement discovered by means of the new tools at our disposal.

This concerns the rotational axes of the inner planets. It has been found that over the period researched (18 million years) these axes could have fluctuated by 60–100° owing to the long-term gravitational influences of the other planets. The fluctuations calculated occurred non-periodically but in large steps over a few million years, behaviour which has been interpreted as being chaotic and which also deserves this epithet even in the non-mathematical sense. Abrupt changes in the tilt of an axis would have disastrous consequences for the climate of the planet concerned. On Mars there are irregularities in the polar deposits which some researchers interpret as the concrete consequences of short-term movements of the axis. Possible fluctuations of up to 85° have been calculated for Earth. Although we know from geology that there have been ice ages and meteoric hits in the past, nothing has come to light in the way of events so revolutionary that they might have affected evolution itself or even prevented the development of higher forms of life altogether. If that had occurred, there would presumably be no one here to keep a lookout for fluctuations. And finally, inclusion of the Moon (omitted initially) in the calculations has shown that the torque it exercises stabilizes the axis of the Earth.[11] We can indeed be thankful that a mysterious event in the misty distances of the past has left us with such a satisfactory companion.

But let us now turn our attention again to looking forward in time. How can we explain why on the one hand the solar system behaves chaotically although on the other hand no indications have been found to show that its unpredictability is what is leading to behaviour such as is normally concomitant with the concept of 'chaos'? As far as mathematical chaos is concerned, it would after all be perfectly possible for the planets to depart from their accustomed places and cause mayhem. Certain complicated resonances are what have hitherto been discovered regarding the cause, or at any rate the partial cause, of the unpredictability, at least for the inner planets. These concern the periods of the precession of the apsides and the nodes (i.e. where the orbital planes cross the ecliptic) between Mars and Earth or respectively Mercury, Venus and Jupiter. There is also another idea, as yet only a supposition, namely that other subtle mutual effects could be responsible for the long-term stability, similar to the 3:2 ratio between the orbital periods of Pluto and Neptune which helps to prevent their paths from crossing even though Pluto sometimes comes closer to the Sun than its inner neighbour. There may be other similar as yet undiscovered resonances that are keeping the planetary system '*trapped in its present configuration*' (IP, 264)—a formulation that conjures up an image of the system longing to break out and escape into the arms of the original complete and utter chaos. The '*stubborn regularity*' of the orbits and what has thus far actually been discovered about its origins is perhaps best illustrated by the following description of a scene in a film. Jack Wisdom, one of the pioneers in the field of long-term calculation, has tried to make the results of his research more accessible by means of a film showing how the orbits develop, compressing calculations covering 60,000 years into a single second:

> *The rings representing the orbits of Jupiter, Saturn, Uranus, Neptune, and Pluto seemed to be in continuous motion, restlessly jiggling to jittery, complex rhythms. Though the rings never quite crashed together, they seemed to knock one another around. For instance, as the orbits of Pluto and Neptune bounced back and forth against each other, there were times when Pluto's orbit was entirely outside that of Neptune. At other times, the two orbits crossed. Uranus acted as if it were being erratically kicked around by its neighbors. Jupiter plainly had a strong influence on Saturn's path.*
>
> *A closeup of the inner planets showed a similar restlessness. Earth certainly wobbled and wandered in its orbit, but its mildly erratic motion paled in comparison with the wild gyrations and vibrations of Mars. It was also easy to see the rotation of Mercury's elliptical orbit, along with a host of more subtle changes. The entire solar system seemed to vibrate with a mesmerizing energy—an energy that freely sloshed back and forth among the various planets. What Johannes Kepler would have made of such startling complexity no one can guess.*
>
> *When Wisdom showed his videotape at a meeting early in 1993, he reflected, 'I think if you were given this movie and then asked to prove the stability of the solar system, you would right away say: Hummmm. I'm not so convinced the solar system* is *stable.'*
>
> *Curiously, Wisdom's movie showed no collisions. Despite their erratic vibrations, the rings stayed apart from one another. No one is really sure why.* (IP, 268)

Having already deduced the existence of a danger threatening stability, Newton saw the need for an occasional divine intervention to maintain the planets in their orbits. Laplace, on the other hand, when asked by Napoleon why he never mentioned the Creator in his monumental work about the universe (*Mécanique céleste*), replied: 'I have no need of that hypothesis.' Scientific results have thus far given no answers as to which of these two was closer to the truth.

Perhaps too much is made of the chaos among the planets, of the 'Celestial Disharmonies' (a chapter heading in Ivars Peterson's book) that people imagine they have to deduce from such chaos. Perhaps to call it the spice in the soup would be a better description. Johannes Kepler would have been more inclined to pay attention to the no doubt also perceptible and indeed possibly even heavenly beauty of the interplay. A rhythmic beat of 60,000 years might be inclined to lead to some exaggeratedly abrupt movements when one considers that the rhythms of the fluctuations of eccentricity are of similar dimensions. The almost visionary description given by Thomas Ring during the 1930s of what we are now shown by computer images surely makes the point equally well (see the end of Appendix 1.1). Then Kepler might have begun to try and understand what geometrical order and which harmonies lie hidden in this long-term interplay of the celestial bodies. We may assume them to be the ones that have to do with the stability of the planetary system, surely a subject for future research.

# 6. The Journey Continues: Rhythms and Eccentricities

In Chapter 2 we introduced three examples of star-figures and polygons which demonstrate geometrically the relationship between the orbital periods of two planets. These depictions showed the heliocentric view: the Sun occupied the centre of the Jupiter/Saturn triangle, the Jupiter/Uranus hexagon and the Venus/Earth pentagram. That type of diagram can only be used to show the moments of the conjunctions or other regularly recurring planetary positions. However, the path followed by a planet between two consecutive points joined by a straight line always lies on its elliptical orbit. So if we place a planet at the centre of the view it is possible to depict the spatial relationship between the two planets continuously. Here is the figure that arises in the interplay of Venus and Earth:

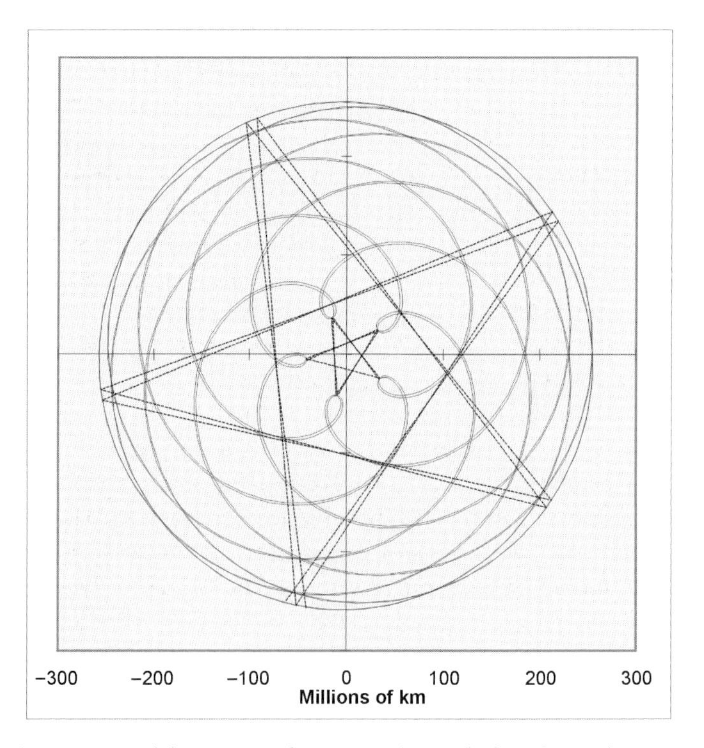


**Fig. 6.1** *Venus/Earth loops as seen from Venus (situated where the coordinates intersect). Period: 15.987 years (10 conjunctions). The pentagrams show the conjunction and opposition (outer) positions*

From the viewpoint of a planet, the loops arise as seen against the static sky of the fixed stars on account of the varying velocities. The inner planet moves round the Sun more quickly, so that for an observer on its surface the outer planet, as it comes closer, first appears to slow down; and then, as it is overtaken, i.e. slightly prior to the moment of conjunction, it changes direction before returning to its usual movement as the distance between the two increases. From the viewpoint of an outer planet, loops are also seen as the inner planet overtakes it.

The picture shows two cycles each lasting the time taken by 5 conjunctions. The loops and the pentagrams are situated very close together, so we shall also have to find out what the probability is for a resonance that is so precise. The pentagrams are not identical to those in Fig. 2.8, as is obvious from the scale. The pentagrams here are the ones that an attentive observer on Venus would be able to recreate in his imagination during just under 26 Venus orbits. The conjunctions and oppositions are temporally displaced and occur at intervals of almost 4 years at the same point on the horizon. The same picture arises as seen from Earth, except that it is rotated by 180°. Of course one cannot observe the extension of the figures on a plane, but the loop formations can be seen in a restricted area around the ecliptic. In order to explain this when they imagined the Earth to be stationary at the centre of the universe, Greek astronomers in the third century BC at first thought that the planets moved in small circles of which the centres orbited on larger geocentric circles. The small circles are called epicycles and the loops they thus form geometrically are epicycloids. The formation of epicyclic and planet-centred (centred on any planet) loops leads to similar figures. So the epicycle theory of antiquity regarding the movement of the planets is not an erroneous fantasy but the most logical explanation mathematically when one believes the Earth to be at the centre. As the model became increasingly refined the larger circles were organized eccentrically which in addition helped to explain a number of planetary movements that were not uniform. This made it possible for the epicycle idea to lead to relatively accurate calculations of the planetary orbits, so that when the heliocentric model of Copernicus came on the scene it yielded predictions which were no more accurate than those made possible by the old theory. All this is discussed in detail in the relevant literature, so there is no need for any more to be said here.*

* The geocentric figure shown in Fig. 6.1 is also occasionally seen, though relatively rarely. Joachim Schultz provides a beautiful compilation of all the loops visible from Earth, including those seen at the ecliptic, in his *Movement and Rhythms of the Stars*.

Once the geocentric view of the universe had been superseded, the planetary loops soon came to be regarded as meaningless or indeed ignored entirely by the science of astronomy. Yet as already mentioned, these figures are not a peculiarity of Earth alone, for all planetary configurations can be displayed in this way. The continuous lines shown in Fig. 6.1 are based on an arithmetical stepping interval of 6.5 days. This interval is small enough to enable a continuous rounded connecting line to be drawn to link the separate points. The same interval is now used to show the connecting line—here termed linkline (see Glossary)—arising in each case between Venus and Earth, whereby the Sun is once more placed at the centre. For greater clarity, drawing was terminated after the 500th line, i.e. shortly after completion of the first conjunction cycle.

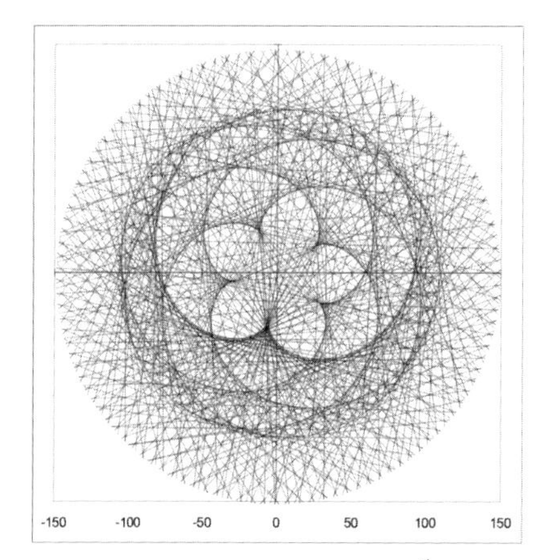


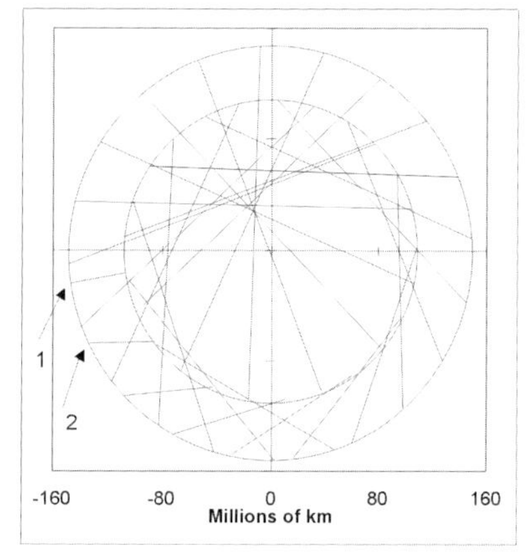


**Fig. 6.2** *Venus-Earth linklines,*[1] *500 times, stepping interval 6.5 days. The left-hand drawing uses a larger interval of 18 days to show how the figure develops (27 steps beginning with the conjunction position on 30 March 2001)* → Plate 1

Other stepping intervals (below a borderline which even the loop figure in Fig. 6.1 would not permit because the points would lie too far apart) would not show any difference except that the lines would lie somewhat closer together or farther apart. The geometrical figure here combines the loops and the inner pentagram of the conjunction positions that appear separately in Fig. 6.1 or, one might say, the flower and the star. It contains, in principle, all the possible configurations between the two planets and thus provides a total picture of the flow of energy in space or of the gravitational interplay.

The depiction of the linklines is now extended by adding the respective middle points to these lines:

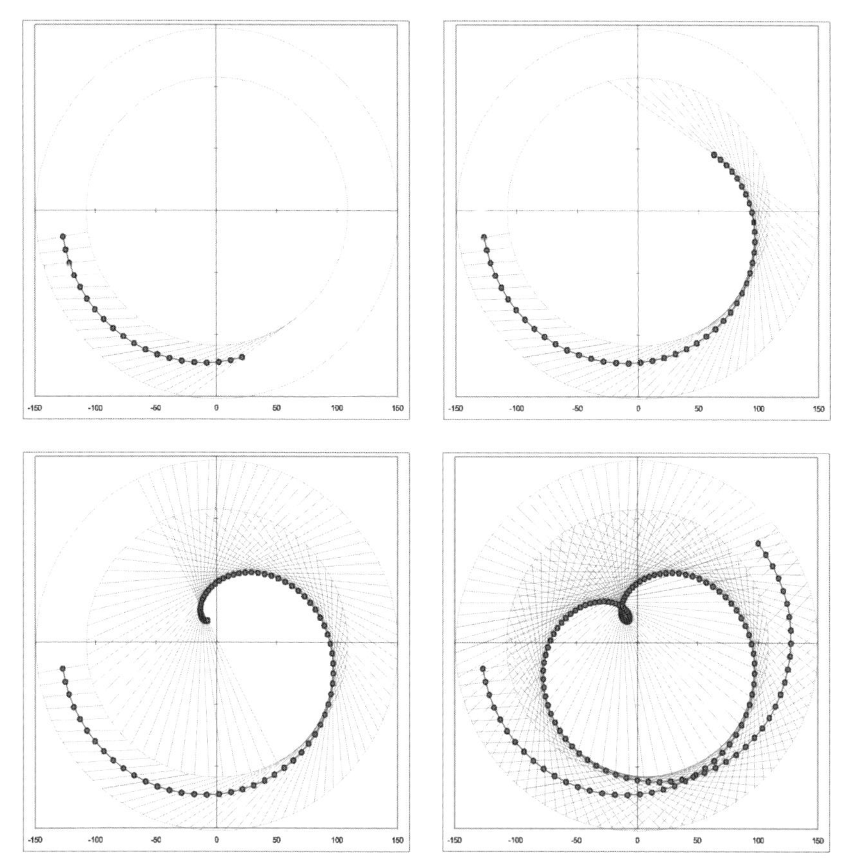


**Fig. 6.3** *Venus-Earth linklines with added middle points beginning with the conjunction on 30 March 2001, stepping interval 3.9 days. Top left, 20 times; top right, 45 times; bottom left, 76 times (opposition); bottom right, 151 times (next conjunction). Inside, Venus orbit; outside, Earth orbit. Scale in millions of kilometres*

At the moment of opposition—Earth having completed 0.8 and Venus 1.3 orbits, i.e. half an orbit more—the route followed by the middle points changes direction and makes a loop in front of the Sun situated in the centre (Fig. 6.3, bottom left). Venus at its closest does the same from the viewpoint of Earth. The cycle of 5 conjunctions is correspondingly rounded off to form the following picture, i.e. Fig. 6.4.

Comparison with Fig. 6.1 shows that the mid-points of the linklines draw exactly the same looped figure, only half the size, as that which arises in the geocentric (or, more generally, the planet-centred) view of the universe—and, I stress, now with reference to the Sun. Is this merely a game, perhaps quite a pretty one, that has nothing whatever to do with

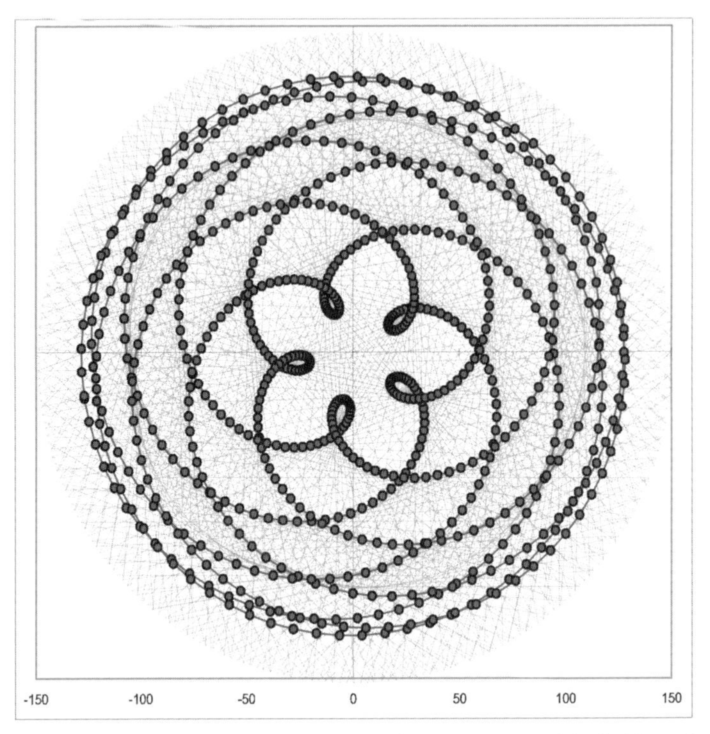


**Fig. 6.4** *Venus-Earth linklines and the route of the mid-points of the linklines during a cycle of 5 conjunctions. Stepping interval 3.9 days, 750 times, 2925 days*

astronomy? In trying to find an answer to this question, let us look at the centres of gravity between the two planets instead of the mid-points of the linklines. We should also note that the mass ratio of Venus to Earth is 0.815, i.e. although Venus is somewhat less substantial the two masses are not so very different (see Fig. 6.5).

The difference between the paths of the mid-points and of the centres of gravity is minimal (except, perhaps, that the inner part brings out an even clearer flower shape). So in principle these seemingly long-outdated geocentric loop figures turn out to be nothing other than the path followed by the centre of gravity of two planets round the Sun. The common centre of gravity of Earth, Venus and Sun also yields the same picture, except that it is 200,000 times smaller because the mass of the central body is correspondingly larger. If our solar system contained only the two planets Venus and Earth, the solar mid-point would move in exactly this pattern around the centre of gravity. (Chapter 9 will deal in more detail with the actual movement of the Sun around the centre of gravity of the overall system.)

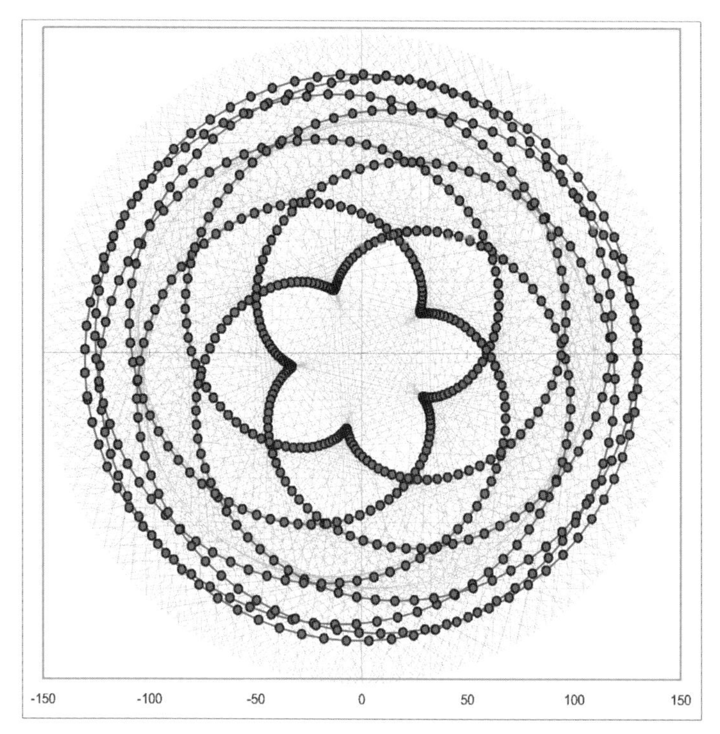


**Fig. 6.5** *Venus-Earth linklines and the route of the centres of gravity*

**Fig. 6.6** *Star-flower 1*

What comes to expression here (speaking mathematically and astronomically, the 13:8 resonance of the orbital periods of Earth and Venus) appears to be one of the stabilizing factors of our solar system. This should certainly be correct regarding the relationship between the planet of humanity and the celestial queen because we can assume that the figure shown has continued to be drawn in the firmament for a very, very long time. The eccentricities have relatively little influence on the creation of this type of figure; in the case of larger eccentricities, e.g. from Mars outwards, the figure would simply be distorted to a greater or lesser degree. We shall be getting to know such effects in other illustrations later. So the star-flower would not fade on account of the fluctuations possible even over the course of millions of years since, as we said in Chapter 5, these would remain within specific bounds which, for Earth and Venus, lie somewhat below the present eccentricity of Mars.

Let us now examine a ratio in the outer planetary system that appears to relate in a special way to that of the Venus/Earth constellation, namely that of Jupiter and Neptune. Their synodic period is 4668.694 days, which is 7.995 times that of the two inner planets. After the figures of triangle, hexagon and pentagram mentioned in Chapter 2, we have here, in addition, the geometric realization of the next 'small whole number' to be found in conjunctions of two planets.

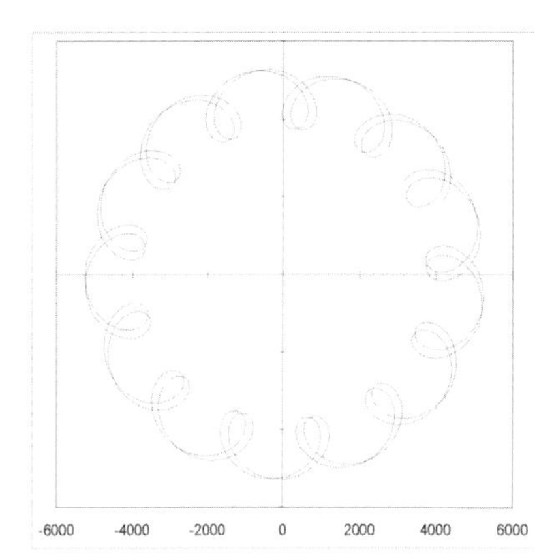


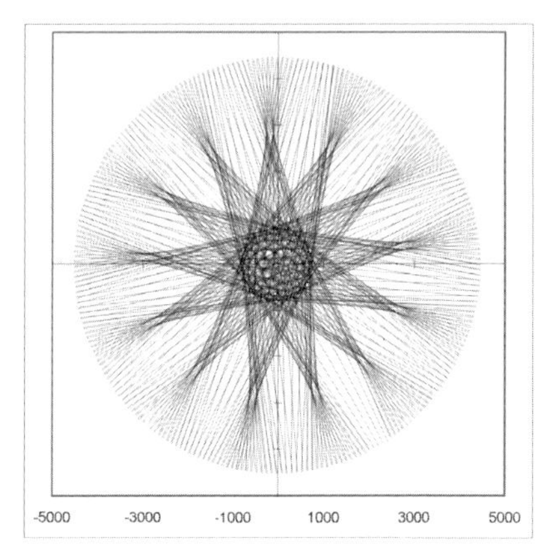


**Fig. 6.7** *Jupiter/Neptune, stepping interval 240 days, 500 times, period c. 328.5 years. Left: Jupiter in the centre. Right: linklines, heliocentric. Scale in millions of kilometres*

The stepping interval here is 240 days. This fits the very much slower movement of those two distant planets. The difference in the distances arises because the maximal distance on the left is derived from the distances of both planets from the Sun added together, whereas on the right the distance of Neptune from the Sun sets the upper limit. The two lines drawn in the left-hand depiction are relatively close together so that it might still be just possible to talk of a resonance (since there is no defined limit for this). The value of the cycle-resonance (see Appendix 3.5 and Glossary) is 3.06°. This means that every time the two planets once again reach almost the same spot after completing one cycle of 13 conjunctions (average 166.17 years) they are further on in their orbits by this amount from their previous meeting. However, the star-figure of the linklines gives a much clearer statement and is much more impressive since it represents the overall image of the interplay between these two planets.

In the following we show a small selection of further loop formations or, respectively, linkline-configurations. First we have the example we already know: the Jupiter/Saturn triangle which is not all that resonant but which nevertheless produces a very nice threefold loop figure. Then comes a somewhat diffuse blossom of 13 petals which is brought into flower after a time by the linklines of Saturn and Uranus. The cycle-resonance of its 20:7 or 13-fold ratio (20 Saturn years are approximately 13 synodic periods or 7 Uranus years) is 6.6°. The next best approach arises through the numbers 77-50-27 with a value of −2.32°. Then we have the 14-fold configuration of Mercury and Venus and finally a very clearly-formed 23-point star-figure formed by Saturn and Neptune. A summary of all relationships between planets that are not too far apart is included in Appendix 6.2.

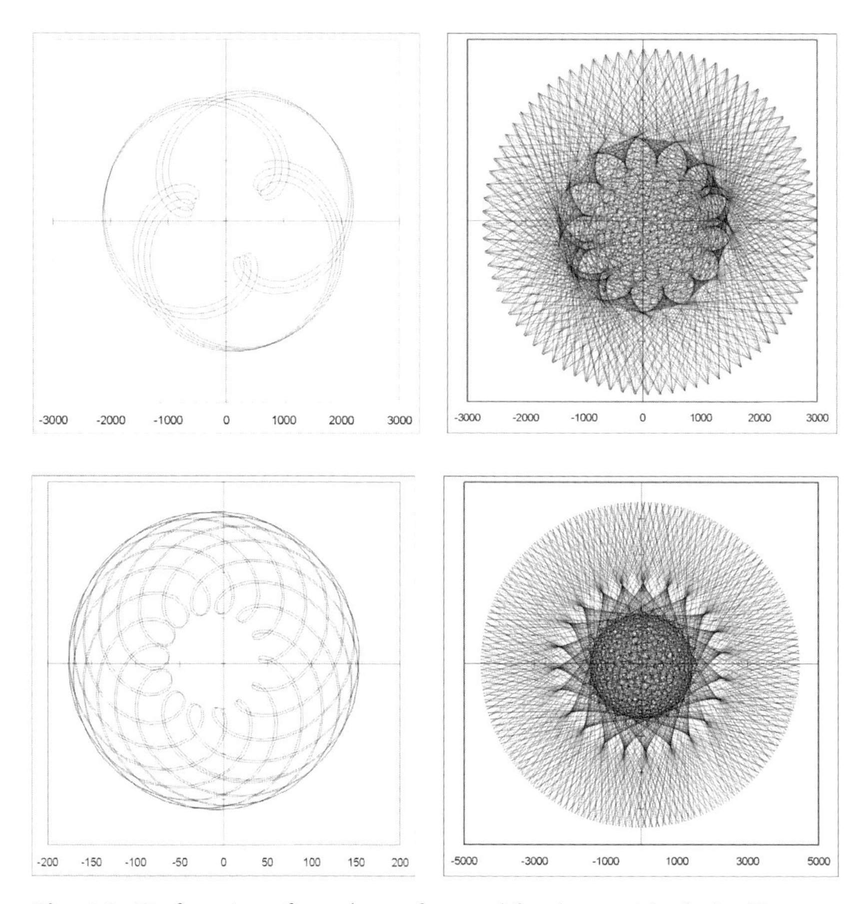


**Fig. 6.8** *Configurations of two planets, from top left to bottom right: Jupiter/Saturn loops, stepping interval 120 days, period* c. *197 years; Saturn-Uranus linklines, stepping interval 365.25636 days, period* c. *750 years; Mercury/Venus loops, stepping interval 4 days, period* c. *11.2 years; Saturn-Neptune linklines, stepping interval 400 days, period* c. *821 years. Scale in each case in millions of kilometres*

(When they are too far apart dozens or hundreds of loops are formed depending on the ratio of the orbital periods.)

We chose the sidereal Earth year as the stepping interval for the Saturn/Uranus figure. The formation of the inner 13-petalled flower is independent of any exact size of stepping interval—as we saw above in the example of the Venus/Earth relationship, which also holds good for the other relevant illustrations. The outer star, however, only shows when this interval is applied. It is an image of the almost precise 84:1 ratio of the orbital periods of Uranus and Earth.

We can now take a further look at the Jupiter/Uranus hexagon also introduced in Chapter 2, or rather at what becomes of it when the linklines are also given an opportunity to play their part:

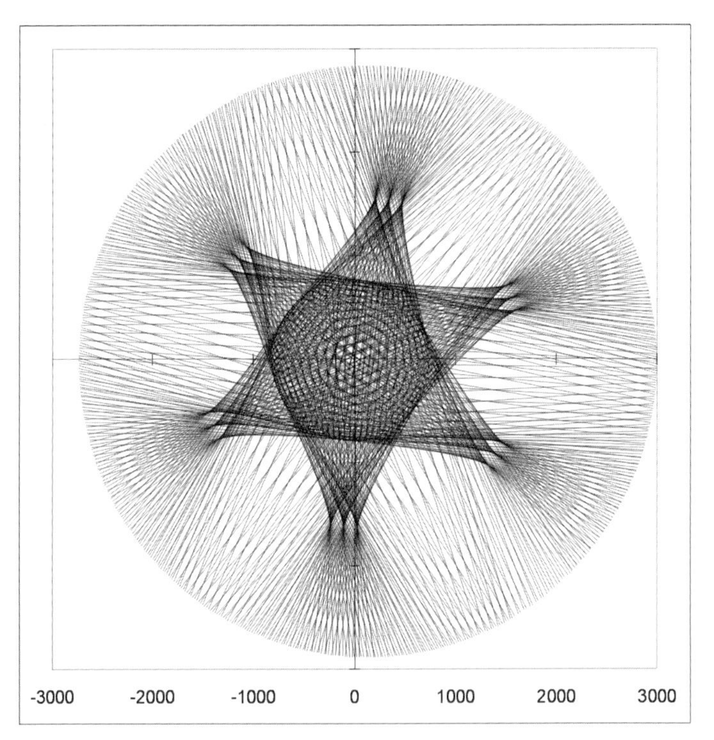


**Fig. 6.9** *Jupiter-Uranus linklines, stepping interval 121.562 days, period c. 248.61 (3*82.87) years* → Plate 2

This figure, which in its entirety turns slowly around the Sun, is probably rather a surprise. Because of the number of lines that make up the whole picture, the stepping interval was set to depict just three sequential hexagrams, whereby in this example, too, the formation as such is independent of the exact stepping interval. The period of 82.87 years needed for one hexagram corresponds to six Jupiter/Uranus conjunctions. The formation of such a hexagram is indeed astonishing since rather than being drawn by a continuous line this figure is composed of two equilateral triangles. Neither the continuous planetary movements nor the depictions of the constellations at specific intervals of time, such as the conjunctions, can lead to the formation of a hexagram.* So the overall geometrical picture of the interplay between Jupiter and Uranus reveals something that would be very difficult, if not impossible, to detect solely by means of an arithmetical consideration of the resonances.

It is also worth noting that after the pentagram in the inner region, which frequently used to be associated with the human being, we here have another sign, now in the outer planetary system, which was formerly given a mystical interpretation expressing, for example, the interpenetration of two polar principles. These two star-figures have probably been the most frequently used to symbolize the order in the cosmos (hexagram) and the position of the human being within it. In the planetary system these are the only resonant (or should we say clearly recognizable and also physically provable) images of a relationship between two planets up to the formation of the 12-pointed star, where a limit is reached even from the purely geometrical point of view.†

We shall now add to the linklines their middle points and their centres of gravity. We saw in Figs 6.4 and 6.5 that a good correspondence arises between the mid-points and the centres of gravity of the linklines when the masses of the two planets are more or less equal. So now let us investigate what arises when the difference in the masses is more pronounced, as shown in Fig. 6.10.

Here, too, the six loops of the mid-points show an exact although smaller image of the planet-centred movement figure. In the case of the centres of gravity, Jupiter's mass (22 times that of Uranus) causes the lines of the mid-points to be drawn a long way inwards. Because of this the path of the second cycle appears to coincide almost with that of the first. However, after six conjunctions these, too, turn by almost 5°, just as do

---

* In some instances the literature speaks of a hexagram which is supposed to result from the positions of Mercury and Earth (or of Mercury and Sun seen from Earth) at conjunctions and oppositions. In this case the oppositions are displaced by approximately 180° from the conjunctions, which leads to the formation of the two triangles of the hexagram. But the triangles are so imprecise (see Appendix 3.5, Fig. 14.22) that I do not want to investigate this further here.

† This limit will be discussed shortly.

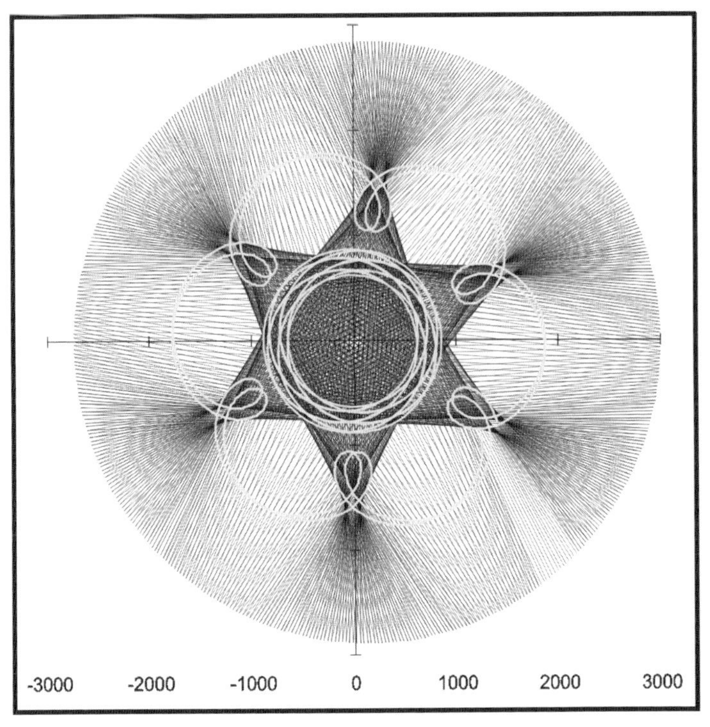


**Fig. 6.10** *Jupiter-Uranus linklines and the path of the mid-points and centres of gravity (inner) through two conjunction cycles, period 165.74 years*

the hexagram and the loop figures (see Fig. 2.6). The sixfold aspect is retained in the same way.

The formation of these two figures, the pentagram and the hexagram, which, as mentioned, is barely influenced at all by the eccentricities, can perhaps be regarded as being a stable fixture in the solar system, a kind of spinal column consisting of small whole numbers. Perhaps these two conjunction periods, which are geometrically so conspicuous, are tuned to one another in some mysterious way. One hint of this being the case could be the fact that $25 = 5^2$ Jupiter/Uranus encounters correspond almost precisely to $216 = 6^3$ Venus/Earth synodic periods (the deviation being only 0.005% or 6.7 days in 345.3 years). And continuing with this reckoning, this means that with the same percentage difference over a period that is 30 times longer, $5^3$ hexagrams and $6^4$ pentagrams are formed. However, I don't wish to read too much into this numerical correspondence as that might make these numbers appear almost mystical. We shall be returning in later chapters to the relationship between these four planets which is astonishing in other ways too.

Let us now turn to the eccentricities which—even though they are barely noticeable in the above images of planetary movements—most certainly play an important part in the interaction of the celestial bodies. This showed in the harmonies of the velocities; and it was also made clear in the discussion of future developments in Chapter 5. Figure 5.5 gave an example of how the rhythms of the conjunctions are moulded by the interplay of the eccentricities and orbital periods of the two planets involved. Here are two more diagrams in connection with this:

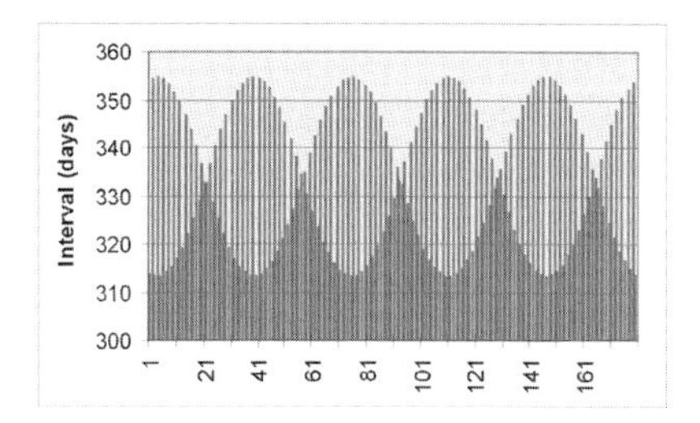


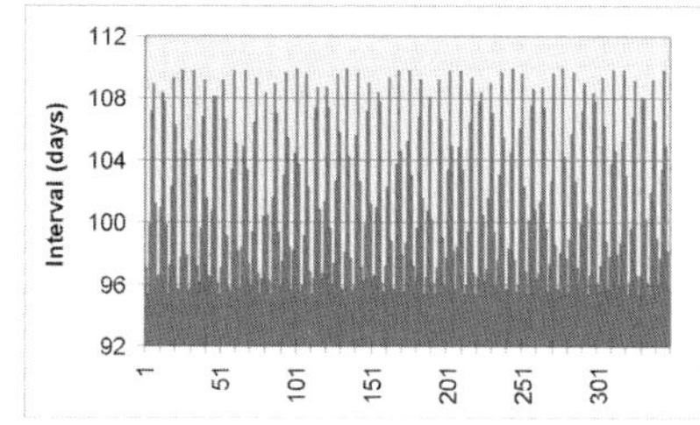


**Fig. 6.11** *Intervals between sequential Venus/Mars conjunctions (left, 200 times) and Mercury/Mars (350 times)*

An oscillation of the intervals around a mean value (i.e. the relevant synodic periods of 333.92 or respectively 100.89 days) in the ratio of small whole numbers becomes very obvious even though the resonances calculated from the orbital periods (see Table 6.2 in the Appendix) are only approximately close to the number two (35:17; Mars: Ve/Ma) or respectively to the number seven (34:5; Mars: Me/Ma). The same rhythms appear when parameters other than the intervals of time are applied, i.e. the distance of the two planets from the Sun, the lengths of the linklines, or the velocities at the relevant points in time. The two most prominent ratios in our planetary system are shown in Fig. 6.12 using a different mode of depiction.

The periods of time here occur in one complete turn of the pentagram or hexagram in the ecliptic. The former was already mentioned in Chapter 2 as being approximately 1195 years; the hexagram turns once on itself in about 6061 years, i.e. about five times as long.* The slight irregularities in the diagram are the result of the perturbations. In the description of the film towards the end of Chapter 5 (p. 98), Uranus behaves on occasion as though it were being 'erratically kicked around',

---

* The cycle-resonance is 4.922°. The turn is complete after an average of 360°/4.922° = 73.14 hexagrams, i.e. 438.85 conjunctions. The synodic period of Jupiter and Uranus is 5044.81 days = 13.812 years. Hence the period of *c.* 6061 years arising from 438.85*13.812.

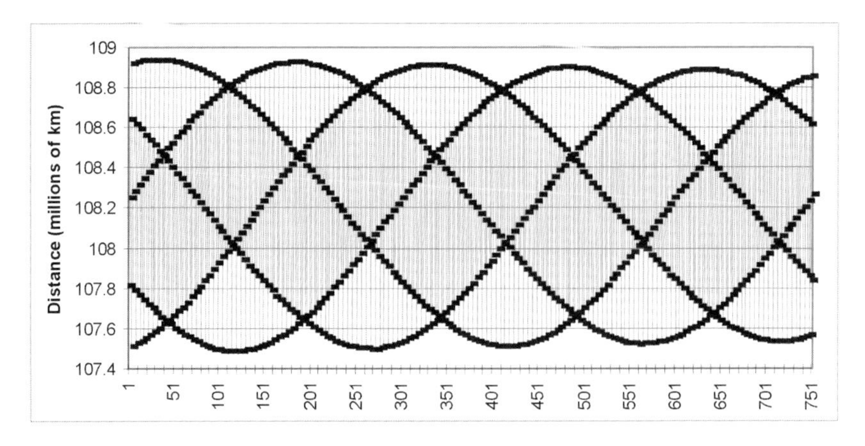


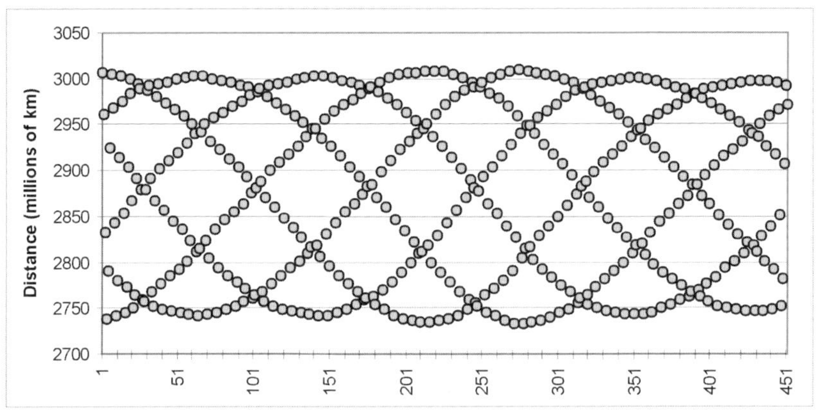


**Fig. 6.12** *Distances from the Sun: top, Venus at Venus/Earth conjunctions (750 times); bottom, Uranus at Jupiter/Uranus conjunctions (450 times)*

but in the timescale shown here this amounts to it enduring no more than gentle nudges. The Venus/Earth oscillation, on the other hand, is of a unique and almost entirely undisturbed elegance.

The wealth of oscillations in the solar system is, however, not restricted to the conditions pertaining to the orbital periods or, respectively, the conjunctions. In another way it finds expression in the mutual gravitational effects. The minimal deviations of the elliptical orbits are the cause of the long-term dynamic that exists in the planetary system. So the next aspect for us to observe with the help of some more detailed depictions is the way in which the planets exercise mutual influences on one another. In order to portray this we shall add the difference in the positions of the planets that arises from calculations either with or without the perturbations at the same points in time:

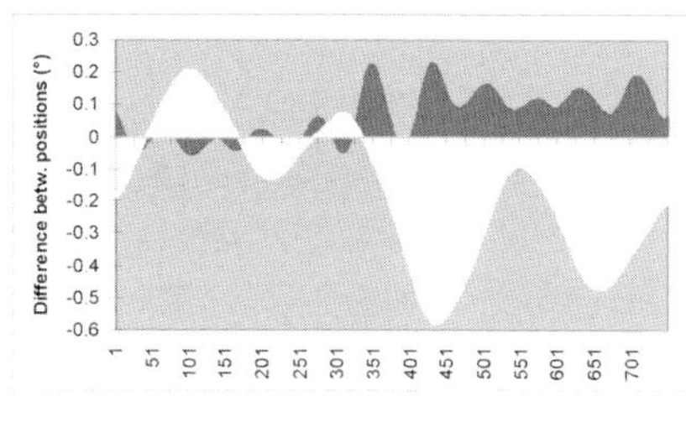


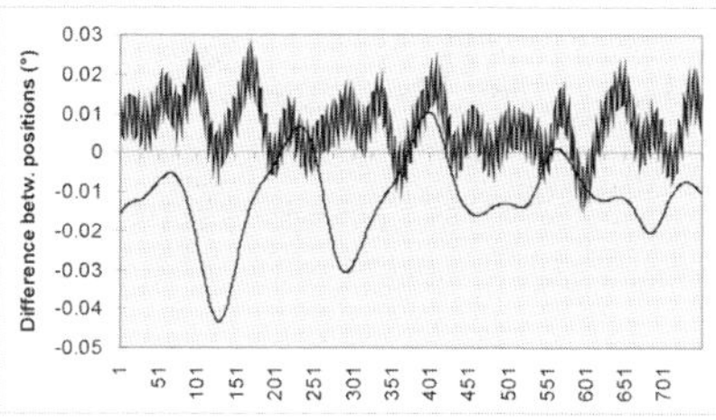


**Fig. 6.13** *The influence of mutual gravitational effects. Top: Jupiter (black) and Saturn (white), 750 times, corresponding to c. 9 Jupiter and 3.5 Saturn orbits, stepping interval 50 days. Bottom: Earth (upper) and Mars (lower), 750 times, corresponding to c. 10 Earth and 5.5 Mars orbits, stepping interval 5 days. The deviations in respect of Earth are shown magnified three times. Beginning in each case: 1 January 2000*

In every case, these wild fluctuations include the influences of all the others on one specific planet. In the case of Earth, the vibrations superimposed on its curve are brought about by our all-important companion. Compared to the overall movements of the planets the deviations are very small. Even in the case of the gas giants—in the space of the period investigated, namely several thousand years—they only amount to a little over 1° (for Saturn *c.* 1.25°, for Uranus 1.05°, for Neptune 0.7°, and for Jupiter about 0.5°). In the inner region the maximum differences of *c.* 0.04° for Venus and Mars are distinctly smaller, while for Mercury they rise to *c.* 0.08°. In the case of Earth the deviations are at most merely 0.01° so that, strangely enough, in this respect at least, we have here the greatest degree of stability. Nevertheless, it is these very minimal perturbations that bring about the long-term changes in the eccentricities, in the precession of the apsides and nodes, etc. of the planetary orbits. In view of the periods of time involved this once more shows the truth of the saying that small causes can lead to great effects.

The impression easily gained is that one cannot take it for granted that the solar system remains stable for millennia or indeed for millions of years, so that since Isaac Newton the elite among astronomers have been devoting their greatest attention to this. The manner in which the fluctuations of eccentricities and orbital inclinations run their course for untold millions of years reminds us strongly of the oscillations of Earth's position shown above.[2] The equalization that has always been calculated for all timescales can, though, only be shown here for a few thousand years.

Here once again we see how our planetary system is interwoven with many and varied rhythms superimposed on one another which are then further bound up in yet other, larger oscillations. The mutual per-

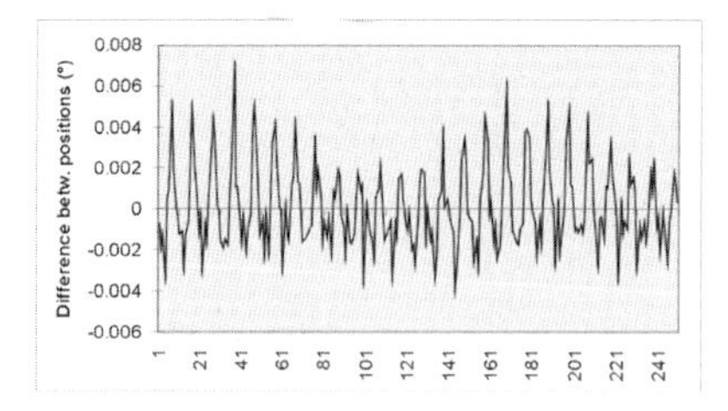


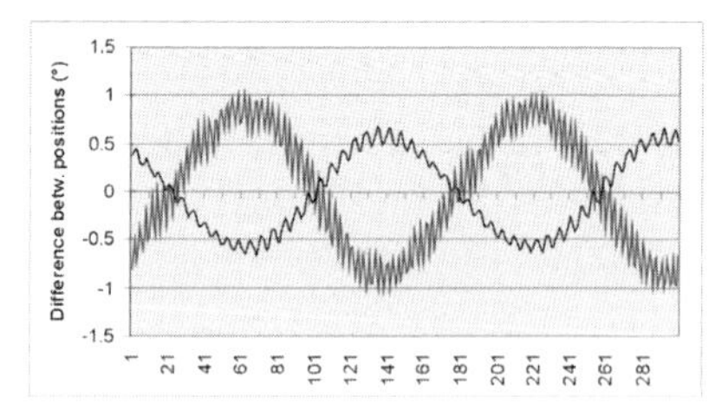


**Fig. 6.14** *Equalization of mutual gravitational effects, beginning 1 January, Year 0. Left: Mercury, 250 times, stepping interval 6000 days, period* c. *4100 years. Right: Uranus (larger curves) and Neptune, 300 times, stepping interval 10,000 days, period* c. *8200 years*

iod of Uranus and Neptune of just under 4300 years coincides with the resonance of their orbital periods of 51:26 (51 Uranus orbits take *c.* 4285 years and 26 Neptune orbits take 4284 years) or at least comes very close to this. However, the oscillation of perhaps 2100 years seemingly superimposed in the case of Mercury is not real. It is the product of the stepping interval chosen; in the case of other intervals of time the periods would differ. But a long-term rhythm which seems to be making its appearance may be showing up in the very slight decline in the overall tendency. Whatever the case may be, as far as the perturbations are concerned, the circumstances in the inner planetary system are more modest but very much more complicated, so that one can understand why chaos found a wider target here.

All the examples introduced here on the theme of gravitational interaction may, when viewed as a whole, serve to give us an idea of what some scientists suspect, namely that in the interaction of the planets there may still be some as yet undiscovered resonances that could be playing a part in creating the mysterious stability of the system as a whole. So let us now turn our attention to this matter. To do so we shall depart from the realm of perturbations which only become interesting in geological periods of time and turn our attention to the effects of the eccentricities in more manageable dimensions.

The eccentricity of a planet is usually shown as the numerical eccentricity $\varepsilon$ (epsilon). The linear eccentricity e, the distance in space, then arises as $e = \varepsilon \star a$ (a: semi-major axis). Let us remind ourselves of the relationships for an ellipse:

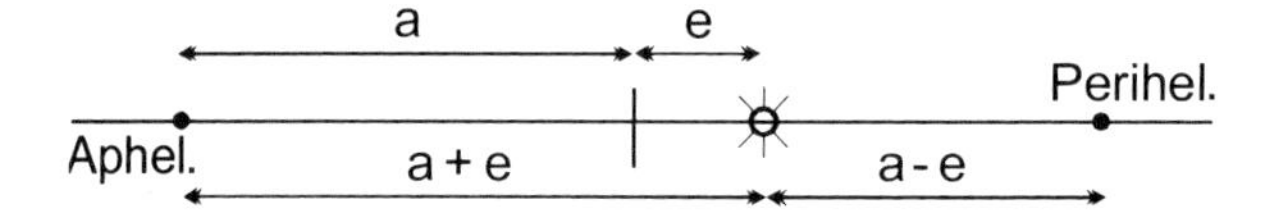

From this we can easily deduce that the ratio of the aphelion distance (a + e) to the size of the semi-major axis is 1 + ε.* By using this version it is easy to depict the order of the eccentricities very clearly:

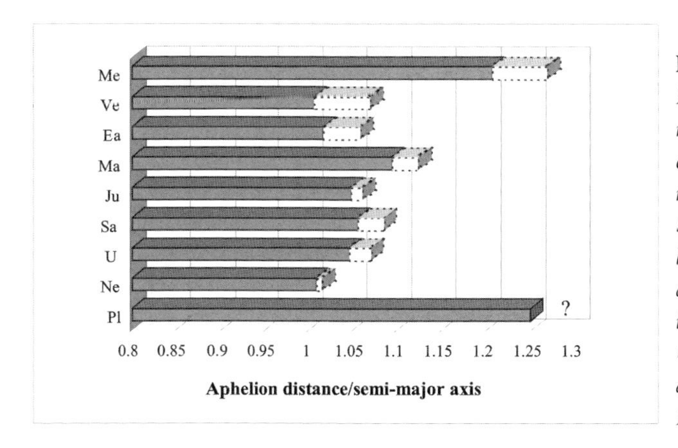


**Fig. 6.15**
*Eccentricities as the ratio of aphelion distance to semi-major axis. Shaded: the upper limits calculated in a numerical integration covering 6 million years, according to Jacques Laskar et al.*[3]

This depiction shows us clearly that—in relation to present values and compared with musical intervals—we have almost three semitones (Jupiter, Saturn, Uranus), one whole note (Mars), a minor third (Mercury) and a major third (Pluto). Perhaps this is also the hidden reason why the harmonic ratios in the planetary system are anchored in the aphelion velocities and not in those of the perihelion.

One must take into account, however, that the eccentricities change over the course of very long periods of time. This does not occur arbitrarily but in a way that preserves the characteristics of the present order. Every planet has upper and lower limits, and as regards the former there exists in the literature a numerical integration covering 6 million years, except for Pluto (see Fig. 6.15). So in the current thousands of years what we are able to observe in the heavens is the middle section of the range. We have also found a reference to Mercury covering a period of 400 million years showing the numerical eccentricity fluctuating between 0.1 and 0.3. At present it is 0.206.[4] We have heard that no one really knows the reason why the planets remain within their ranges. And why each one of them has one specific eccentricity or another is also unknown. In a similar way an individual has characteristics at birth and others that come to light later in life. Whereas some of these can be traced back to heredity and environmental influences, they still do not explain the person as a

---

* $e = \varepsilon \star a \Rightarrow a + e = a + \varepsilon \star a = a \star (1 + \varepsilon) \Rightarrow (a + e) / a = a \star (1 + \varepsilon) / a = 1 + \varepsilon$

whole. Ever since the death of the man who discovered that the eccentricities are caused basically by the elliptical orbits, astronomy has no longer asked *why* these eccentricities are as they are. Perhaps we shall discover that this has been an erroneous attitude for more reasons than solely that of musical harmony.

Let us look first at Mercury. In the inner realm, which is more threatened by instability, this planet has the greatest eccentricity. It is also the one which might in the distant future more easily extend its orbit beyond its limitations and cross that of Venus. Perhaps it is for this reason the planet most in need of protection, like an over-excitable child needing gentle restraint by unseen fetters. The conjunction period of Mercury and Venus is 144.566 days, but this is an average. The actual intervals vary between approximately 126 and 158 days. In Fig. 6.8 (bottom left) we saw the two planets forming 14 loops caused by the 23:9 resonance of their orbital periods. In the inner part of the loop figure it would be possible to add the star that also arises from the heliocentric view when one adds the planetary positions at sequential conjunctions and joins these together. In the figure below this is shown firstly for the positions that would arise using the mean value and secondly for the actual positions (here of Mercury) when all other conditions are the same.

Apart from the eccentricity of Mercury shown in two ways on the right, the fluctuations around the mean value strangely enough lead to the star being reversed by almost 180°. Although this may not be too important physically, it does reveal the existence of effects which would

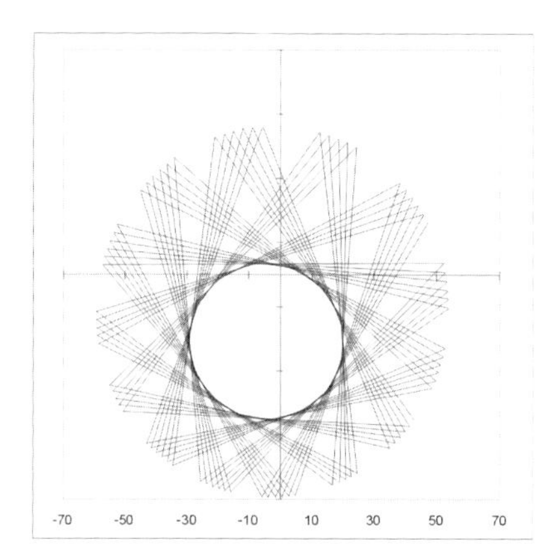


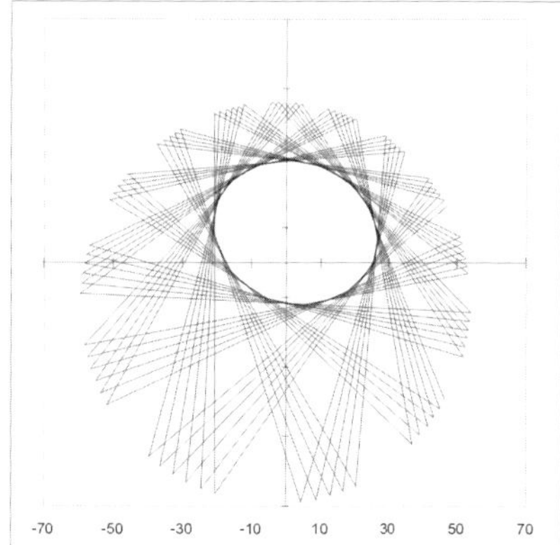


**Fig. 6.16** *Mercury in its orbit at Mercury/Venus conjunctions, 100 times, period c. 39.6 years, beginning 3 May 2000. Left, positions in the rhythm of the mean value of 144.566 days; right, actual conjunction positions*

not show up in an examination of the orbital periods and their resonances alone. As with other planets, Mercury is affected not only by the gravitational influences of its closest neighbour but, in principle, albeit in varying degrees, also by all the other planets. When two of these, let us say Venus and Mars, are in conjunction, the joint effect they exercise is at its strongest, pulling at that moment, in relation to the Sun, in the same direction. So it is possible not only for direct resonances between two planets but also perhaps for resonances between three or more of them to play a part in holding the whole system together. As already mentioned in Chapter 5, the long-term unpredictability of movements in the inner solar system is thought to be due partly to a complicated resonance between Mercury, Venus and Jupiter. Let us now have a look at the figure arising for Mercury when we plot its positions at Venus/Mars conjunctions. Once again the result is a geometrical overall picture, this time of the effects on Mercury caused by the two planets. We show this again for the mean value of 333.922 days and for the actual positions:

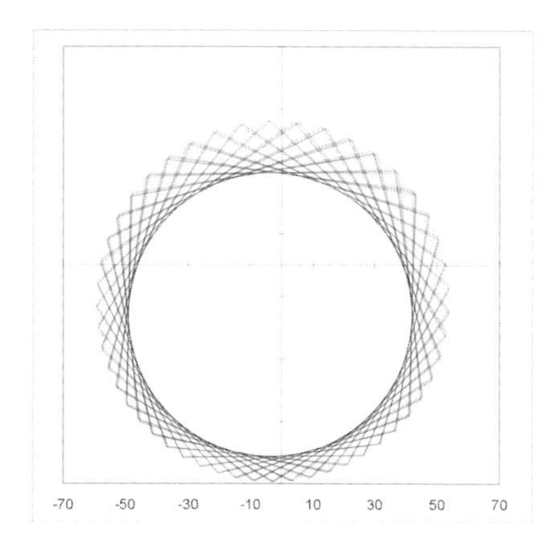

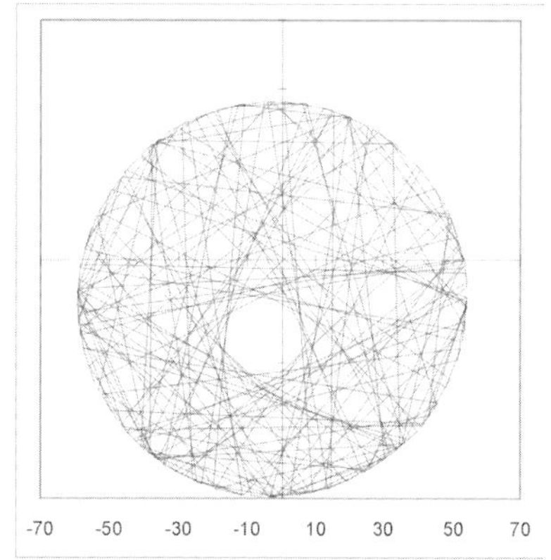


**Fig. 6.17** *Mercury at Venus/Mars conjunctions, 200 times, period c. 182.8 years, beginning 19 June 2000. Left, positions in the rhythm of the mean value of 333.922 days; right, actual conjunction positions*

Mercury's orbital period is in a very precise ratio of 49:186 to the conjunction period of Venus and Mars, expressed by a 49-pointed star-figure. However, this only exists on paper. The figure that arises in reality is certainly not resonant; indeed, the sequence of lines would be better described as chaotic. We are not, however, looking for chaos here, but for a hidden order which it can sometimes take patience to find. So let us send the messenger of the gods off on a few more trips under the joint influence of the goddess of love and her martial colleague.

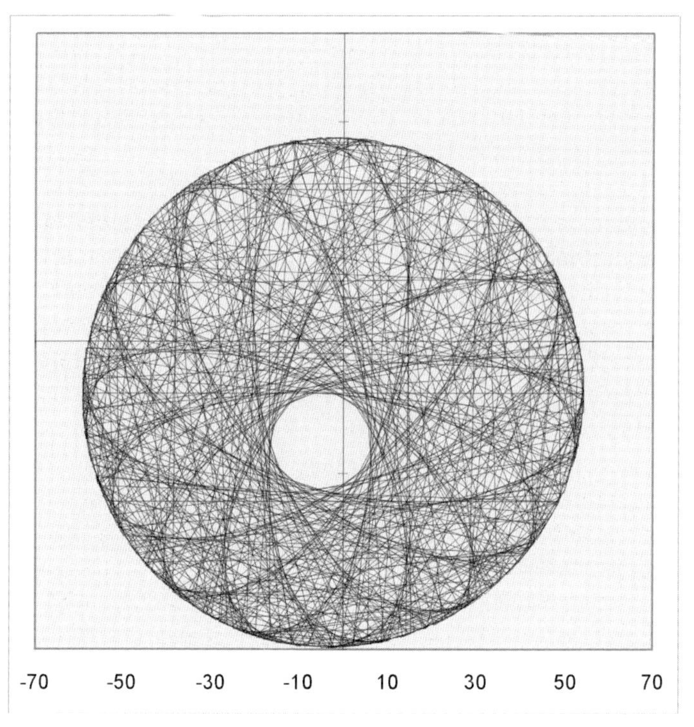


**Fig. 6.18** *Mercury at Venus/Mars conjunctions, 650 times, period* c. *594.2 years, beginning 19 June 2000*

What we find is that after about 600 years the calming twofold rhythm of Venus and Mars introduced in Fig. 6.11 (left) made it possible for a flower with 13 petals to emerge out of the tangle within Mercury's orbit. It would have been almost impossible to foresee this. And this flower is by no means short-lived, for another calculation beginning, for example, 4500 years later brings it to light again, though it has rotated a little meanwhile. Perhaps this flower is one of those invisible bonds that play a part in keeping the eccentric Mercury true to its task.

We can now turn to look at Venus itself. What is the number that might accord with the planet which humans have so often linked to beauty and love? And where should we seek it? Please believe me when I say that clear and expressive geometric figures which reveal themselves in this way are quite a rarity. Although it is not unusual for star-figures to appear in this type of depiction (showing the positions of a planet at the conjunction of two others), the lines mostly shift relatively quickly and thus render the figure they have formed unrecognizable. But before we look any further I must first interpolate something else.

The moment of a conjunction, i.e. of an encounter of two planets at the same spot on the ecliptic, is as a rule only approximately the moment when they are closest and thus exercise the greatest influence on one another or, respectively, when the greatest combined force field arises. Resulting from the eccentricities of the orbits and—to a lesser extent—from the varying small inclinations to the ecliptic, there can be a difference of several days between the date of least distance between them and that of the conjunction. (Fig. 14.20 in Appendix 3.4 shows an example of this.) The differences between the points of conjunction and those of minimum distance vary periodically around a zero value at which both coincide exactly. Therefore the mean value for the temporal intervals of sequential smallest spatial distances corresponds to that of the conjunction period. In most cases there is no or almost no difference between the figures shown for 'normal' conjunctions (also termed conjunctions 'in ecliptic longitude') and for minimal distances (or distance conjunctions) of the same constellation because the rhythm of the temporal intervals remains the same, except that the fluctuations are somewhat greater in the former.* In the previous example, in cases of distance conjunctions the 13-petalled Mercury flower would merely have somewhat broader petals than those in Fig. 6.18. Because other differences can for the most part be ignored geometrically, in what follows I shall mostly work with conjunction data in ecliptic longitude.

There are, however, a few cases that lead to obvious differences. Take, for instance, the Venus movement at Mars/Jupiter conjunctions. These occur every 816.435 days on average, with the fluctuation range of approximately 37 or 20 days respectively (with minimum distances). The difference between the two types of conjunction at the same moment in time can be up to 8.5 days during which Venus moves on by an average of 13.6°. This shift of position could of course noticeably influence the geometrical figure which takes longer to develop. Fig. 6.19 shows those that arise from the position of Venus at the two different types of conjunction.

Once again one wonders whether this was to be expected. It is an outcome that is probably the most impressive result to arise from this way of depicting the harmony of eccentricities and orbital periods. The rhythm of the Mars/Jupiter conjunctions which relates in this unexpected way to the orbital period of Venus is quite complicated and multi-layered, so I shall not attempt to depict it here. Nevertheless it would be possible almost immediately to expect and then follow the development of these

---

* Appendix 6.3 lists the values for all planetary constellations.

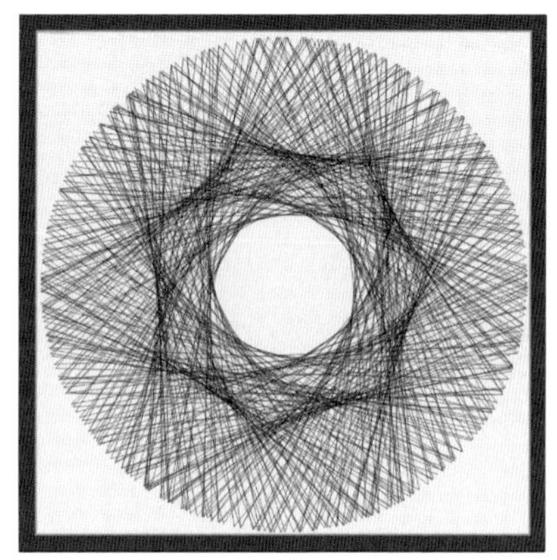

**Fig. 6.19** *Venus at Mars/Jupiter conjunctions, 400 times, period* c. *894.1 years. Left, at conjunctions in ecliptic longitude, beginning 13 March 2000; right, at minimum distances, beginning 21 March 2000*

figures that are ordered in accordance with the number seven, something that would have been impossible in the case of the initial tangle of lines leading to what subsequently became the 13-petalled flower. Although the Venus star-figure loses its points in distance conjunctions, nevertheless the structure of the inner heptagon remains and continues to demonstrate the basic principle quite clearly. Once more we gain the impression that a mysterious order is woven into our solar system.

Initially all we can say for certain is that the way in which the eccentricities and distances of the planets are arranged is indeed mysterious and that this arrangement involves geometrical and harmonic principles which in many ways are profoundly obscure. As far as the harmony of the velocities is concerned, we showed that this is scarcely likely to have come about by accident. But what is the situation with regard to the geometrical figures? Even if such a thing were meaningful or indeed possible without making too many assumptions, to assess the probability of creating for example a 7-pointed star out of variously interwoven rhythms, or even a 13-petalled flower out of seemingly utter chaos, would go far beyond the bounds of our present considerations—as also, I have to admit, beyond my own capabilities in this respect. By comparison, it is relatively easy to assess the probability of the star-figures, e.g. the Earth/Venus pentagram, that are formed by two orbital periods. As shown in Appendix 4.3 this is *c.* 1:20 to 1:25.

The pentagram may appear to be something special as far as human beings are concerned. And it is, moreover, the first in the series of infinitely many theoretically possible star-figures. Purely mathematically, however, for an analysis of probabilities it is irrelevant which figure is formed. But since it is always a star-figure* that arises on the basis of any degree of accuracy chosen, it is necessary for a further assessment to set a second limitation, namely concerning which star-figure should be the final one in the consideration. Geometrically, the most useful boundary mark seems to me to be the 12-pointed star. Up to here, including this one, there are 12 continuous star-figures, i.e. those that can be drawn with a continuous line, 5 slender ones and 7 more corpulent ones like that on the left in Fig. 6.19. According to this (see Appendix 4.3 again), we ought to find in the planetary system just under one half of a star-figure arising directly from the synodic periods and three quarters of a polygon, which together amount to approximately one resonant geometrical formation with a maximum deviation of 3°, just as is indeed the case (see Appendix 6.2). In other words, the number of resonances arising from the orbital periods of any two planets corresponds very well to a random distribution.

Are there, then, further aspects of order elsewhere in our solar system? In bringing this chapter to a close let us now ask whether the multiplicity of rhythms and figures arising from eccentricities and distances can be incorporated within an overarching structure. The eccentricities are primarily expressed in the form of ellipses; each one presents its planet with an aphelion and a perihelion while also giving it a spatial sphere by means of the precession of the line of apsides as illustrated in Appendix 1.1, Fig. 14.2. We showed in Chapter 1 that the architecture of the distances of the planets is determined by the semi-minor axes b. These represent the geometric mean between aphelion and perihelion, which signifies that the ratio of aphelion distance (ap) to the semi-minor axis equals that of semi-minor axis to perihelion distance (pe). Using Mercury as an example:

$$\frac{ap}{b} = \frac{b}{pe};\ \frac{69.8167}{56.6717} = \frac{56.6717}{46.0017} = 1.23195\ \left[\frac{km * 10^6}{km * 10^6}\right]$$

We saw that the planetary system gains a threefold structure through the ratios of the semi-minor axes, with two somewhat smaller circles outside and inside and a larger one in the middle (see Fig. 1.5). A glance at the ratios arising between aphelion and perihelion distances

* Or polygon, such as the Jupiter/Uranus hexagon, which is only transformed into the hexagram in the depiction involving the linklines. In principle, a similar probability calculation with a similar result can also be undertaken in respect of the polygons (see Appendix 4.3).

(all the ratios are given in the table in Appendix 6.6) shows that here again there is a division into three regions, but now of (almost) equal sizes:

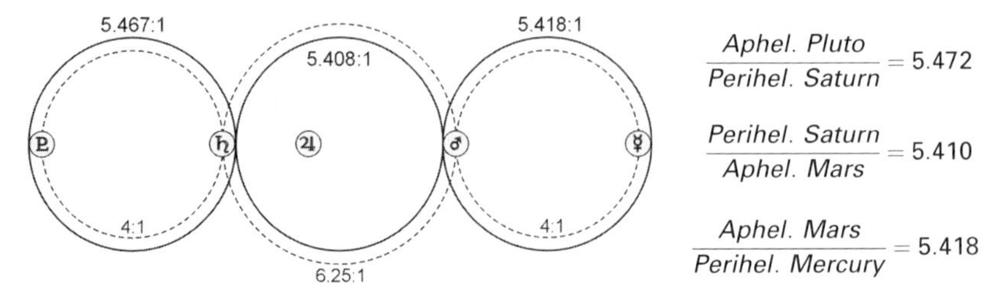


**Fig. 6.20** *Threefold structure of the aphelion and perihelion distances. From left to right: Pluto, Saturn, Jupiter, Mars, Mercury. Dotted circles show the structure of the semi-minor axes*

Here again we see that the eccentricities are not allocated to the planets haphazardly but that they bring about an at least reasonably accurate distribution of a logical and thus all the more convincing kind. The special case of Jupiter also becomes even more clearly visible in that it takes up the interval space otherwise occupied by four other planets. One question to ask is whether there is anything special about the value of a mean of 5.43:1, particularly since the Pluto perihelion and the Neptune perihelion, which are almost identical, stand in a ratio of 5.450 to the Jupiter aphelion (mean value of 5.437 and 5.463); and whether there is at least a geometrical correspondence on which Fig. 6.20 is based.

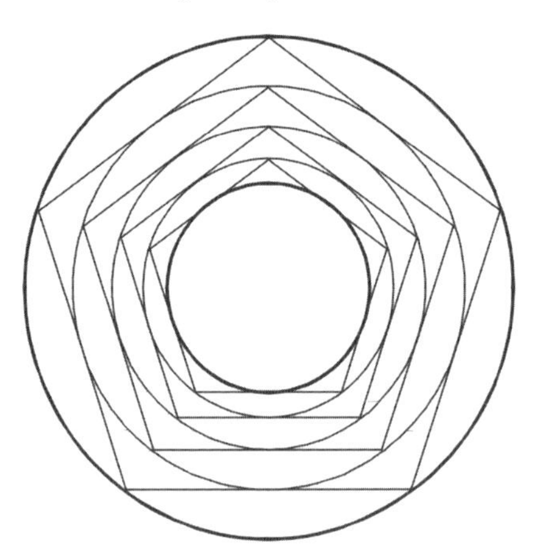

We find it in the pentagon, i.e. in the ratio of the areas of the outer and inner circles of four pentagons one inside the next. It is 5.449:1.* This results in the overall structure of the aphelion and perihelion ratios of 3⋆4 = 12 pentagons. Although this approximation is

* The ratio of the areas of the outer and inner circles for one pentagon is 1.52786:1, see Appendix 2.1. For the figure shown this amounts to $1.52786^4 = 5.44928$. The ratio of Pluto aphelion to Mercury perihelion is 160.378. The coincidence of the overall ratio is thus: $160.378/5.44928^3 = 0.9911$ or 99.11%.

only 99% correct (see footnote on p. 122), it does appear that the numbers five and twelve play a not entirely insignificant part in the structure of the solar system.

Whether this is the case or not, it is obvious that eccentricities and distances are closely linked by their ratios which have such a common measure. This inner connection emphasizes the importance of an overall view for reaching an inkling of the being and nature of the solar system. The Titius-Bode law, for example, deals only with the distances. There are a number of contemporary scientists who are investigating the consequences of resonances also from viewpoints different from those we have been discussing here, for example their connections with the rhythms of solar activity and/or their effects on the Earth (which we shall look at more closely in Chapter 9). The orbital periods, too, involve only the mean distances, since without taking the eccentricities into account both are related via Kepler's $T^2 = a^3$.

The Third Law can also be interpreted as an indication that relationships in the solar system are determined at a higher level which we have termed 'quadratic time' (see Glossary). The obvious correspondence between the intervals of the *distances* and the geometrical ratios of the *areas* is, from the mathematical point of view, a linking up of the planetary ratios with a superior dimension. The incalculability dealt with in Chapter 5, and the mystery connected with it, can be seen as a further indication that something has been or still is at work in our cosmic home which is more akin to the mystery involved in every genuine creative act than exclusively to the consequences of blind physical forces.

Perhaps it is in music that this mystery—in relation to the working of the human spirit—is most obviously and yet also most unfathomably made manifest. Music is forever dependent upon a brimming over from, and access to, a hidden source that exists no one knows where, so that we shall never be able to fathom completely the creative depths of works like the symphonies of Anton Bruckner or the piano concertos of Amadeus Mozart. All we can do is endeavour to draw closer to them. One way of attempting this is to gain an inner image of their architecture, not in the musicological sense of studying their form but by gradually finding a way as a listener into what is going on inside the music, how its themes are formed and how they hang together. An overall spatial picture of the musical structures existing in time can then gradually be painted. Such a picture will always be an individual thing, just as our emotional reactions to the notes must be, yet it is not something arbitrary. In the end this is what is needed if one is to comprehend something of the spiritual content of a great work, to gain what might appropriately again be termed 'an inkling' of what it contains. In the same way our search for the mystery of our planetary system can only ever amount to drawing gently somewhat closer to it.

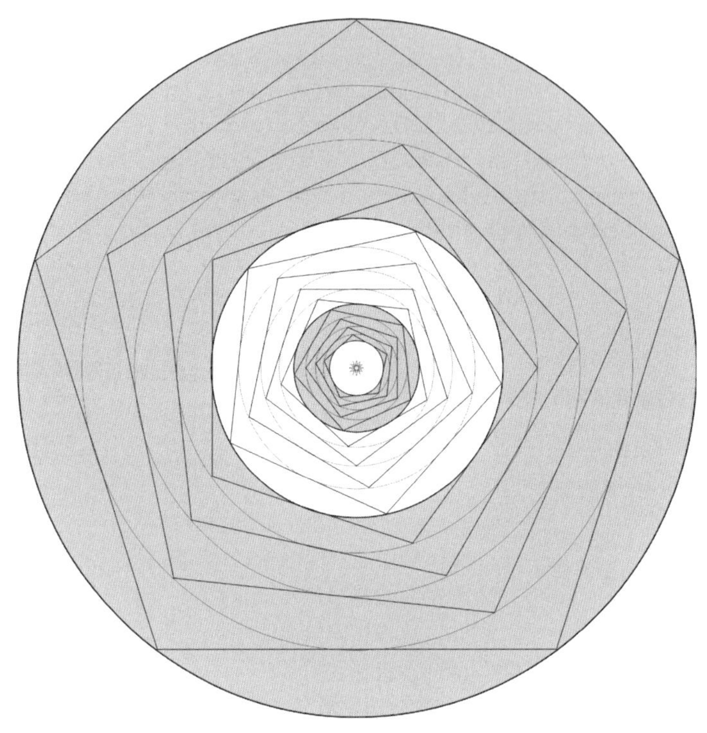

**Fig. 6.21** *Approach to the superior ratios of aphelion and perihelion distances (Pluto aphelion—Saturn perihelion—Mars aphelion—Mercury perihelion) via the area ratios of 12 (3*4) pentagons*

# 7. Venus, Earth, Mars

Ancient mythological associations of the visible planets with various divinities who possessed many and varied characteristics may have been based, perhaps not exclusively but indeed for the greater part, on observation of the heavens since time immemorial. The familiar Graeco-Roman ideas had their roots in Babylonian astronomy and astrology, and one assumes that it was Pythagoras who, returning from his travels, brought with him the stellar religion that gave such impetus to Greek astronomy.[1] Mars was identified with the god of war. The degree of fluctuation in its orbit in relation to Earth (approximately 48 days) is higher than that of all the other planets by a considerable margin. The figure it traces in the firmament over decades possesses nothing of the beauty we see in the loops of Venus. Viewed heliocentrically, the following image arises after two conjunction periods:

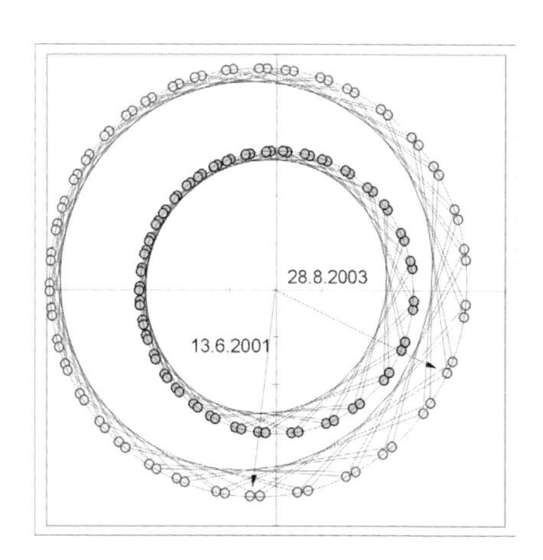


**Fig. 7.1** *Earth/Mars conjunctions, 74 times, period c. 158 years, beginning 13 June 2001*

The orbital ratio of the two planets is 79:42, so that with a cycle-resonance of approximately 2.35° a 37-pointed star-figure is formed. The disturbances caused by the eccentricity of Mars are mirrored correspondingly in the loops perceptible from Earth. The distance between the two planets at conjunction varies between nearly 56 million and 101 million kilometres. This also causes the noticeably greater difference in brightness compared with that of the other planets. At the time of the conjunction in 2003, Mars adorned the night sky in maximum brightness.

According to another idea from the world of Greek mythology, the union of the eccentric god of war with the lovely Venus brought forth the legendary figure Harmonia who thus personified the blending or balancing out of opposites. With this in mind we can see our human planet existing in a tension field between polar forces which it may be possible to reconcile here. The Pythagoreans widened the original concept of harmony to include an ordering of all things on the foundation of harmonic number ratios. Such ordering has been shown to exist in the proportions of music and of geometry, and its existence among the interplay of the planets has at all times also been surmised. So let us begin by taking a look at geometry, and at the traditionally revered archetypal triangle of Pythagoras which was also known to the Egyptians whence it derived its other name: the Egyptian triangle. The principle governing one of their great monuments, the pyramid of Khafra (the second of the pyramids at Giza), for example, is based on it.

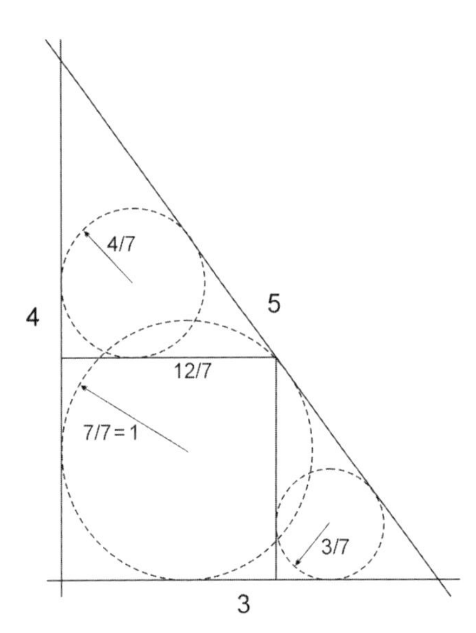


**Fig. 7.2** *Pythagorean triangle and inscribed square with side-length 12/7.**

It is the triangle with the numerical triple that best embodies the theorem of Pythagoras: $3^2 + 4^2 = 5^2$. By inscribing its square we create two smaller triangles of which the sides are in the same ratio as those of the original since they are similar triangles. The radii of the three inscribed circles are in the ratio of 7:4:3. The same ratio is shared by the sides and the sections divided off by the inscribed square—a wonderfully simple construction. Even though it cannot be described as harmonious in the musical sense, since sevenths do not form musical intervals, it nevertheless embodies all the magic of elementary geometry that speaks so clearly to our feeling for harmony. Of course only the basic triangle played a part in the construction of the pyramid. Turn it on its side by 90° and you have before you one half of its profile aspect.

In measuring the orbital periods of the three planets with which this

* The side-length x arises from: $\frac{4-x}{x} = \frac{4}{3} \Rightarrow x = \frac{12}{7}$
The radius of the inscribed circle (= 1) is determined via Heron's formula.

chapter will be chiefly concerned, only the ratios created by the square with side length 12/7 are relevant. The three synodic periods Earth/Mars, Venus/Earth and Venus/Mars are 779.936, 583.921 and 333.921 days. The mutual relationships of the three periods are:

$$\frac{Ea/Ma}{Ve/Ea} \cong \frac{4}{3} \qquad \frac{Ea/Ma}{Ve/Ma} \cong \frac{7}{3} \qquad \frac{Ve/Ea}{Ve/Ma} \cong \frac{7}{4}$$

This yields the following full equation:

$$3\ Ea/Ma \ \cong\ 4\ Ve/Ea \ \cong\ 7\ Ve/Ma$$

The correspondence is fairly good, with the almost accurate values being 3–4.0071–7.0071. An even more accurate match with the actual situation can be attained by means of the numbers 140–187–327.* Another way of depicting the threefold proportion is by means of a ratio equation:

$$Ea/Ma : Ve/Ea : Ve/Ma \cong 4 * 7 : 7 * 3 : 4 * 3 =$$
$$28 : 21 : 12 = 4 : 3 : 12/7$$

This is how the ratio of the three synodic periods can be directly deduced from the Pythagorean triangle shown above. It appears that the movements of the planets in the way they interact are ordered very simply. But we easily come up against our limitations when we try to imagine the three bodies moving around the Sun. Over the course of (several) of the cycles of 3, 4 and 7 conjunctions the star-figures already introduced arise: the Venus/Earth pentagram and the star-figure with 37 points shown in Fig. 7.1. In addition there is the figure created by Venus and Mars in their 107:35 ratio.

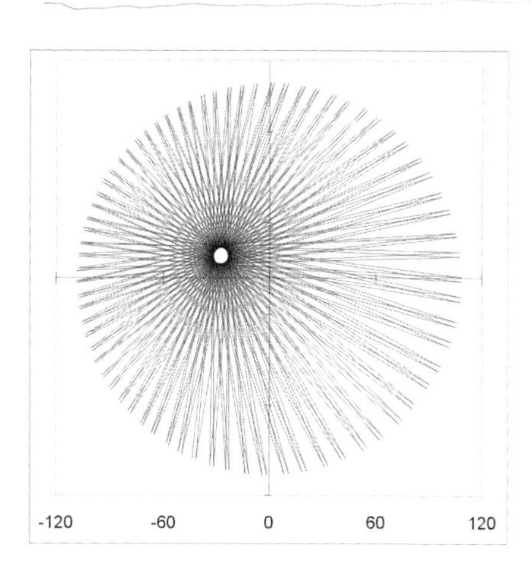


**Fig. 7.3** *Venus/Mars conjunctions in the orbit of Venus, 144 times, period c. 131.6 years, beginning 19 June 2000, scale in millions of kilometres*

This eccentric 72-pointed star-figure is the one that leads, for example, to the 13-petalled flower of Mercury shown in Fig. 6.18. In spite of the simple basic

---

* These three values—and also further number pairs and triples in other planetary constellations to be mentioned later—were obtained with the help of computer programs.

relation of 3–4–7, we might be inclined to regard the interaction of Venus, Earth and Mars as being highly complicated or indeed almost inconceivable. So let us begin by extricating the essential elements of the mutual interaction of these planets. When all three are on one line at the beginning of a cycle, i.e. when they form a threefold conjunction, this will be the same again at the beginning of the next one. In this period of time, during which Earth's two neighbours meet seven times, the Earth itself encounters the outer planet three times and the inner planet four times. After half the period of time, i.e. after two Venus/Earth conjunctions, the other constellations have occurred 1.5 and 3.5 times respectively. This means that at every second Venus/Earth encounter in a cycle Mars is situated exactly opposite them, i.e. in opposition.

This is the ideal basic pattern which, however, very rarely occurs in reality. The small deviation from whole numbers brings it about that the positions are correspondingly displaced each time. Because the difference is so small, it will take a relatively long time before another threefold conjunction can occur. The whole pattern of the movement shifts in the same way without, though, losing its character. Before examining this in a graphic depiction we should first consider something else in connection with simultaneous conjunctions of three planets. Such a conjunction will never be 100% exact, since this could only happen if the ratio of the orbital periods were that of proper whole numbers. So the concept of a threefold conjunction always requires the statement of a limitation up to which the positions deviate from a straight line between the three planets and the Sun. The next very similar constellation between Venus, Earth and Mars will not take place until 14 December 2101. On that day these three will approach one another to within 1.2°. Beginning on that date, Fig. 7.4 shows the rhythmically changing distances by means of the angles formed in each case by two of the planets.

The angular separations are shown as cosines. This makes it very easy to detect the conjunction and opposition positions, namely at +1 or −1 respectively. After 4 and 8 Venus/Earth conjunctions all the curves are at +1, and after 2 and 6 such conjunctions Ea/Ma and Ve/Ma are at −1, i.e. Mars is on the opposite side of the Sun from the other two which are once more in conjunction. The order thus becoming visible also signifies that a Venus/Mars or an Earth/Mars conjunction at which the third planet in each case is in opposition is not possible, at least theoretically. The reason for this is as follows. If the whole cycle is set at being a single unit of time, the intermediate values of all the conjunctions and oppositions fall at fourteenths, eighths and sixths of the total. The fractions that thus arise

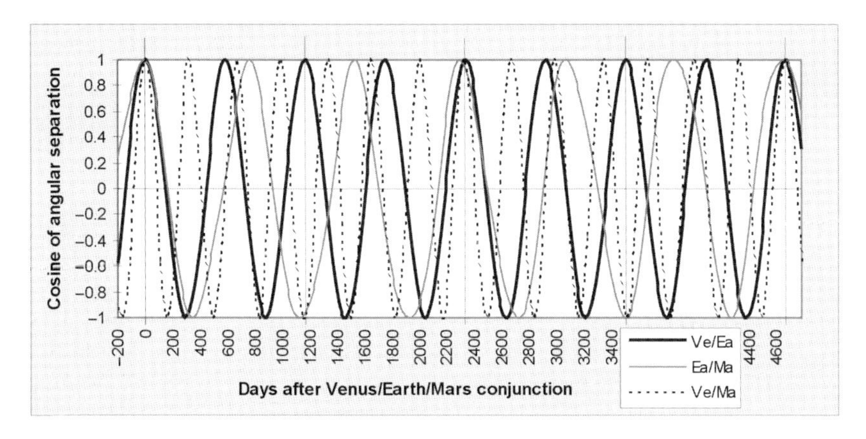


**Fig. 7.4** *The 3–4–7 rhythm of the synodic periods of Venus, Earth and Mars, beginning 14 December 2101 (zero), 2 cycles, c. 12.8 years (= 4675 days)*

only coincide at 0.5 ($= \frac{7}{14} = \frac{4}{8} = \frac{3}{6}$), i.e. after two Venus/Earth conjunctions. This applies to the ideal situations, but the pattern shift and the deviations caused by the eccentricities nevertheless allow for approximations in the constellations under discussion.

So in summary and without exaggeration one can say that the three planets move in a rhythm that follows the beat of the Venus/Earth conjunctions. It is a 4/4 beat with a strong accent on 1 and a somewhat lesser emphasis on 3. This means that every second and even more so every fourth encounter of our planet with that of the goddess of love is likely to have an added significance which will presumably also be expressed geometrically. But first let us look at movement diagrams of the three celestial neighbours documenting sequential Venus/Mars and Venus/Earth conjunctions respectively (see Fig. 7.5).

The conjunction star-figure, which for the planets involved turns out to be similar, has been shown only once in each case. The same goes for the linklines. Their other depiction, not shown here (Earth-Venus or Mars-Earth respectively), would in the left-hand picture have to be imagined mirrored inwards from Earth towards Venus rather than towards Mars; and in the right-hand picture from the Mars position, turned slightly, running towards Earth rather than towards Venus. The striking figure of Earth's movement on the left is one of those which only arise on the basis of a special tuning of the rhythms formed by the eccentricities, as explained in Chapter 6 with reference to the 13-petalled flower and the 7-pointed star-figure. The ratio of Earth-year and Venus/Mars conjunction is *c.* 35:32, so that if these rhythms did not fit together a

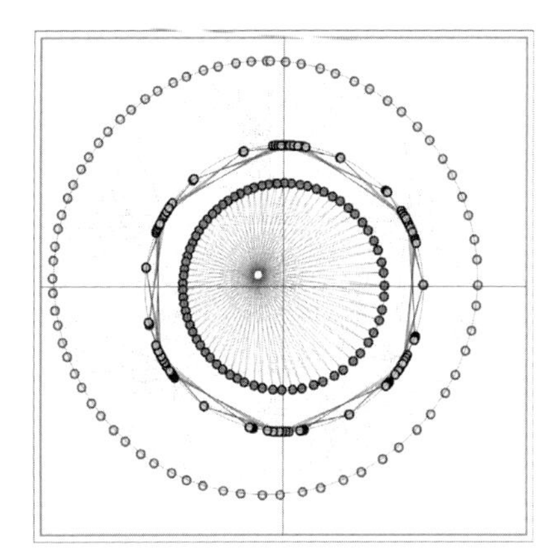
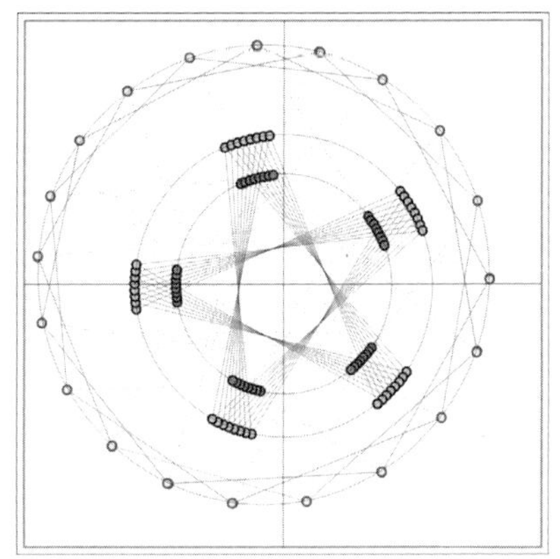

**Fig. 7.5** *Movement diagrams of Venus, Earth and Mars. Left: Ve/Ma conjunctions, 72 times, period c. 65.8 years, beginning 19 June 2000. Right: Ve/Ea conjunctions, 40 times, period c. 64 years, beginning 30 March 2001. The dotted lines show the linklines between Mars and Earth and, respectively, Mars and Venus* → Plate 3

35-pointed star-figure would arise. Instead, Earth mostly bounces around in a hexagon when Mars and Venus embark on a rendezvous. The figure is very resonant. We have actually shown two cycles in the illustration, i.e. 144 Venus/Mars conjunctions, but it would still be easily recognizable even after four such cycles. It is as though the number six, otherwise not particularly concerned with the inner planetary system, were wanting to make its influence felt by this means. And as in the minor third (6:5), it is here also allied to the number five in accordance with which the lines connecting Earth to Mars (and similarly to Venus) are organized. After several hundred conjunctions this would become even more apparent, but I shall not go into this here since we encounter the number five elsewhere.

And here it is, within the 20-pointed star-figure of Mars in the right-hand diagram. This 20-point figure stands in the sky, immutable, like a commander-in-chief in the midst of battle—to use a designation befitting the fourth planet—firmly restraining the pentagram as though with invisible fetters. Once again the 20:17 ratio between the orbital period of Mars and the Venus/Earth synodic period mentioned in Chapter 2 comes into play. With a cycle-resonance value of 0.12° this is the strongest resonance of small whole numbers in the whole planetary system, apart from the moons (and some even more concealed harmonies which we shall meet in later chapters). Perhaps we should once more point to the fact that the above diagrams are not just more or less pretty pictures. They provide a geometrical method of illustrating gravitational interactions—

and who knows, perhaps also other as yet undiscovered forces or aspects of gravity. The purely arithmetical approach has so far failed to solve several riddles connected with the way the planets hold together. Perhaps it will never find solutions, just as measurement and calculation of the pitch and duration of notes, however accurate, will never fully embrace the inner nature of music.

Let us now see how the interaction proceeds when the 4/4 beat commences:

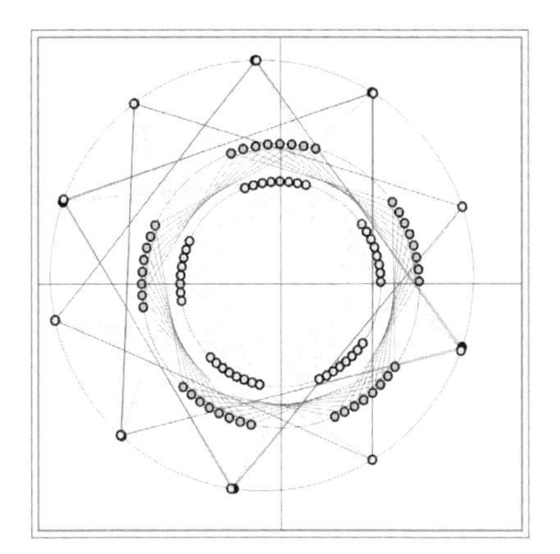

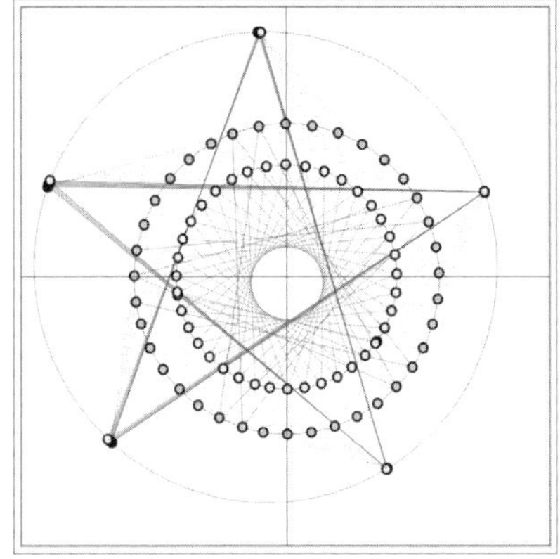

**Fig. 7.6** *Movement diagrams of Venus, Earth and Mars. Left: at every second Ve/Ea conjunction, 40 times, period c. 127.9 years; dotted Mars-Venus linklines. Right: at every fourth Ve/Ea conjunction, 40 times, period c. 255.8 years; dotted Mars-Earth linklines. Beginning in both cases: 30 March 2001* → Plate 3

**Fig. 7.7** *Star-flower 2*

It is a truly shrewd move to combine the three crucial time-spans with the help of Kepler's Third Planetary Law in the manner suggested by circle, square and triangle. The 10-pointed star-figure here demonstrates the full power of Mars' steadfastness. And the 5-pointed star-figure of the Venus/Earth relationship has turned into a pentagon held by cross-over lines which also take on this shape (certainly at this moment; at other times another figure is possible, namely a 10-*rayed* star). And when the pentagram, having hitherto belonged only to Venus and Earth, has virtually dissolved and become transformed into a 38-pointed star-figure (Fig. 7.6, right) it is then gloriously reborn in far greater splendour in the harmonious interaction of the 3:4:7 ratio of all three planets.* Just as Johannes Kepler would have been delighted about the correspondence between his two stellar children and the ratios of orbital period, velocity and distance which give a boundary to the overall structure, so surely would Pythagoras have been pleased to discover the correspondences between his archetypal triangle and the rhythms of the conjunctions in their combination with the resulting movements of Venus, Earth and Mars.

One might also, of course, interpret this merely as an expression of the simple mathematical fact that doubling the interval of observation can logically only lead to this development. Initially alternate points of the 20-pointed star-figure are skipped so that a 10-pointed star-figure is formed. Then a further multiplication by 2 of the duration brings about a similar doubling of the determining angle of the geometrical figure, thus resulting in the pentagram. Yes, that's how it is. And we can also find this logic and this geometrical archetypal image in the sky of the solar system because those two so very different planets framing our own have a unique relationship not only with one another but also with our Earth in all its vitality. Over and above this, as we shall see, there are in relation to other members of the planetary system correspondences in the longer-term movements which are so astonishing that it is impossible to imagine them taking place without some kind of inner coherence.

So let us now look at the superior rhythms. One cycle, for example, of 7 Venus/Mars conjunctions takes *c.* 6.4 years. All three oscillations expressed in the exact durations, which are different each time, are ordered according to the number five. Entering these durations consecutively results in oscillation diagrams similar to the top diagram in Fig. 6.12. The next higher mutual temporal measure is thus 6.4★5 = 32 years, corresponding to 1 Mars pentagram and 4 Venus/Earth pentagrams. The table opposite shows all the periods discussed and others that correspond with them.

Thus after 5 cycles the orbits of the three planets also again nearly coincide with the synodic periods, so that after this period of time they can once more be found almost at the same spot on the ecliptic from

* The first two appearances are shown in the Earth orbit. With the fourfold period of time during which the pentagram is formed, the cycle-resonance is 9.63°, so that one must now speak not of a 5-pointed star-figure but of one with 38 points.

**Table 7.1** *Periods of conjunctions and orbits*

| ① | ② | ③ | ④ | ⑤ |
|---|---|---|---|---|
| | Days | ② ★7; ★4; ★3 | ③ ★5 | Mean ④/② |
| Ve/Ma | 333.9215 | 2337.4506 | 11687.2531 | 35.0030 |
| Ve/Ea | 583.9214 | 2335.6855 | 11678.4274 | 20.0168 |
| Ea/Ma | 779.9361 | 2339.8083 | 11699.0414 | 14.9862 |
| | *Means:* | *2337.6481* | *11688.2407* | |
| | | Mean ③/② | | ③ ★5 |
| Venus | 224.70080 | 10.4031 | | 52.0169 |
| Earth | 365.25636 | 6.4000 | | 32.00010 |
| Mars | 686.97985 | 3.4028 | | 17.0139 |
| | | | | |
| Me/Ve | 144.5662 | 16.1701 | | 80.8504 |
| Me/Ea | 115.8775 | 20.1734 | | 100.8672 |
| Me/Ma | 100.8882 | 23.1707 | | 115.8534 |
| Ju/Ne | 4668.6936 | 0.5007 | | 2.5035 |

*'As an aid to memory* [Kepler is referring to the idea in his *Secret of the Universe*] *I give you the proposition, conceived in words just as it came to me and at that very moment: "The Earth is the circle which is the measure of all [the other circles] ... by the Earth I understood the orbit on which it travels, called the Great by Copernicus ..."'*

Johannes Kepler[2]

which they started. Column ⑤ in the Table shows the ratio of the orbital periods of *c.* 52★Ve ≅ 32★Ea ≅ 17★Ma. In addition the mean values of the three conjunction periods have been calculated. Although one has to be quite careful in astronomy as far as averages are concerned, in the case of the 3:4:7 constellation one may perhaps assume that this can be regarded as a genuine superior measure. The cycle of Earth with its inner neighbour is perhaps two days shorter while that with Mars, which lies further out, takes *c.* two days longer. The average of 11,688.24 days in the next column corresponds almost 100% to 32 Earth orbits, for in this period of time the difference is less than 1 hour.

It is rather remarkable that in the 32-year period as many Mercury/Earth conjunctions occur as there are days in one Mercury/Mars synodic period (or vice versa). Furthermore, the synodic period of Jupiter and Neptune appears to be very closely linked with the pentagram rhythms. We shall soon be discussing this in greater detail. But first I want to show how a diagram of the positions of two planets at the moments of their conjunction produces two other star-figures. They arise when one plots the position of the one planet whenever the other has completed exactly one orbit. In the case of Venus and Earth the result is shown in Fig. 7.8.

The 13:8 resonance of the orbital periods of the two planets taken here as an example is expressed geometrically in figures that accord with this ratio, whereas the difference between the two numbers results in the number of points in the conjunction-star. As the current scientific view puts it, *'the orbital period of any planet ... although an important observational quantity, is meaningless in dynamic terms, whereas the periods of the conjunctions document useful parameters in celestial mechanics.'*[3] For the most part this statement is in agreement with the point of view expressed in the present

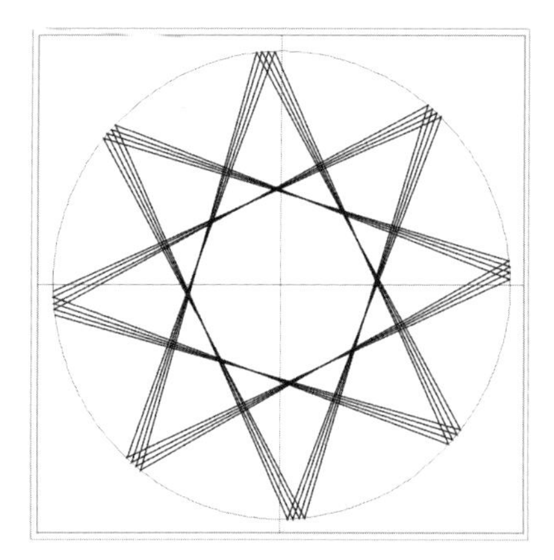
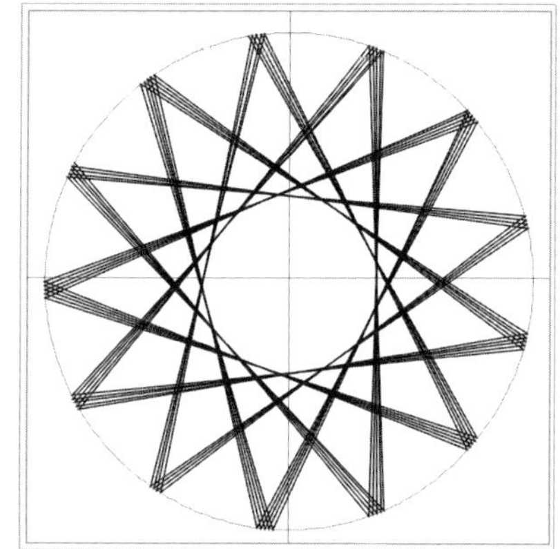

**Fig. 7.8** *Position diagrams, period c. 32 years. Left: Venus after in each case one Earth orbit, 32 times. Right: Earth after in each case one Venus year, 52 times*

work, so I shall continue essentially to investigate movement processes related to conjunctions. However, it is appropriate to have mentioned at least once what could be referred to as the 'chemical components' of the pentagram, which is so important from our point of view, and also those of the star-figures in general.

Let us now consider how all this relates to the outer planetary system. Fig. 6.7 showed the relationship between the conjunctions of Jupiter and Neptune by means of the 13-looped figure and respectively the corresponding star-figure of the linklines. The stepping interval of 240 days was chosen since this showed the continuous looped movement and the clearly defined star-shape. If the stepping interval chosen is too great the outcome is of course different. In the case of 600 days, approximately the duration of the Venus/Earth conjunction, the figures are still recognizable but have become much more sharp-edged or, respectively, more blurred. In the centre of the 13-pointed star-figure in Fig. 6.7 more or less all that was visible was a dense tangle of lines. But never mind; we did not miss much. Things look quite different there, however, when we plot the Jupiter-Neptune linklines at the conjunctions just mentioned or else in the rhythm of the two or, respectively, four encounters.

A kaleidoscope, a unique kaleidoscope for the gods! Whether one views the pictures at every second or at every fourth conjunction, the images vary over the millennia as shown in the examples in Fig. 7.9. This spectacle only occurs in the interaction of the two outer planets with Venus and Earth. Just the two-day difference of the sevenfold conjunctions of Venus and Mars suffices to do away with all of it. Four Earth/Venus encounters correspond to 0.5003 Jupiter/Neptune conjunctions;

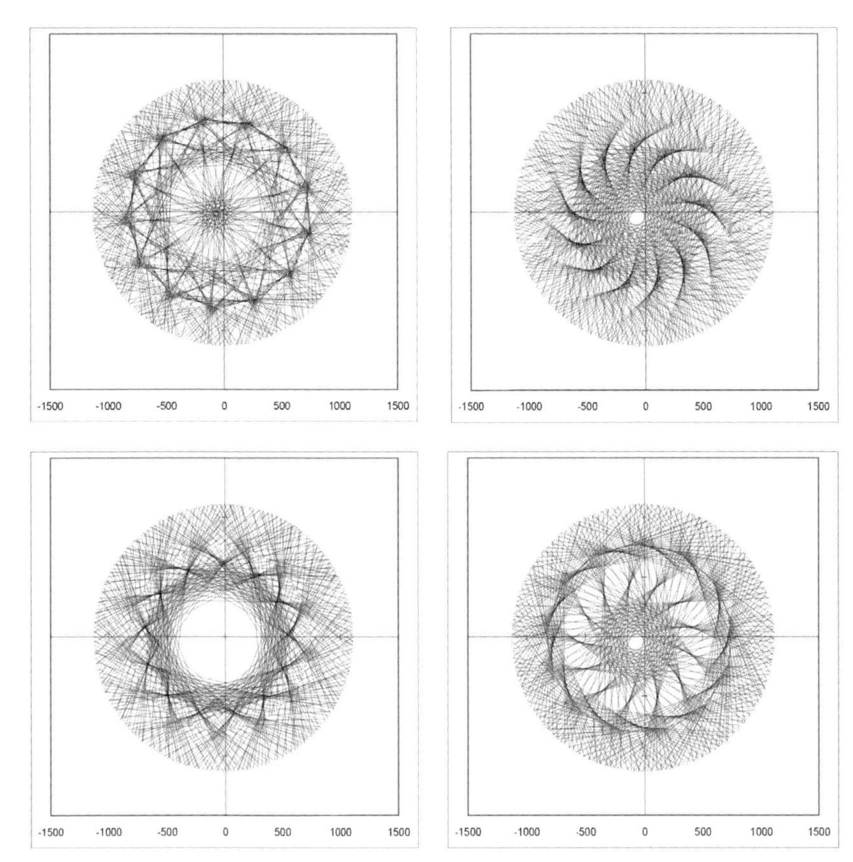


**Fig. 7.9** *Jupiter-Neptune linklines at Venus/Earth constellations, in each case 500 times. Top left: at every conjunction, period 799.3 years, beginning 30 March 2001. Top right: at every fourth conjunction, period 3197.3 years, beginning 30 March 2001. Bottom: at every second conjunction, period 1598.7 years, left beginning 30 March 2001, right beginning 3 October 3596. Detail enlargements 4:1*

the value for the other constellation is 0.5007. We see here what can genuinely be called a true resonance. The supreme divinity of Graeco-Roman mythology, lord of light, lightning and thunder, and his colleague who rules the fluid element appear to attach especial favour to the link between the planet of human beings and the celestial symbol of love. We may assume this approval to be an integral aspect of the framework of factors in our solar system which are arranged in a way that for millions of years hold the great orbit of Copernicus firm amid its companions while it weaves its signs into the ether.

But the longer-term interplay as such of the trio with which we have been concerned in this chapter also challenges us to investigate further. In order to do this we shall introduce another method by which longer-lasting movement sequences among the planets can be made visible. Actually this is merely an extension of the method with which we are already familiar, namely that of recording the path of one celestial body from the perspective of another which we place in the centre of the picture. In our customary use of the method we apply relatively short periods of time, and this yields the looped figures introduced in Chapter 6. But there is nothing to prevent us from taking the points of conjunction instead and thus forming, for example, a diagram in which Mars is seen from Venus at moments of closeness to Earth.

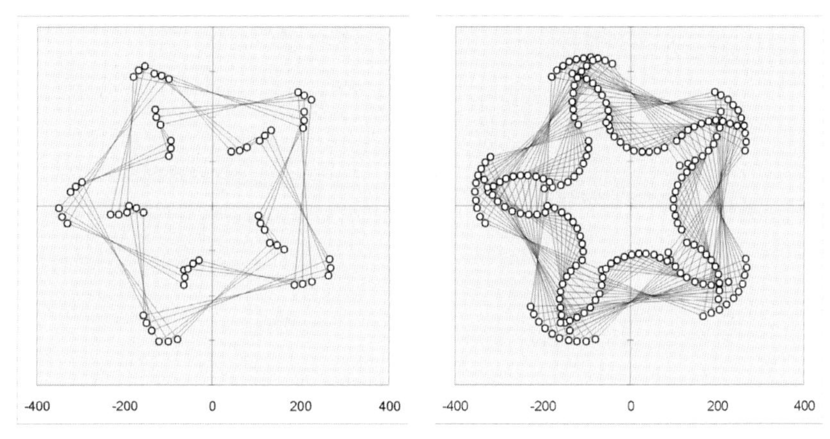


**Fig. 7.10** *Venus-centred view of Mars at Venus/Earth conjunctions, beginning 30 March 2001. Left: 60 times, period* c. *95.9 years. Right: 200 times, period* c. *319.7 years. Scale in millions of kilometres*

The view of Venus from Mars at the same points in time would yield the same image rotated by 180°. We see not only the fivefold structure of the Earth/Venus relationship but also the superior rhythm of all three planets which in this case shows up as a 20-fold movement sequence. After 200 conjunctions the pentagram-like structure is dominant but the figure is still in process of being formed. The overall picture only shows after *c.* 750 'snapshots', thus corresponding to the period of one rotation of the pentagram by 360° in just about 1200 years.

In some of the positions we can see that the points following the finished figure are slightly offset, on average by *c.* 4.5°. This is due to the slight shift of the Mars square as shown in Fig. 2.10. The 20 loops or, more accurately, circles, which have been formed during the course of 12 centuries are due to the 17:20 resonance of the conjunction period with the orbital period of Mars.

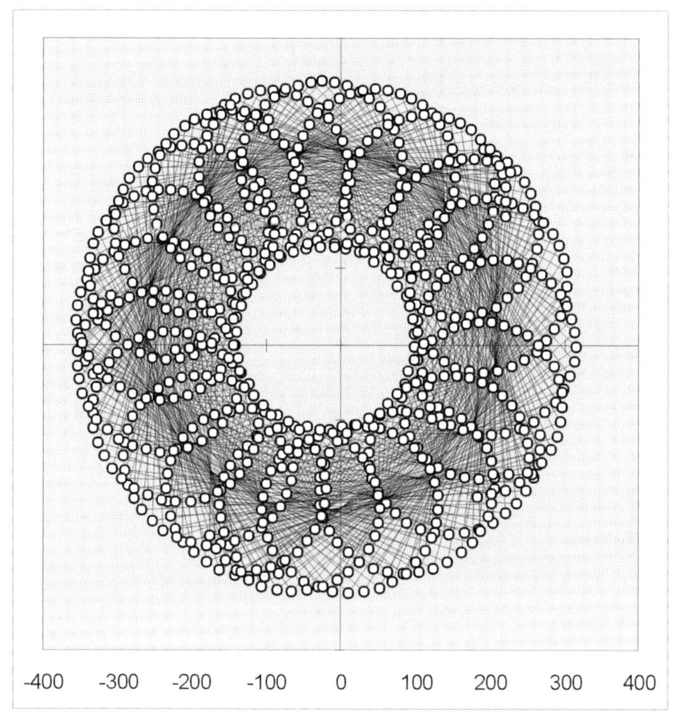


**Fig. 7.11** *Venus-centred view of Mars at Venus/Earth conjunctions, beginning 30 March 2001, 800 times, period c. 1278.9 years*

Fig. 7.12 shows the movement images that are formed when the time period is based on the beat of two or respectively four Venus/Earth conjunctions. The first two pictures show that both in this planet-centred method of depiction, and also in the corresponding method related to the Sun, the basic number on which the formation of the figure is founded remains the same. This is at any rate the case in most instances. The main advantage of the planet-centred movement images is that they combine two aspects: the patterns that arise from the positions of the planets and the fabric woven by the paths between them. Depending on the constellations, the one or the other type of diagram can have more to say or, respectively, can uncover more interconnections.

So we see the interaction of the three planets being formed not only in relatively brief spans of time (see Figs. 7.5 and 7.6) but also in the longer-term movement forms of the numbers five, ten and twenty (4*5)—with some of the resulting pictures being almost unbelievably beautiful.

In addition, the frequency of 4 Venus/Earth conjunctions appears to have a resonant relationship with a further constellation of the outer

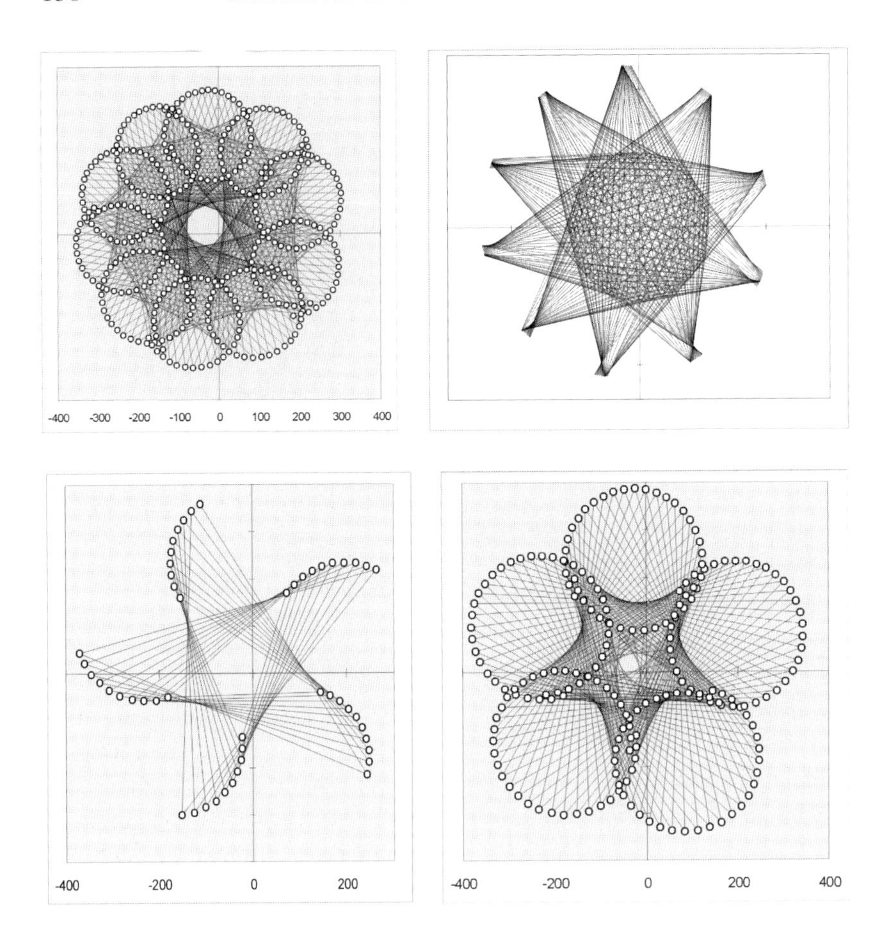


**Fig. 7.12** *Venus, Earth, Mars constellations. Top left: Venus-centred view of Mars at every second Venus/Earth conjunction, 400 times, period* c. *1278.9 years, beginning 30 March 2001. Top right: depiction of the same constellation using the Venus-Mars linklines. Bottom: geocentric view of Mars at every fourth Venus/Earth conjunction—left 50 times, period* c. *319.7 years; right 200 times*

planetary system. Once again this involves Jupiter, this time together with Uranus, i.e. the two planets which in their movements bring about the second figure formed by direct conjunction positions: the hexagon and the hexagram respectively. The two mystical signs, if one wants to call them that, are linked at least geometrically over the millennia by an invisible bond. It takes fairly accurately one half of the rotation of the hexagram (see Chapter 6) for the threads to be woven in the next depiction, i.e. *c.* 3030 years.

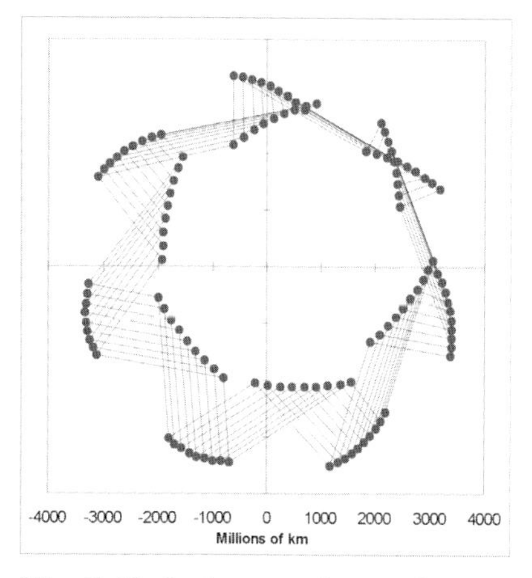


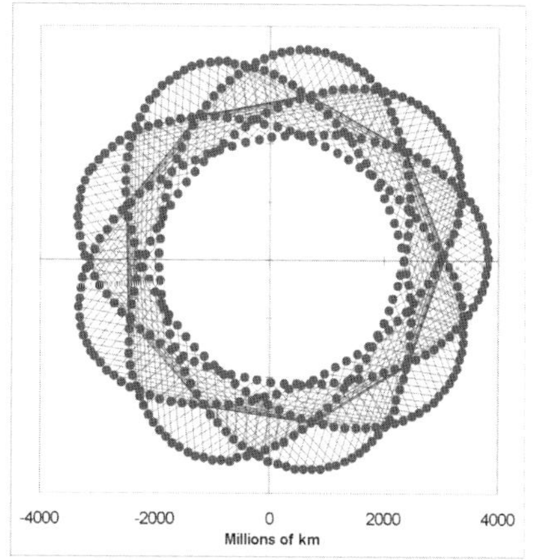


**Fig. 7.13** *Jupiter-centred view of Uranus at every fourth Venus/Earth conjunction, beginning 30 March 2001. Left 120 times, period* c. *767.4 years; right 473 times, period* c. *3024.7 years*

The formation of the figure on the left takes place in a periodic sequence of 13 steps. The basis of this rhythm is that 13★4 Venus/Earth conjunctions coincide approximately with 7 Jupiter and 1 Uranus orbit, i.e. when the two large planets have approximately reached their starting point again. That the 13 slowly forming sections of adjacent points subsequently come to form a coherent figure of 9 elliptical loops seems to me to amount to a minor planetary miracle of geometry. There is, however, also a mathematical reason for this which can be found in what we shall term a fractional resonance (see Glossary). We saw in Chapter 6 that $25 = 5^2$ Jupiter/Uranus encounters correspond exceedingly precisely with $216 = 6^3$ (= 54★4) Venus/Earth conjunctions. In this overall period of time Jupiter completes 29 $\frac{1}{9}$ orbits and Uranus 4 $\frac{1}{9}$. The ninths (i.e. the fraction after the decimal point) then bring it about that the number nine steers the whole development of the movement form with, as it were, invisible reins.

We can surely sum up by saying with some justification that even from a purely aesthetic point of view the interplay over specific spans of time between Earth and its two companions with all their symbolic overtones, or indeed between Earth and other planets as well, is something quite unusual. From the scientific point of view, too, I find it highly unlikely that what we have seen depicted geometrically here can have no bearing on the way the planetary system holds together. What cannot be deduced from the calculations, however supposedly objective, and what remains untouched by one's sense of beauty alone, is the question as to whether an intention to communicate some kind of message might lie hidden within the interplay of the planets. Think of the music brought into the world by some composers. Here, too, there seems to lie in its depths a spiritual core that escapes an adequate description in words or indeed any description at all. Those whose souls are touched by a true work of art know they are being addressed by something real even though they cannot claim any general validity for their experience.

We shall bring this chapter to a close with a small collage depicting real planetary constellations on three different though interweaving temporal levels. The starting point for this may be found in Fig. 2.10: the square drawn by Mars at Venus/Earth pentagrams. The pentagram is shown as it were in real time within the Earth orbit (a). The first time-lapse leads to the Mars square and the positions of Venus with the addition of the linklines between these two planets (b). Up to here the figure as a whole is not, as hitherto, orientated in a way that places the vernal point on 1 January 2000 at 0°—i.e. at the right-hand end of the horizontal axis—but rotated (clockwise) by 52°. At the third level, which in addition is offset spatially, Jupiter and the as yet absent large planet Saturn are active. Their interplay is shown from the planet-centred viewpoint, i.e. either Jupiter's view of its neighbour or vice versa. The temporal interval here is the superior rhythm of 20 Venus/Earth encounters or 4 pentagrams (c). The quotations are from Johannes Kepler's *The Secret of the Universe.*[4]

> *Bear with me now, patient reader, if I trifle for a moment with a serious subject, and indulge in allegories a little. For I think that from the love of God for Man a great many of the causes of the features in the universe can be deduced. Certainly at least nobody will deny that in fitting out the dwelling place of the universe God considered its future inhabitant again and again. For the end both of the universe and of the whole creation is Man.*
>
> *. . . Certainly God did not start the motions [of the planets] at random, but from some single definite starting point . . . since every thing is for the sake of Man.*

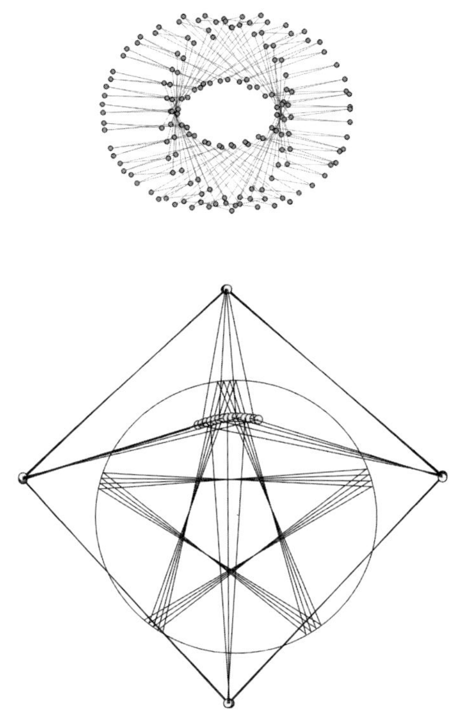

**Fig. 7.14** *(a) Earth in conjunction with Venus, 20 times, 15 July 2127 to 5 July 2159; (b) Venus and Mars at every fifth Ve/Ea conjunction, 15 times, 6 March 2081 to 1 February 2193; (c) Saturn as seen from Jupiter at every fourth pentagram, 160 times, beginning 30 March 2001, period* c. *5116 years*

# 8. Underneath the Clouds of Venus

## *Prelude*

According to a tale passed down by the Roman architect Vitruvius (first century BC), the Greek philosopher Aristippus (*c.* 435–356 BC) and some of his companions had once managed to swim to an unknown island after being shipwrecked. There they found geometrical patterns drawn in the sand, whereupon Aristippus declared: 'Friends, we may permit ourselves to entertain some hope, for here are signs of human beings.'[1]

**Fig. 8.1** *Aristippus*

In the twenty-first century, meanwhile, we have come to see nothing unusual in geometrical figures 'traced', as it were, by nature. Such shapes are present in the configurations of molecules, in the ordering of crystal lattices, where soap bubbles meet and in many another domain. Those of a scientific turn of mind, however, are not inclined to conclude from the geometry they find in nature that a creative spirit might be the source of such things. They regard the laws of nature per se as sufficient explanation.

But what if such individuals were to discover the following figure drawn in the sky?

Having so often felt marooned in a universe measured in billions of light years, might they not hope, as did Aristippus on finding signs of the human spirit in the sand, that they are, after all, not entirely abandoned? Might they not hope that this cosmos out of which they have been born may be the product of something more than unseeing and impersonal natural laws alone? Might they not hope that spirit and soul, instead of having accidentally evolved on this planet, could actually fill the whole universe in an infinitely more lofty form?

## *The 'Lady of Heaven'*

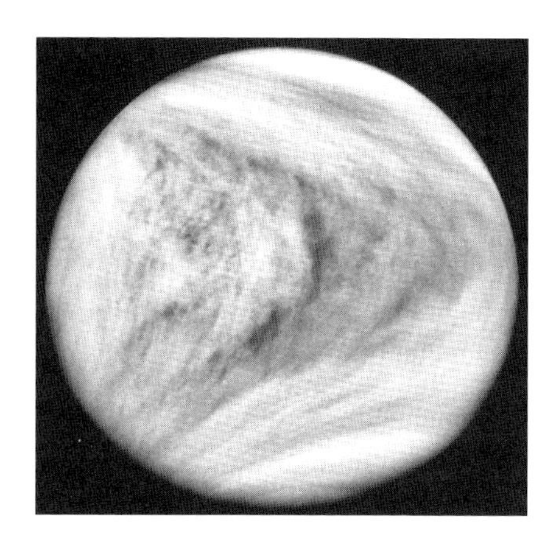

**Fig. 8.2** *Venus (NASA)*

In 1964 humanity succeeded for the first time in espying what lies beneath the dense and ubiquitous veil of cloud that envelops Venus. The Babylonians referred to this planet as the 'Lady of Heaven' whom they also worshipped as the goddess of love. Elsewhere in the literature she is also mentioned as being the goddess both of love and of war. Being probably the most advanced astronomers of antiquity, those Babylonians knew the identity of the morning and evening star as long ago as two thousand years before Christ. Cuneiform characters on a tablet of that time give observation dates for the heliacal risings and settings of Venus covering a period of 21 years[2] (heliacal meaning the first rising after and last setting before a period of invisibility). Venus also played a very significant role in the eyes of the Maya who were similarly able to carry out exceedingly accurate observations of the skies. Their calendar was based on the synodic period of around 584 days. By applying cunning corrective measures comparable to our introduction of extra days in leap years they were able to calculate the positions of the planet with an error margin of just two hours in 500 years.[3] So it is likely that both those cultures would have been familiar with the *Pentagramma Veneris*. But it was to take many millennia before it became known how the 5-pointed star-figure—or should we rather say the mythological principle embodied in it—brought into being by collaboration between the 'Lady of Heaven' and planet Earth constructs conditions in the inner solar system in quite another and far more subtle and all-embracing manner. The basis for this was created in 1964 when radar made it possible to define the period of rotation of Venus as being almost precisely 243 days.

This is by far the slowest rotation of all the planets. And more astonishing still is the fact that the rotation of Venus is retrograde, i.e. the direction of its rotation is opposite to that of all the other planets. (Uranus and Pluto are the only ones with what might be regarded as a similar retrograde rotation, although the inclination of their axes falls close to the planetary orbital plane.) The observation which puzzled researchers most of all was that, apart from a very small deviation, this period has a synchronous resonance with Earth's orbital period. The resonance period relating to the movement dynamic of Venus and Earth would be 243.165 days. (The source of this figure will be explained shortly.) For such a link between these two celestial neighbours to be caused by gravity would be very puzzling because their mutual gravitational influences would be too weak. Yet as late as 1980 Carl Sagan, for example, was still maintaining: '*However the Earth's gravity has managed to nudge Venus into this Earth-locked rotation rate, it cannot have happened rapidly*.'[4] And if indeed the rotation of Venus were found to have been caused by Earth's gravity, one would then have to ask why the opposite had not occurred, since the mass of Venus and therefore its gravitational pull amounts to 0.81 times that of Earth's gravitation, i.e. it is roughly the same and ought therefore to have exerted a similar effect.

In the meantime the initially not quite accurate measurements have been made more precise, so that an exact figure of 243.019 days* has been determined for the rotation of Venus. Today, no doubt, this apparent and inexplicable resonance is regarded as a coincidence since, after all, the calculations do show a small deviation from the linked period of 243.165 days, a difference '*well outside the error bounds of the present determination*'.[5] Or else, in other instances, no mention at all is made of it any longer. Nevertheless we shall soon see that in this matter, too, a number of very clear indications have something quite different to say about many of the harmonies existing among the oscillations in the inner planetary system. And these will show that it is precisely this small deviation from exact

---

* In the 1997 *Encyclopedia of Planetary Sciences* the values 243.025 ± 0.001 (p.887) and 243.01852 ± 0.014 days (p. 595) are given. The 1999 *Astronomical Almanac* gives 243.0185 without any indications of error limits. All in all we must assume that the rotation of Venus can as yet not be determined as accurately as those of Earth or the Moon. The calculations leading to the results shown in this chapter are based on 243.019 days. For all the diagrams, some of which cover longer periods of time, in which a lack of accuracy could be cumulative and thus lead to deviations, the calculations have also been checked using the value 243.025. No differences worth mentioning occurred.

synchronicity which enables some exceedingly astonishing phenomena to occur.

The first consequence of the slow retrograde rotation is that one day on Venus is slightly longer than one half of its year. In relation to any point on its surface, the Sun rises again after a period of 116.75 days. Another way of putting this would be to say that the same spot on the surface of Venus once again faces the centre of the planetary system; or—more poetically—that the 'Lady of Heaven' once again turns her gaze upon the Sun. In a more comfortable rhythm she would also be looking at Earth once in a while during the course of her orbit. The following sketch shows the main features of these two movements combining orbits round the Sun and the planet's own rotation:

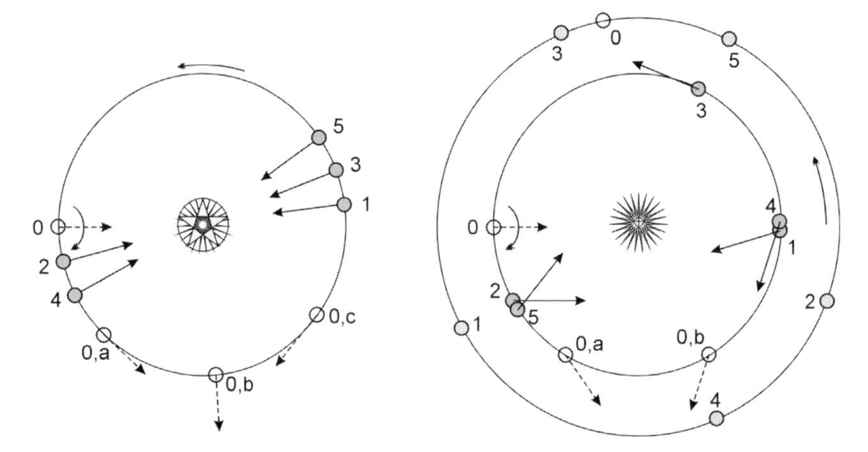


**Fig. 8.3** *Venus rotation relating to the Sun (left) and Earth (right). Beginning 1 January 2000 with a 'Venus axis' directed, at this zero point in time, towards the vernal equinox (0°, at the right-hand edge of these diagrams)*

Here and in what follows the term 'Venus axis' refers not to the rotational axis but to an axis orthogonal to this in the equatorial plane of the planet. This corresponds to a point on the surface which is crossed by the horizontal axis. By definition the 'Venus axis' is here so determined that at zero hours on 1 January 2000* it is directed towards the zero point

* The times are given in TDT on which astronomical calculations are usually based. This is absolutely regular, so that small differences arise in relation to Universal Time (UT) which is oriented according to Earth's slightly irregular rotation. If necessary, TDT can be converted to Universal Time (related to the Greenwich longitude) or to Central European Time (CET).

of the reference system. Any other angles would be just as good and would—in this and the following depictions—lead to the same results, merely displaced by a corresponding measure of time. Venus is at 181.79° at the starting point, i.e. the 'Venus axis' was directed to the Sun several hours ago or, respectively, it was midday at the relevant point on the surface. The positions 0, a–c show the forward movement against the retrograde direction; the points 1–5 show the sequence of positions at which the 'Venus axis' once again points directly towards the Sun. The period of time during which this takes place is 116.7506 days. (See Appendix 3.4 for the calculation.)

This, too, is a mean value; but owing to the small eccentricity of Venus the variation is only about one day. In relation to the orbit, the duration of a Venus day amounts to *c.* $\frac{13}{25}$ (12.9896:25) which leads to a not very resonant 25-pointed star-figure. This arises by plotting the ensuing positions of Venus at exactly midday—or at any other fixed time such as midnight or the onset of twilight. In the following I shall use the term 'Venus-Sun-View' (VSV).

The interplay with Earth turns out to be more varied. Venus appears a number of times at almost the same spots when it turns its glance onto Earth. And the distance between positions 3 and 4 (Fig. 8.3, right) is also considerably larger than in the other instances. It amounts quite accurately to four fifths of its orbit or, respectively, one half of Earth's orbit. This lengthening arises because during the conjunction which occurs between positions 3 and 4 Venus has, as it were, its back to Earth and therefore needs correspondingly more time to turn once more towards its partner. On average the period of the 'Venus-Earth-View' (VEV) is 145.9277 days. (For calculation see Appendix 3.4.) However, the value obtained is very rarely exact; the shorter periods take on average 133.77 days and the longer one 182.41 days. However, for the purely mathematical resonance the average is decisive, namely when one relates the Venus-Earth-Views to the conjunction period of both planets.

The result obtained shows that during one synodic period Earth is viewed exactly four times by Venus (583.921/145.9277 = 4.0014). With a rotation period of 243.16498 days (see above) the periods would be exactly synchronous. Since the difference is almost indiscernible or, respectively, since 20 years ago it could not yet have been determined with such accuracy, the conclusion reached in various quarters[6] was that in conjunction positions Venus always turns exactly the same side towards Earth. In other words Venus would be as rigidly fixed to the five points of the pentagram as is a moon to the body around which it circles. This, however, would not be at all appropriate for a 'Lady of Heaven', and indeed the actual situation is almost the complete reverse. Because of the tiny difference of the periods amounting to 583.921 or respectively 583.711 days (= 4 ⋆ 145.9277), any assumed fixed point on Venus directed towards Earth during a specific conjunction would take about 1110 years to reach the same orientation towards its neighbour once again, i.e. it would be turned away from Earth for the far greater span of time. We shall soon throw more light on these long-term aspects.

But first I want to show that a good though not perfect superimposition of one conjunction and four Venus-Earth-Views is not a matter of what is in reality a meaningless average value. After all, the sum of *c.* 583.7 days arises from alternation between three of the shorter and one of the longer distances mentioned (3 ⋆ 133.77 + 182.41 = 583.72). As could only be hinted at in Fig. 8.3, we shall see in a moment that this rhythm is founded quite regularly on the sequence of movements. The 4/4 beat of the interplay of conjunctions of Venus, Earth and Mars as shown in the last chapter thus appears in a remarkable way to be mirrored in a somewhat modified form in the rotation of Venus when this is shown in relation to the orbits. So let us now take a look at the way in which the 'Lady of Heaven' moves over a somewhat longer period of time. The positions plotted show the moments when she turns her attention to the planet of human beings:

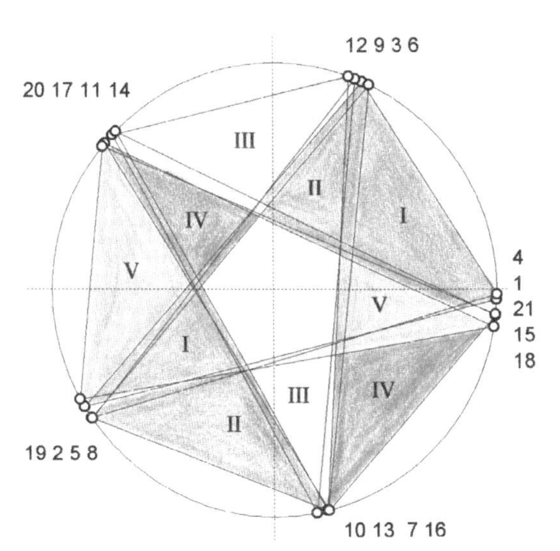


*'. . . the truth about the gods is adapted to mathematical entities, and the Creator of the whole universe used mathematical archetypes, coeternal with Himself, in the construction of the world.'*

Proclus (410–85)[7]

**Fig. 8.4** *Movement diagram showing Venus at Venus-Earth-Views, 21 times, period 7.99 years. The sequence of VEVs is numbered. The Roman numerals show a triangle formed by four sequential Venus positions* → Plate 4

This is utterly breathtaking . . . and the calculations are indeed entirely correct.* Five times four Venus-Earth-Views form an overarching movement sequence just as five times four Venus-Earth conjunctions belong together in a 32-year rhythm (see Table 7.1). At the same time the alternation between three shorter and one longer distance causes a triangle to form and also brings it about that the next cycle of four creates a triangle that is rotated by 216°, which is the amount of the angle of the pentagram. The sides of these triangles could be thought of as the points of a correspondingly larger pentagram. With those of the pentagram shown above their ratio is the square of the golden section (1.618.. ⋆ 1.618.. = 2.618..). One might call this an almost perfect cosmic display of the principle of the golden section. Taking the average of 20 ⋆ 145.9277 = 2918.55, the period of time needed for the figure to take shape is the same (except for 1.05 days, or 0.036%) as that in which the *Pentagramma Veneris* is formed. And, by the way, Venus performs very nearly 12 (12.0096) rotations during this same period. Even though there is no synchronous resonance between the synodic period and Venus' period of rotation, nevertheless the ratios are exceedingly well attuned to one another. But there is no need to discourse at length about this, as Johannes Kepler himself remarked after summarizing his thoughts on the origins of geometry and of human perceptions of geometrical order in the world, whereupon he reached the conclusion that his ideas were very like those of Proclus.

What, meanwhile, is Earth up to in all this? Let us now consider its movements over the course of 32 years together with those of Venus, while also including Mars, the third member of this confederacy (see Fig. 8.5).

The figure drawn by Venus shows that the movement sequence ordered according to the number twenty repeats itself while progressing by an average of 4.1°. As three pentagrams and one pentagon are formed during the course of 20 VEVs, each figure undergoes a deviation of *c.* 1°, so that here, too, one can speak of a relatively exact resonance. This is very much less the case with the figures traced by Earth. Although these are also fivefold they are not pentagrams but in each case approximately 5 lines of a 30-pointed star-figure. In the centre there is also a rather more precise 10-rayed star-figure which can be attributed to the longer period of 182.41 days covering half an Earth orbit. Mars, finally, appears to remain true to its 20-pointed star-figure which resonates extremely accurately with the conjunction period of Venus and Earth. It even forms

* The position data leading to this depiction are given in Table 6.7 in the Appendix and can be checked without recourse to a computer.

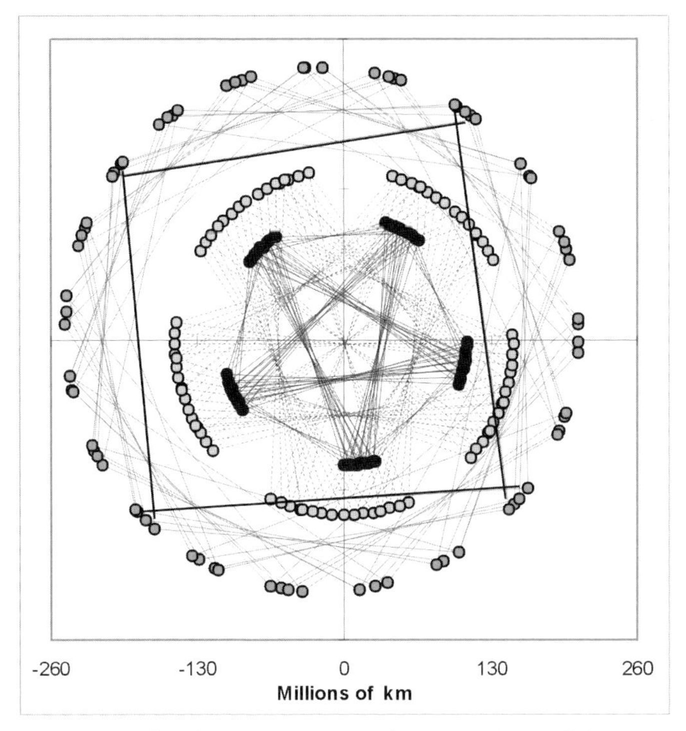


**Fig. 8.5** *Venus, Earth and Mars at Venus-Earth-Views, 81 times (80 connecting lines), period 31.96 years*

two sub-types of this figure, skipping every three corner points and, three times less often, every four corner points. So actually it forms, three times over, four pentagons and five squares. One of the latter is here emphasized for greater clarity. The whole figure of Mars, then, is only complete after all four cycles of 20 Venus-Earth-Views are finished; only then has every line been traced at least once. So in the overall interaction of these three planets we are right to deduce that the Venus rotation, or rather the movement sequences caused by it in relation to Earth, is also built into the 32-year rhythm.

Eighty Venus-Earth-Views, the 20-pointed star-figure of Mars with its five- and four-part divisions, Earth's 10-rayed star-figure and the Venus pentagrams harnessed to the pentagon pertaining to it—all these are linked to one another in such an inconceivable manner that one cannot help but speak of a geometrical and architectural work of art in movement. So perhaps the time has come to call upon a human artist to speak to us not in words but through one of his sculptures. I refer to the sculptor Otto Flath who died in 1987 and who in my opinion is one of the greatest artists of the twentieth century. A visit to the museum dedicated to his work at Bad Segeberg, Germany, is most warmly recommended. In the picture below we have a motif from the twelfth chapter of the Book of Revelation which tells of the appearance in the heavens of a woman upon whose head rests a crown of 12 stars. A link with what we have just been discussing is provided by the fact that the heavenly queen among the planets traces 12 pentagrams in the firmament as she turns her glance onto the planet of human beings in that 32-year rhythm.

**Fig. 8.6** *Altar of the Revelation by sculptor Otto Flath, Church of St Anschar, Neumünster, Germany (detail)*

The figures depicted thus far show the situation arising when Venus and Earth are far away from a conjunction position at Venus-Earth-View. We can recognize it in Figs 2.8 and 8.4, where the pentagrams are rotated against each other by almost the greatest possible amount. As we have already pointed out, when oriented towards Earth during a conjunction the 'Venus axis' will only again point to its neighbouring planet at a conjunction after about 1100 years.* On account of the variations in the

* One can calculate this interval of time by ascertaining the more exact correspondence of a ratio of whole numbers (as 4:1) with the ratio of synodic period and VEV. It comes to 2773:693, corresponding to 1107.87 years. With this degree of accuracy the figure is of course only valid for the basic rotation period of 243.019 days.

timing of synodic periods and Venus-Earth-Views around specific mean values, the transitions are variable, i.e. one can speak of a near conjunction for some time in relation to the VEV. The approximation to this phase is accompanied by a very slow though continuous change in the distances between the separate Venus-Earth-Views. These changes are also noticeable in the movement-figures of the planets. In the case of Venus this means that the pentagon and one of the pentagrams are transformed into a 10-pointed star-figure which is, however, only roughly discernible. The lines of Earth's 10-rayed star-figure diverge and thus merge into a pentagon. However, I want to point out these metamorphoses where their documentation shows them up in the most impressive degree. For this purpose let us recall the period of the Venus day, i.e. the Venus-Sun-View, and compare it with the interval of the rhythm of 80 Venus-Earth-Views:

$$145.9277 * 80 = 11674.22; \quad \frac{116.7506 * 100}{11674.22} = 1.000072$$

corresponding to a difference of 0.0072%

During this period Venus looks at the Sun almost one hundred times. The ratio between VEV and VSV comes fantastically close to that of the major third: 4 Venus-Earth-Views correspond to 4.9996 Venus-Sun-Views. In consequence, 5 VSVs also occur during one synodic orbit of the two planets (583.921/116.7506 = 5.0014). The period of the VSVs is consequently also linked with all the other time periods oscillating in the 32-year rhythm. Initially this may seem surprising, but it must be the case since there is a mathematical dependency between the periods of Venus/Earth conjunctions, VSV and VEV (see Appendix 3.4). It has to be the case in principle anyway, but it appears in this harmonious manner because the rotational and orbital periods are as they are. That the rotation of Venus is so closely linked with its own orbit as well as with that of Earth and, as we shall see, with those of other planets too, is utterly mysterious. By its very unusualness it is just as puzzling for the scientists. Current theories seek to explain this extraordinarily astonishing rotation as being caused—at least partially—by chaotic effects that cannot be calculated in advance and thus produce results on the basis of chance. I shall go into this more accurately in the next chapter. Whatever the actual cause may be, let us have a look at the effects it has produced. Having gained an overview of the long-term tendencies we shall then go into more detail.

Depicting the lines linking the two planets is perhaps the clearest way of expressing the more long-term tendencies. The second of these two diagrams showing the wondrous dynamic of the movements begins close

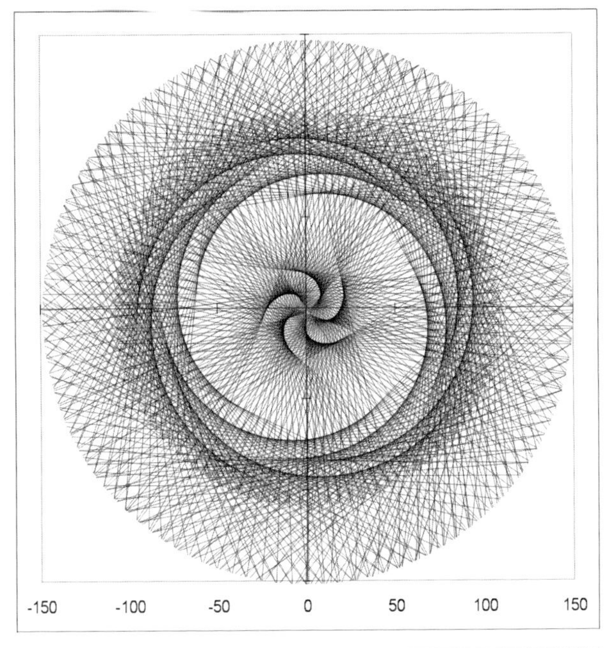


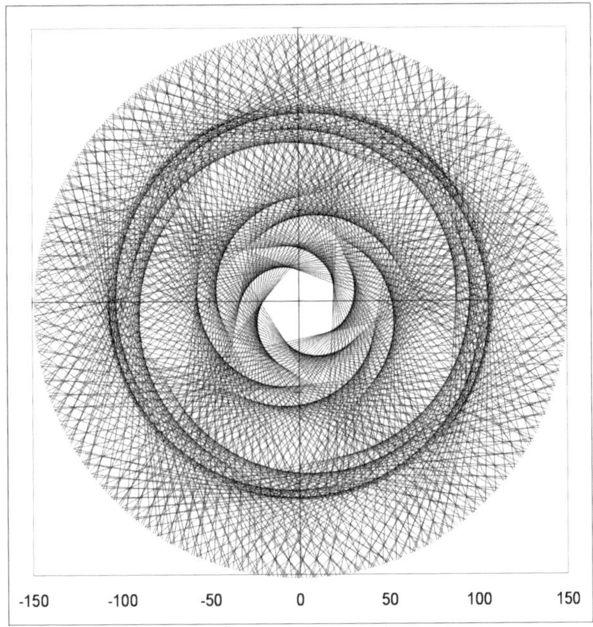


**Fig. 8.7** *Venus-Earth linklines. Top: at Venus-Earth-Views, from 15 January 2411, 700 times, period c. 279.7 years. Bottom: at Venus-Sun-Views, from 15 July 2205, 750 times, period c. 239.7 years (the dates relate to fixing the 'Venus axis' as in Fig. 8.3)*
→ Plate 16

to the conjunction. The first ends there. When investigating other periods of time one finds that by directing the 'Venus axis' to the Sun and to Earth—in relation to the linklines but also in a planet-centred depiction—almost the same long-term movement patterns arise, although shifted along time-wise. In a suitably adapted form this also goes for the other two inner planets when they are viewed by Venus. The changes in the geometric figures formed in shorter periods of time such as those arising in the relationship between Venus and Earth in the 1110-year cycle are, however, unique. Overleaf we show them in a sequence of six details using the linklines at Venus-Sun-Views. Since this is equivalent to one Venus day we might imagine a long-lived astronomer on Venus making a systematic record of Earth's position as it appears in the sky when night falls (or, respectively, where it is when he cannot see it, just as Venus is not always visible to us in the night sky). Our astronomer is sufficiently advanced to calculate his findings as heliocentric coordinates. (Later on we shall also make the acquaintance of an earlier planetary researcher on Venus who persists in adhering to what we might term a Venus-centred view of the world.)

What we see in this picture is utterly sublime.

Beginning with the double pentagram or the inner decagon, a metamorphosis of contracting and expanding pentagons and pentagrams runs its course until in the final picture (where the views, i.e. every fifth view, of Venus to the Sun coincide with its oppositions to Earth) the number ten appears once more, dominating the image with its rays. The pentagram has disappeared, but it will reappear again, as surely as inhalation follows exhalation. After *c.* 550 years half a cycle has been completed; it will now be repeated in the opposite direction until it once more comes close to conjunction. One cannot help comparing this with living, rhythmical processes such as the beating of a heart or the opening and closing of a flower. And this play of forces is made possible because the rotation period of Venus differs in small measure from a resonance tied to Earth. If full synchronization were present, a specific point on the surface of Venus plotted in relation to Earth would always produce an identical pattern.

In his book on the golden section, Walther Bühler introduces us to a series of metamorphoses of the pentagram.[8] The points of the pentagram are bent over inwards until they reach the opposite corner of the inner pentagon. This causes the original pentagram to disappear while two smaller ones come into being. It takes quite an effort of geometrical imagination to picture this metamorphosis to oneself, but a great deal more mental agility must have been needed to create this development

**Fig. 8.8** *Venus-Earth linklines at Venus-Sun-Views, 75 times in each case, corresponding to 23.97 years. From top left to bottom right: beginning close to conjunction on 17 April 2133 and then, relating to the initial date, 58.5, 126.9, 225.7, 414.6, 543.1 years later*

out of the geometrical possibilities inherent in the pentagram. Bühler sees this process as being symbolic of the mysterious ability of living things to divide like individual cells or to proliferate like the higher organisms, to pass away while leaving descendants to follow, thus maintaining the chain of life. The 10-pointed/5-pointed star-figure metamorphosis that rules in the interplay between Venus, the symbol of love, and Earth, the human planet (whereby Venus plays the greater part), seems to me to be showing us not only a geometrical symbol of life but also a picture of the way in which celestial bodies, geometry and the characteristic marks of life are all inextricably interwoven with one another.

Immersing oneself in these movement sequences is akin to undertaking a kind of geometrical meditation. The same goes for the corresponding picture of the 'Venus axis' when Venus turns its gaze to Earth. We shall be coming to this towards the end of the chapter. For me the discovery of these archetypal images was among the most wonderful experiences I had while working my way towards the conclusions put forward in this book. When the mysteries connected with the rotation of Venus were beginning to dawn on me I did not yet have at my disposal the computer programs capable of entering hundreds of planetary movements in circles on the monitor within a few seconds as in the standardized, square graphics shown here. But observing how a figure begins to take shape as one works with compasses, ruler and protractor and immersing oneself in the thoughts that arise in consequence (for the subtleties of its form are not always immediately recognizable even once it has been put down on paper) is something far more humanly personal than whatever the computer presents in the twinkling of an eye, however essential it may be as a tool. I mention this in order at least to give the reader one opportunity to participate in this experience. If your curiosity has been aroused you might like to draw the figure that arises through the rotation of Venus in relation to Earth yourself, before looking at it in the coming pages. The necessary data are given in the table in Appendix 6.7.

The following two diagrams show the longer-term figures that arise between Venus and Mars or, respectively, Venus and Jupiter (Fig. 8.9).

The periods for the Venus views are on average 179.515 or 230.112 days respectively. With regard to Mars the linklines at Venus-Sun-Views were plotted. When Venus is looking towards Mars a very similar 10-pointed starlike movement picture arises, although a shade less resonant. In the planet-centred mode of depiction ten inward-facing loops arise. In the case of Jupiter and also of the other planets the period is longer than one Venus orbit. Presumably this is the reason why the similarity of the figures is no longer, or only approximately, present. In the case of the

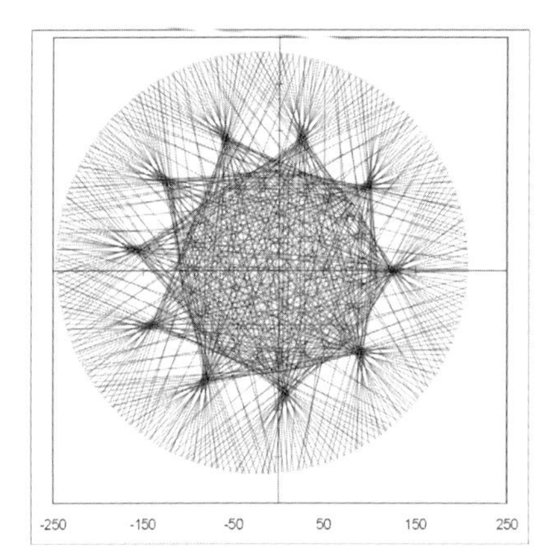


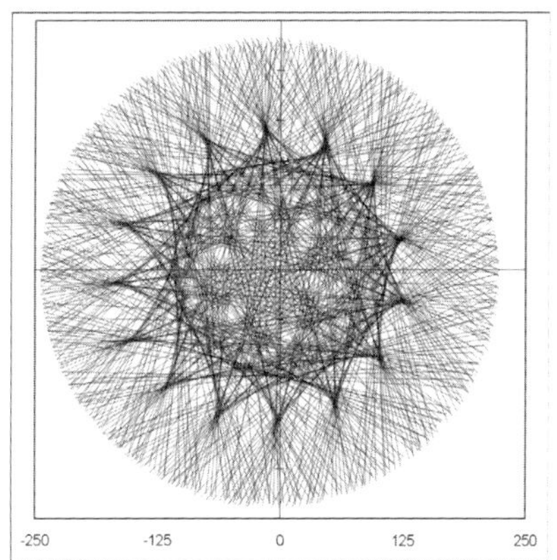


**Fig. 8.9** *Linklines. Left: Venus-Mars at Venus-Sun-Views, 500 times, period 159.8 years. Right: Venus-Jupiter at Venus-Jupiter-Views (detail enlargement 10:3), 750 times, period 472.5 years. Scale in millions of kilometres*

Venus-Jupiter-View, the Venus-Jupiter connecting lines form a clearly visible 15-pointed starlike figure.

To sum up we can say that the order formed primarily by the number five and its initial multiples, which arises through the rotation of Venus, reaches up to Jupiter in a very clear sequence facing outwards. In this sense the Babylonians were entirely right in giving the second planet in our solar system the name Ishtar, the 'Lady of Heaven'. Again and again we cannot but marvel at the appropriateness of many of the ideas—translated into modern concepts—developed before the days when telescopes, radar equipment or computers began to dominate astronomy. But of course these instruments are essential in helping modern humanity to recognize those ideas. Until the 1960s the foundations of what was going on remained totally inaccessible to human view underneath the impermeable cloud cover of Venus. And then it took another 35 years before a firm belief in a celestial order, and in the fact that men like Proclus and Johannes Kepler were not mistaken, made it possible to find the signs of the geometrical harmonies expressed through rotation.

We must now turn inwards towards the place where the messenger of the gods pursues his orbit. Or we might express our question in a different way by asking what is revealed in the inmost part of the pentagram or of the figures and their metamorphoses related to it. What is the spiritual context of the images that are coming into view? Are they merely the consequence of an accidental rotation period and brought into existence by geometrical constructs in accordance with purely mathematical laws? Or is there a kernel of truth in the mythological notion of Mercury being the bearer of a message?

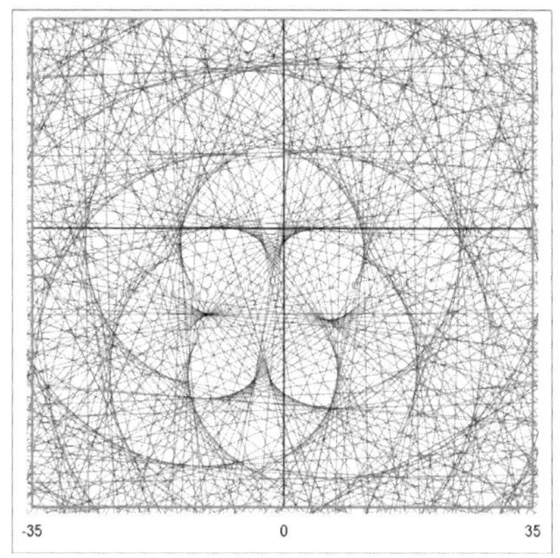


**Fig. 8.10** *Venus-Mercury linkline at Venus-Mercury-Views, 800 times, period c. 255.7 years. Right: detail enlargement. Scale: millions of kilometres*

So the cross and the pentagram are the geometrical principles, brought down to the simplest denominator, on which the movements of the inner planetary system are founded. In its interplay with the other planets, that very unusual rotation of Venus thus reveals the symbol of the Christians as well as that of the Pythagoreans. These are the symbols that stand for thoughts through which their spiritual originators gave to humanity what is probably the greatest moral and religious impulse on the one hand and the greatest intellectual and scientific impulse on the other. To combine these two seemingly, or often also actually, contradictory orientations was and is still one of the most arduous spiritual and cultural tasks with which humanity has to grapple. Hermann Hesse claimed that this has already been successful in one particular field. In his novel *The Glass Bead Game*, he wrote: '*We consider classical music to be the epitome and quintessence of our culture, because it is that culture's clearest, most significant gesture and expression. In this music we possess the heritage of classical antiquity and Christianity . . . Classical music as gesture signifies knowledge of the tragedy of the human condition, affirmation of human destiny, courage, cheerful serenity!*'[9]

For long ages, in the sky of our solar system, both those symbols—that of antiquity, in general terms, and that of the belief in a force that is more powerful than death—have together been sharing in the task of ensuring that human beings shall be provided with a dwelling place.

Of course one could say that the thoughts expressed in the previous paragraphs amount to reading more than is justified into the images created by the planets through their movements. But an astronomy that has withdrawn from any link with humanity and with the spirit is just as certainly a science that has become quite radically impoverished. And even from a not utterly one-sided scientific point of view such a reduction is moreover not at all desirable and lacks any justification. Surely this is made obvious by the extraordinary improbability of our universe coming into being at all, quite apart from the tremendous steps it has taken in its evolution (among which one must of course also count the development of human consciousness). And further evidence is perhaps also provided by at least some of the research results presented here. However, since words so easily lead to arguments we shall once more call upon the sculptor Otto Flath to speak in his own way as we draw near to the end of this chapter. He is one of those who place the heavenly queen shown in Fig. 8.6 within a wider context involving both those special symbols.

**Fig. 8.11** *Altar of the Revelation by Otto Flath, Church of St Anschar, Neumünster*

We can now once again imagine the situation of that astronomer on Venus. His observatory is positioned at the spot where we have fixed the planet's axis, i.e. the spot which points towards other celestial bodies at specific moments in time. Whenever the two planets closest to his own reach the position along the extended axis which corresponds to the Venus View, what he sees from his standpoint is shown in the following two illustrations.

The period of time is the same in both pictures. Of course the Venus-centred astronomer cannot discern the distances involved in the diagrams, with the exception, perhaps, of the difference in the planets' brightness. As explained in Chapter 6, the planet-centred view leads to a depiction that is valid for both the planets involved except for being rotated by 180°. Thus the right-hand picture in Fig. 8.12 also shows the image traced by Venus in the firmament as seen from Earth during Venus-Earth-Views.

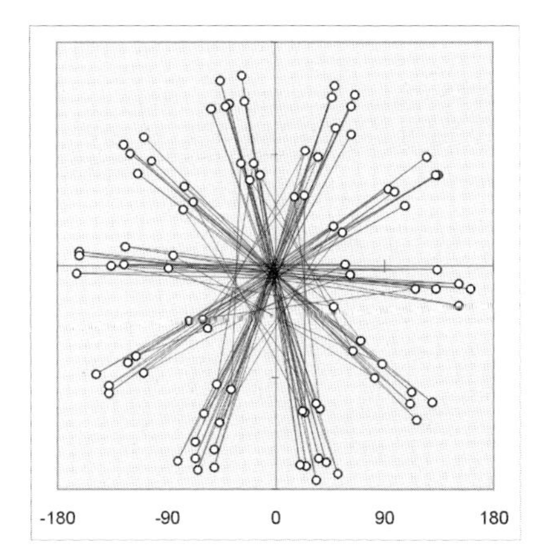


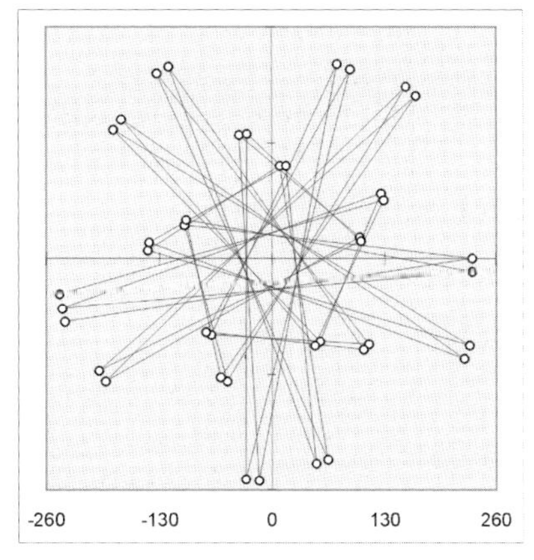


**Fig. 8.12** *Venus-centred views of Mercury and Earth. Left: Mercury at Venus-Mercury-Views, 90 times. Right: Earth at Venus-Earth-Views, 41 times*

Finally we have the geometrical figure that arises directly from the rotation of Venus when Venus looks at Earth, i.e. the sequential points arising when Venus' horizontal 'axis' points towards Earth are plotted. These points are entered on a circle that represents the rotating surface. Putting ourselves in the situation of the observer who wants to find out the movement figure traced by his home planet whenever his cosmic neighbour appears over the horizon, we discover that the resulting rotation figure (which corresponds to the right-hand illustration in Fig. 8.12 except that it is disengaged from the distances between the planets) is the very one which can actually be observed in the sky or, rather, which can be constructed from longer-term observations.

As in Figs 8.4 and 8.12, the movement sequence comprises 20 steps which are then repeated with, here again, a slight displacement. The formations based on the numbers four and five, to be found in almost all the figures shown in this chapter, appear here almost like a synthesis of a small work of art. Since the figure is not dependent on distance, one could say that as simply as possible it reveals an archetype which might be the foundation of the concrete phenomenon that is actually worked out on a spiritual plane. Or, as Johannes Kepler put it: '*Geometry, which before the origin of things was coeternal with the divine mind and is God himself . . . supplied God with patterns for the creation of the world.*'[10]

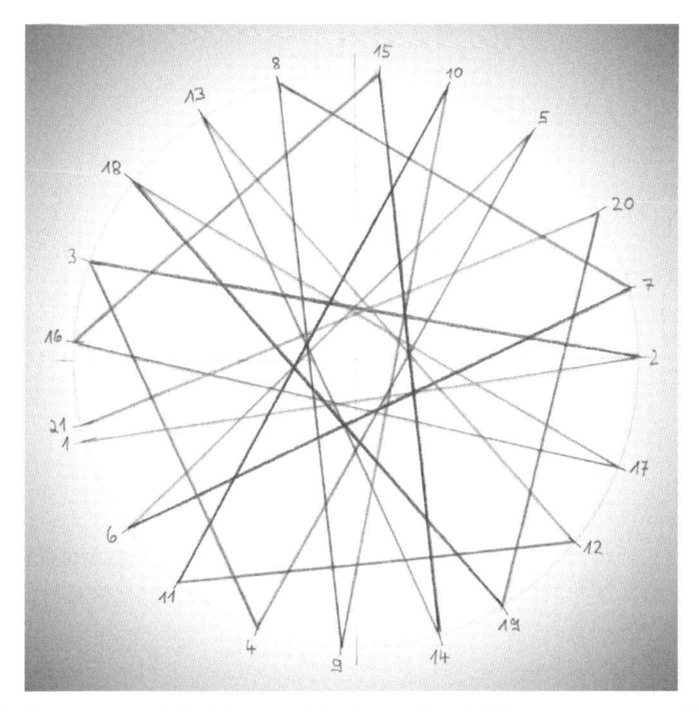


**Fig. 8.13** *Rotation of the 'Venus axis' at Venus-Earth-Views* → Plate 5

# 9. Rotations

'One of the most profound mysteries of the universe consists in the fact that everything in it rotates,' wrote the English astronomer Sir Arthur Eddington.[1] And the way in which the celestial bodies turn on their own axes is, surely, a mystery of similar magnitude. We saw in the previous chapter how closely the rotation of Venus is linked with the orbits of the inner planets. That particular section of the universe certainly gives the impression that all movements—rotations as well as orbits—are interconnected in a structured manner. Perhaps relationships of this sort are more akin to those that exist between the various sections of a musical composition than to dependencies which are physically tangible and entirely determined by natural laws. As we gain familiarity with a symphony we begin to discover initially hidden inner correlations which are certainly present even when external scientific analysis of the music can throw only partial light on them. The secrets of the creative process cannot be fully discerned from the outside. Not until we engage in our own inner act of re-creation can we broaden our limitations in this respect. But by doing so we of course also run the risk of reading something of ourselves into the things we discover. Although this can never be entirely avoided, we can endeavour to minimize it by adhering as closely as we can to the notes we hear and by letting the music speak for as long as possible before we ourselves begin to hold forth and express our own interpretations.

The purely scientific view, too, has to make its way along the narrow watershed that runs between understanding and interpreting. And when we reach, or think we have reached, an understanding, this, too, is still influenced not only by our own preconceptions but also by ideas which we do not include in our considerations. It is thus no surprise that I have so far not found in the relevant literature anything that questions whether the way in which the various celestial bodies of the solar system rotate might have something to do with a systematic coherence of the whole. In general the assumption is that over the course of aeons these rotations have either gradually swung into their present measure more or less by chance, and will remain subject to further very slow changes, or that there is a direct gravitational influence at play, as is the case with the moons or Mercury, which I shall come to shortly. One assumes that originally all the rotations in the solar system originated in those of the cloud of gas and dust from which our cosmic home is presumed to have emerged. The rotations of the planets, too, are seen as resulting from that of the protoplanetary disc from which they are derived in accordance with the ideas described in Chapter 5. We shall be considering how it could have been possible for the rotations present today to emerge from the collisions and agglomerations of the planetesimals circling around the Sun. But let us first examine this star at the centre of our system and consider some of its characteristics.

Within the Sun, hydrogen is transformed into helium by means of nuclear fusion. This is what produces the immense amount of light and heat that has been incessantly radiating out into space for billions of years, creating in the process the conditions needed for life on our planet. In view of this, the first surprise we have to consider is why such a mode of energy production does not cause, or has not long since caused, the Sun to explode according to the principle of the hydrogen bomb. Well, our Sun, and with it billions of others, is sufficient proof of the fact that controlled nuclear fusion is possible within stars. In physics the explanation given is, firstly, that the two hydrogen protons needed for fusion into a helium atom only very rarely come sufficiently close to one another to cause a reaction and, secondly, that in every part of the Sun there is a hydrostatic balance between all the forces present, whether they are directed outwards or inwards. But it is this very balance that is so startling since many processes take place inside the Sun, most of which we have scarcely begun to comprehend. Among them are the processes that guarantee the transport of radiation to the Sun's surface, or those that lead to the cyclical increase and decrease in the frequency of sunspots, or to gigantic loop formations of various kinds termed solar prominences and flares. We are quite justified in making a comparison between such balance amid countless dynamic processes and the biochemical processes in living creatures which always remain in balance within a specific range, thus guaranteeing the continued existence of the organism in question.

Another highly astonishing observation in modern solar research is the fact that the Sun fulfils its task of giving us life in a thoroughly musical way. This is not supposed to be an allegorical description (well, perhaps it is, in the first half of the following quotation) but a true statement of fact. '*The Sun is playing a secret melody, hidden inside itself, that produces a widespread throbbing motion of its surface . . . the entire Sun rings like a bell for days and weeks at a time.*'[2] This phenomenon was first observed in 1960 by measuring minimal rhythmical up and down movements of the visible solar surface (termed the photosphere) which pulsate approximately in a five-minute rhythm. Subsequent more accurate observations showed that the vibration in question is composed of millions of different individual oscillations of varying frequencies. Sound waves are generated in the convection zone beneath the Sun's surface. This region comprises about one quarter of the Sun's diameter. According to current research, it is there that the energy is transported to the surface by means of fluctuating gaseous currents. The turbulences that arise in the process are said to cause the inner regions of the Sun to resonate. On reaching the surface these sound waves are reflected back to the inside, so that the Sun is like a resonating space similar to the hollow interior of a guitar contained within its wooden sides. The overall effect resembles the sounding of a bell involving vibrations of the whole body of the Sun which last for days and weeks.

Many of the Sun's activities are linked to magnetic processes. Magnetic fields are constantly generated, altered, shifted and pole-reversed within the Sun. Sunspots, first observed in the seventeenth century, are also a product of solar magnetism. Here very strong magnetic fields arise which interfere with the convection and thus the radiation of energy. Where this occurs the Sun's surface is cooler, so that the spots appear darker. The sunspots which, seen from Earth, move once across the surface of the Sun in about 27 days, also led to the Sun's rotation being discovered. Systematic observations conducted in the nineteenth century then showed that there was something unusual about this rotation of the Sun: it moves more rapidly at the equator and more slowly towards the poles. This is termed differential rotation. However, in the astronomical literature we find that a fixed value is given for mean solar rotation. This was measured accurately in the mid-nineteenth century by the amateur astronomer Richard Christopher Carrington (1826–75) who also discovered the differential rotation. The value is given as 25.38 days,* i.e. a synodic period of rotation of 27.2752 days as seen from Earth, which is the average period of time taken by the sunspots to rotate once around the axis. The appearance of new spots (each one being visible for a maximum of about 100 days) occurs in a cycle of *c.* 11.07 years on average, during which the locations of the new appearances migrate from about latitudes

* The value of 25.38 days is found in most of the literature. This can give the impression that the further numbers after the decimal point have been deleted thus rendering it not all that accurate. The figure of 25.38 days is based on Carrington's measurement of the angle covered by a spot in one day on average. In relation to the sidereal rotation this value amounts to 14.1844° (Waldmeier, p. 44). This leads to a period of 360°/14.1844° = 25.3799949 days, so that the figures shown as we proceed, which are calculated on the basis of 25.38 days, can in this regard be seen as sufficiently accurate.

30° north and south up to the equator. The sunspot minimum (meaning the period during which the fewest new spots appear) occurs with sunspots closest to the equator. The period of the sunspot cycle is actually 22.14 years, since the sunspots in every second 11-year period have a different magnetic orientation.

The magnetic activities of the Sun are also felt, and probably have been felt, on Earth throughout history. For example sediments in an Australian glacier lake dating back 680 million years show a sequence of 'annual layers' corresponding to an 11- or, respectively, a 22-year cycle which is also overlaid by a longer cycle. This points to variations in the strength of solar activity such as can still be measured today. The widths of the various layers indicate the annual amount of precipitation. At intervals of about 11 years there is 'a dark clayey band. Alternate 11-year intervals often show systematic fluctuations between dark bands spaced widely and closely', showing that the 22-year period has also left traces.[3] So we can conclude that the magnetic processes on the Sun have remained at least more or less stable even over long geological periods. Another effect is revealed in 'geomagnetic storms and short-wave radio fade-outs which show approximately a 27-day period'.[4] They mirror the rotation of large, stable gaps in the solar corona which point towards Earth in the rhythm of the synodic period, thus allowing a greater portion of the solar wind to pass through, which causes magnetic storms. The solar wind is a permanent stream of particles varying in strength which emanates from the Sun and fills the whole of the solar system.

In summary we can say that the Sun, and with it every other star, amounts to a great deal more than a nuclear fusion reactor brought into existence by the original agglomeration of a cosmic cloud. It is increasingly being revealed as a great mystery with which we are closely connected. That it shines is as much a matter of course for us as the functioning of our own body; stars come into being and die away just as do living creatures on our planet. In every realm the seeking human spirit has hitherto been denied access to an understanding of the most profound causes affecting the world in which we live, so that we seldom come to realize just how filled with marvels it is.

Before now turning to the actual subject of this chapter I want to mention one further point regarding a movement of the Sun apart from its rotation, namely its movement around the common centre of gravity of the entire solar system. The law of gravity states that the manner in which two bodies attract one another depends on their masses. In the case of the solar system it is generally possible to ignore the force exerted upon the Sun because the mass of all the planets together amounts to only about one eight-hundredth of that of the Sun. In fact, however, the gravitational effect of this causes the Sun to move around the overall system's centre of gravity which is termed the barycentre. In doing so the centre of the Sun moves a maximum of 1.5 million km from this centre, which amounts to 1.1 times the diameter of the Sun itself. In other words, at times of greatest distance from the barycentre, the real Sun is situated only a little removed from the position that would be occupied by an imaginary, fixed sun. At that moment the situation would be almost precisely that of an outer circle rolling epicyclically over an inner one. And this is indeed what is shown when the movement is plotted. The illustrations here give examples of various situations, in each case over a period of just over 60 years, i.e. the duration of three Jupiter/Saturn encounters.

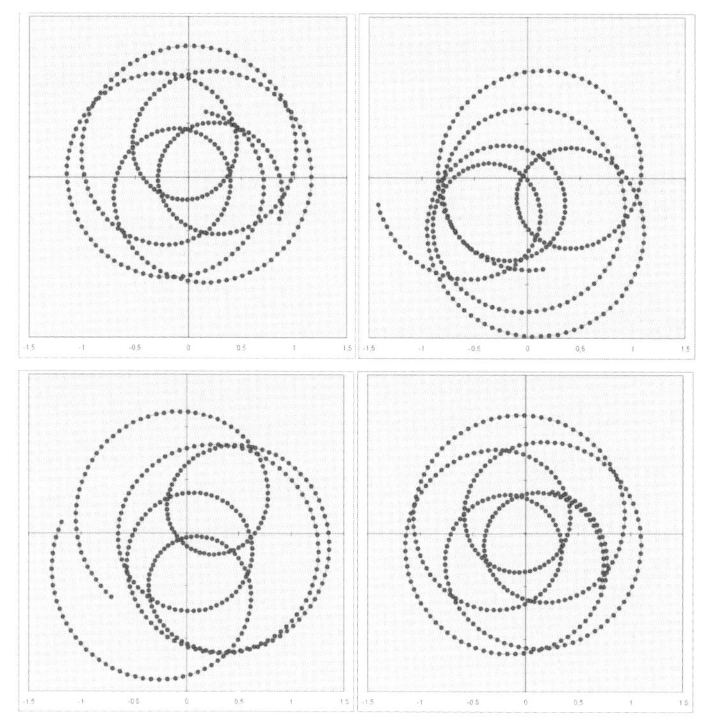

**Fig. 9.1** *Movement of the centre of the Sun around the barycentre of the planetary system, stepping interval 60 days, 380 times, period 62.42 years in each case, scale in millions of km. Top left, from 1 January 1380; top right, from 1 January 1626; bottom left, from 1 January 1980. In the final figure, bottom right, only the gravitational forces of Jupiter and Saturn are calculated (from 1 January 1380, as in top left)*

From 1380 till 1442 (top left) the movement of the Sun follows almost exclusively the interaction of Jupiter and Saturn (bottom right).* During this period of time there is an opposition of Uranus and Neptune in 1393. So these two gaseous planets, with their virtually equal masses, are situated on opposite sides of the Sun, thus more or less cancelling out one another's gravitational influence. The two other time periods show rather more disorderly movement sequences, although the three loops of the Jupiter/Saturn relationship are still recognizable. In these we see Uranus/Neptune conjunctions (in 1650 and 1993), in which the two exert their pull in the same direction and thus disturb the primary loops. In this connection the influence of the much smaller inner planets is entirely negligible.

So the movement of the Sun around the barycentre is influenced primarily by Jupiter and Saturn, the two planets with the greatest mass, while there are also times when the influences of Uranus and Neptune lead to much more irregular pictures. In a later chapter we shall give an airing to the almost unbelievable mystery of how the movements of the four large planets are ordered in the longer term. At first, however, a more or less repeating sequence of 3★3 loops is formed which are combined in a cycle of the Sun's movement. The duration of one such cycle is taken to be on average 178.733 years which corresponds to nine Jupiter/Saturn conjunctions except for minimal deviations (< 0.004%). Various researchers therefore speak of the measure of the Jupiter/Saturn synodic period of 19.859 years as the '*pulse of the solar system*'.[5] The period of just under 179 years is said to be linked to the longer periodic variations of the number of sunspots in that the phases of orderly solar movement coincide with a maximum of activity and those of more chaotic movement with minimum activity. Such considerations show that the search for order in the solar system is still on-going in a number of fields.

## *Sun and Venus*

Arithmetic, however, can only supply us with the scaffolding. Although exact calculations do provide the basic foundation, the results obtained without its banished sisters music and geometry are, in the last resort, merely some not very convincing partial aspects of the mysterious architectural structure that is built into our solar system. This is shown

---

* On a reduced scale this figure corresponds with the centres of gravity on the linklines as shown in Chapter 6 for other planetary constellations.

clearly by some harmonies connected with solar rotation that can only be depicted clearly with the help of geometry. We shall now look at these more closely one by one.

First let us investigate the Sun's rotation when viewed from Venus. As we saw in the previous chapter, the Venus-Sun-View (VSV) relates to all the planetary orbits right up to that of Jupiter in a way that bears the stamp mainly of the number five. As previously in the case of the Venus rotation shown in Fig. 8.3, in order to depict the movement of the solar rotation we shall define an axis in the equatorial plane pointing towards the vernal point on 1 January 2000 (TDT) at zero hours. In the following diagram we have plotted the point towards which the rotating 'solar axis' is directed whenever Venus looks with its 'Venus axis' towards the Sun. Here too the sequentially arising points are then joined up:

*'The sun makes music as of old*
*Amid the rival spheres of Heaven,*
*On its predestined circle rolled*
*With thunder speed: the Angels even*
*Draw strength from gazing on its glance,*
*Though none its meaning fathom may:*
*The world's unwithered countenance*
*Is bright as at Creation's day.'*
J.W. Goethe[6]

**Fig. 9.2** *Solar rotation at Venus-Sun-Views, 250 times, period 79.91 years. The vernal point on 1 January 2000 (0°) is in this instance situated at the top of the diagram*

And, sure enough, as Johannes Kepler might have said, we find a pentagram or, as in the picture, 50 of them lying so close together that their lines are superimposed one upon another. The precision of the ratio of 23:5 (23.0005 solar rotations to 5 VSVs) which is expressed here is phenomenal; the value of the cycle resonance is only 0.184°. This also means that one of the solar pentagrams is almost formed in the period of one Venus/Earth synodic period, and that therefore one Venus/Earth pentagram corresponds to five Sun pentagrams drawn by 25 VSVs (the deviation amounting to 0.84 days in 8 years, or 0.029%). Furthermore, the pentagram of the Sun completes the sequence of movement figures of the planets when Venus is aimed towards them, shown in the previous chapter: 5 (Sun)—4 (Mercury)—4*5 (Earth)—10 (Mars)—15 (Jupiter). So we can really say that it would be impossible to conceive of a more geometrically and architecturally convincing correlation between the Sun's rotation and the interrelationships thus far discovered in the inner solar system.

Furthermore, five Venus-Sun-Views take almost the same amount of time as four Venus-Earth-Views, so that purely arithmetically we find the corresponding figure of a square superimposed on the pentagram being formed during the course of 23 solar rotations. However, since the time taken for the VEVs to be completed varies from about 133 to 182 days, it is not necessarily a matter of course that the rotating Sun would actually trace a regular square if the relevant movements were plotted. It is only after every fourth VEV that a period of time of 583.7 days occurs which is subject only to slight variations, as we can see in Fig. 8.4. In relation to the rotation of the Sun, the intervening intervals could lead to any kind of figure if it were expected to obey only the condition that every fourth point coincides approximately. Yet there is probably no need to be a prophet in order to foretell that, regardless of there being many other theoretical possibilities, in this way too the pentagram is given a frame. The Sun's rotation thus gives us a pretty accurate picture of the Venus-Earth-Mars relationship (see Fig. 2.10). This certainly holds good for most of the time during which the Venus-Earth-View does not take place when the two planets are close to conjunction. In those instances the figure changes for a while.

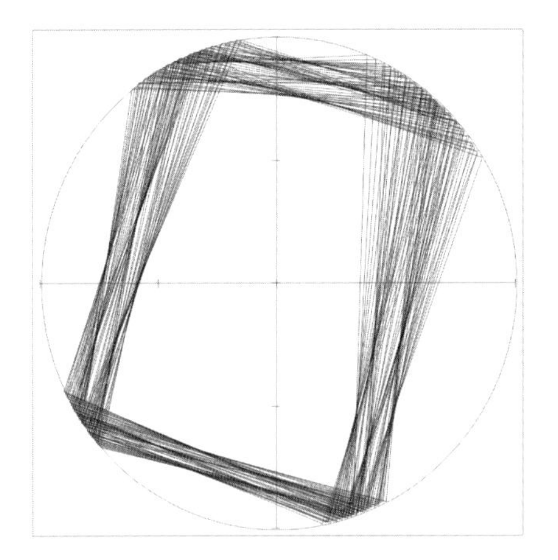

*'Hear the tempest of the Hours!*
*For to spirit-ears like ours*
*Day makes music at its birth.*
*Hear it! Gates of rock are sundering*
*And the sun-god's wheels are*
*thundering:*
*See, with noise light shakes the earth!*
*Hear it blare, its trumpets calling,*
*Dazzling eyes and ears appalling,*
*Speechless sound unheard for dread!'*
J.W. Goethe[7]

**Fig. 9.3** *Solar rotation at Venus-Earth-Views, 250 times, period 99.88 years*

The solar rotation, then, is interwoven with the interactions determined by the Venus-Views and the conjunction periods of Venus, Earth and Mars. This is shown more comprehensively in a table incorporating all the relevant periods of time (Table 9.1). But first let us complete the picture by considering a further constellation, namely one involving Mercury, Venus and Jupiter. This next diagram shows the movement of Venus which arises when the sequence of its positions at Mercury/Jupiter conjunctions is plotted.

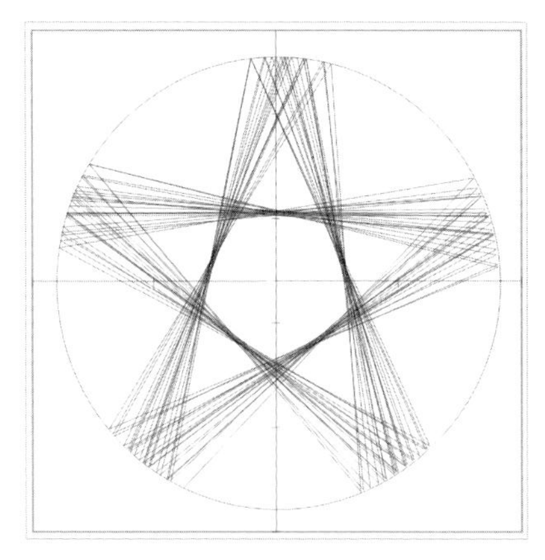

*'How it all lives and moves and weaves*
*Into a whole! Each part gives and*
*receives,*
*Angelic powers ascend and redescend*
*And each to each their golden vessels*
*lend;*
*Fragrant with blessing, as on wings*
*From heaven through the earth and*
*through all things*
*Their movement thrusts, and all in*
*harmony it sings!*

J.W. Goethe[8]

**Fig. 9.4** *Venus in its orbit shown at Mercury/Jupiter conjunctions, 100 times, beginning 9 February 2000, period c. 24.58 years*

Here once again the sign of the inner planetary system makes its appearance. Five conjunctions of the highest divinity with the messenger of the gods bring it about that the goddess of love traces in the firmament the symbol of the human being or, shall we say, the Sign of the Microcosm. Let us call this 5-pointed star-figure the Minor Pentagram. It comes into being during an average period of 448.96 days in a rather resonant manner: 5 encounters of Mercury and Jupiter amount to 1.998 Venus years; the cycle-resonance is $-0.704°$. The synodic period of Mercury and Jupiter is 89.7924 days and thus almost half the period of the Venus-Mars-Views (VMaV)(0.50018 ★ 179.5189). In combination, the kind of star-figure which arises and the exactitude of the additionally appearing octave ratio together indicate that both the Sun's closest companion and the planet which reigns in the middle region of the overall planetary system are harmoniously linked in the web woven by the three planets which lie between them. The number ratios that apply are shown in the following table:

**Table 9.1** *Periods in the inner planetary system*

| | Periods (Days) | Ratio numbers | ① ★ ② (Days) | Deviation from mean (%) |
|---|---|---|---|---|
| | ① | ② | ③ | ④ |
| Ea/Ma | 779.93610 | 3 | 2339.808 | 0.168 |
| Ve/Ea | 583.92137 | 4 | 2335.685 | −0.008 |
| Ve/Ma | 333.92152 | 7 | 2337.451 | 0.067 |
| VMaV | 179.51889 | 13 | 2333.746 | −0.091 |
| VEV | 145.92772 | 16 | 2334.843 | −0.044 |
| VSV | 116.75059 | 20 | 2335.012 | −0.037 |
| Me/Ju | 89.79241 | 26 (2★13) | 2334.603 | −0.055 |
| Sun | 25.38 | 92 (23★4) | 2334.960 | −0.039 |
| | | *Mean* | 2335.878 | 0 |

With small to very small deviations, all the periods involved amount to whole numbers in the fundamental rhythm of four Venus/Earth synodic periods worked out in Chapter 7. The frequency of this synodic period also coincides most accurately with the mean of all the time periods shown. As mentioned at the beginning of this chapter, the purely scientific view of the universe merely concludes that the ratios shown must have been brought about by a mixture of random collisions of blocks of matter and the self-regulation caused by gravity.

Here, however, we are concerned with bringing all the ratios shown in Table 9.1 together in an overall picture, although 'all' here means 'almost all' since the ratios of the periods of the Venus-Mars-Views and of the rotation of the Sun are not included in Fig. 9.5. But as we have seen, the value of the VMaV represents nothing other than the octave of the Mercury/Jupiter synodic period. Thus it would be possible simply by multiplication (or, respectively, division) by the number two to derive the ratios of the former from the ratios of the latter shown in the picture. The ratios arising from the magnitude of the Sun's rotation have also been omitted since they are no longer formed by small whole numbers. If one wishes, one can derive these from the relationships of the Venus/Earth synodic period which is 23 times the rotation of the Sun.

At the top we have the figure of Venus at VEV which consists of 3 pentagrams and 1 pentagon (see Fig. 8.4). The Venus/Earth pentagram, the Minor Pentagram, and the Sun pentagram are united in one of the triangles of the hexagram. The lines which show the most exact correspondence with the relevant number ratios are emphasized; the deviations here are in every case less than 0.05%. The ratio of the three pentagrams is 13-10-2. In other words, 10 Sun pentagrams and 13 Minor Pentagrams

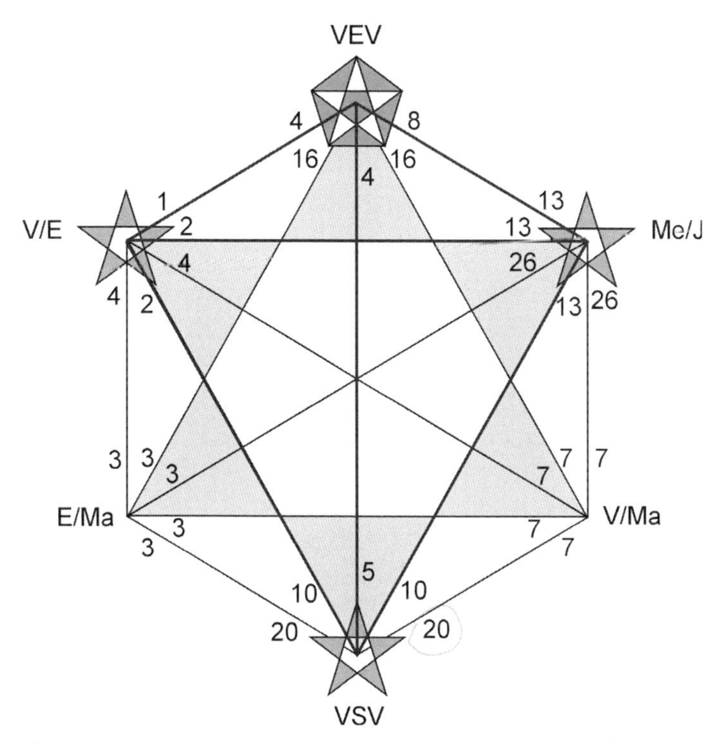


**Fig. 9.5** *Hexagram of the relationships in the inner planetary system. The ratios shown beside the lines joining the points indicate, for example bottom right: 20 Venus-Sun-Views (VSVs) ≅ Venus/Mars conjunctions (V/Ma)*

are formed during the course of 2 Venus/Earth pentagrams. This means that 25 pentagrams encompass the movements in the inner planetary system. In addition, Jupiter is included in what happens via the Minor Pentagram. By using a hexagram to depict all the relationships arising we have probably chosen the most suitable and simplest method. Perhaps we can discern in this picture something like a spiritual message in the widest sense, a message that speaks to the onlooker through the movement sequences of the celestial bodies of our solar system which are so clearly attuned to one another. We hope that this message will be further clarified in the coming chapters.

## *Sun, Mercury, Venus*

Let us now turn to the rotation of Mercury. This, too, was only discovered, at about the same time as that of Venus, once radar became available. It takes 58.6462 days, i.e. exactly two-thirds of the planet's orbital period, a resonance that is attributed to the tidal forces of the Sun which over the course of time brought about the link between the rotation and orbit of the planet. In this sense one might almost speak of Mercury as the Sun's moon. However, because of the relatively high eccentricity of Mercury's orbit combined with the planet's slightly deformed shape, the ratio here is not 1:1 as with most moons, but 3:2, as we have already mentioned. The effect of this for a potential inhabitant on Mercury is that one Mercury day takes two Mercury orbits or years, namely 175.94 Earth days. One could imagine it becoming rather cold during the night and all the hotter during the day on this companion of the Sun.

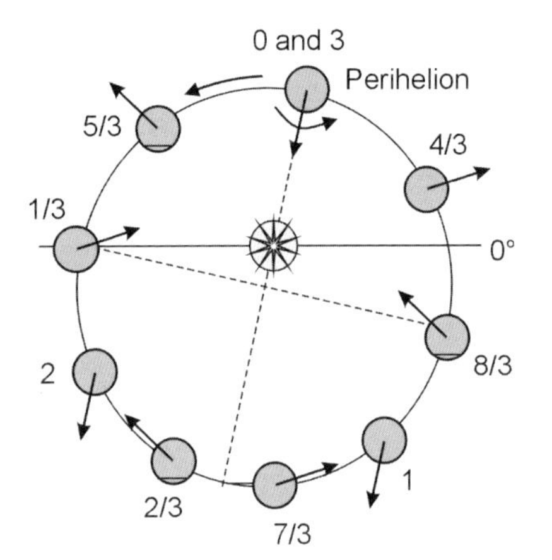


This picture shows the position of Mercury after, in each case, one third of a turn, starting from a position at perihelion. After exactly 3 rotations or 2 orbits the same point once again points towards the Sun.

**Fig. 9.6** *Schematic diagram of Mercury's rotation*

A more long-term depiction of Mercury's position at intervals of a Mercury day, or one could say at Mercury-Sun-Views, would still show only one point because the planet is then always at the same position again. Or perhaps one should talk of half a point, since it only points towards the centre after every second orbit. As in the case of Venus, one can then investigate which configurations arise when the messenger of the gods looks towards his addressee. We shall demonstrate this by means of one example only, since the 3:2 resonance between orbit and rotation always applies. After some time this leads basically to similar movement pictures regardless of which planet Mercury is looking at. The following shows a planet-centred depiction, whereby the horizontal Mercury axis is once again fixed at the vernal point on 1 January 2000, zero hours (TDT).

Although this is a beautiful threefold figure, one cannot attach too much importance to it because, as we have said, it is a direct consequence of the 3:2 resonance. Apart from the fact that Mercury's rotation is bound

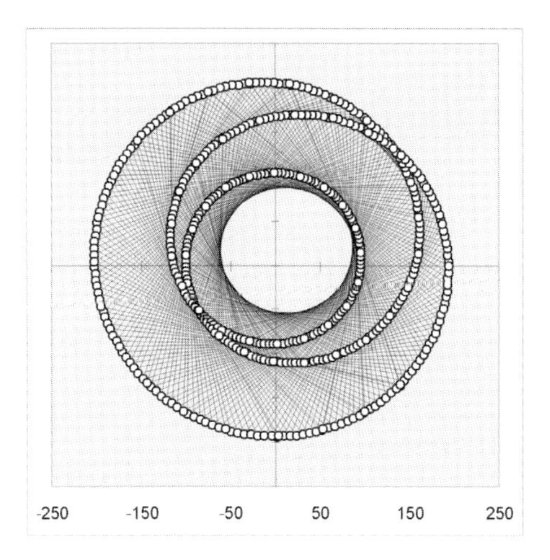


**Fig. 9.7** *Earth from Mercury-centred view at Mercury-Earth-Views, 500 times, period* c. *95.6 years (the average period being 69.8636 days)*

to its orbital period, however, this rotation is woven into a profoundly hidden order which embraces the four bodies in the solar system whose rotation is relatively slow. Imagine the rotation periods of the Sun, Mercury, Venus and the Moon in a ratio of 2:5:6:10. This would make it obvious that there must be some connection here which could hardly be a random one. For the sake of simplicity let us assume that the values chosen show rotation periods in days. This would mean that after 30 days—the least common multiple—all four bodies would show the same directional orientation. We could define four axes which point towards the vernal point at a specific moment, and after 30 days this would once again be the case. Although the four bodies are all at different positions in space their axes, independently of this, are all oriented in the same way, i.e. parallel, after this period of time. This would entirely correspond to four planets whose orbital periods stood in the same ratio; let us now assume we mean years. After 30 years, supposing a constellation in which all four were in a single line with the Sun, they would once again take up the position of a common conjunction.

So with regard to the actual situation, let us begin more modestly with two bodies, the Sun and Mercury. From their rotation periods we can calculate (see the formula given in Appendix 3.4 for synodic periods) that after 44.7433 days they would once again show the same orientation. This period would, so to speak, be the rotation synod. As in the case of the planets it would now be possible to inscribe geometrically a conjunction star-figure for the end-points of the horizontal axes which correspond to the orbiting planets or, respectively, to calculate it by dividing the synodic period by the period of rotation. In this example the result would be a 38-

pointed star-figure because the synodic period is in a ratio of about 29:38 to Mercury's rotation and about 67:38 to the rotation of the Sun. Now let us add the rotation of Venus. Strange to say, its period, too, has a ratio to the period of 44.7433 days which leads to a 38-pointed star-figure, namely 38:7. When we then use the formula for retrograde rotation (see Appendix 3.4) to determine the Sun/Venus and Mercury/Venus rotation synods, we arrive at the values 22.9800 or respectively 47.2449 days. And if we then finally relate all three synodic periods and also all three rotation periods to one another, we find that the following values arise for 38 synodic periods caused by the Mercury and the Sun rotation:

**Table 9.2** *Rotation periods* and synodic periods*

| Rotations | Period (days) ① | Ratio number ② | Exact value ③ | ① ★ ② (Days) ④ |
|---|---|---|---|---|
| Sun | 25.38 | 67 | 66.99161 | 1700.460 |
| Mercury | 58.64617 | 29 | 28.99161 | 1700.739 |
| Venus | 243.019 | 7 | 6.99635 | 1701.133 |
| Synods | | | | |
| Su/Ve | 22.98005 | 74 | 73.98797 | 1700.524 |
| Su/Me | 44.74335 | 38 | 38 | 1700.247 |
| Me/Ve | 47.24488 | 36 | 35.98797 | 1700.816 |
| | | Mean value of synods: | | 1700.529 |

The ratio number in column ② arises in each case in relation to the Sun/Mercury period (= 38), for example:

$$36\ Me/Ve \cong 38\ Su/Me$$

$$\Rightarrow \frac{Me/Ve}{Su/Me} \cong \frac{38}{36} = \frac{19}{18}$$

Seven Venus rotations thus encompass fairly accurately the rotation periods of Sun and Mercury and also all the relevant synodic periods of the three inner celestial bodies. We shall see, however, that the synodic periods in the second section of the table are more significant and that, just as in the case of the planets, the decisive role is played not by the orbits but by the conjunctions. (The latter are, though, dependent on the former.) The ratio numbers of the synodic periods are still divisible by two, so that the result is 37:19:18 together with a mean period of 850.26 days, after which the three periods coincide. We shall see, in addition, that these numbers, too, have no direct significance but that they play as it were merely a mediating role. And this fits well since although an 18-pointed or 19-pointed star-figure may be a thing of beauty in itself it cannot be

* The exact frequency of the Mercury rotation is of course also subject to the present limitations in exact measurement. The 1997 *Encyclopedia of Planetary Sciences* gives a value of 58.646065 ± 0.005 days (p. 595). The 1999 *Astronomical Almanac* and other sources show 58.6462 days without stating accuracy limits. Since there is a high probability that the 3:2 resonance is genuine, the calculations regarding the Mercury rotation use the exact 2/3 value of the orbital period, which can be accurately determined, namely 58.64617 days.

described as being in accordance with a small whole number, and geometrically we are bound to get a star-figure of one kind or another. The rotating celestial bodies of the solar system, however, do not form star-figures of just any kind.

So let us now allow ourselves to be surprised by what actually arises when one shows the rotations of our Sun, of Mercury and of Venus in their mutual geometrical relationship.

The diagrams show the threefold 'conjunctions' and, as usual, the linking of the sequential positions. In the case of the rotating bodies these are—as mentioned above—the points of their axes on the surface

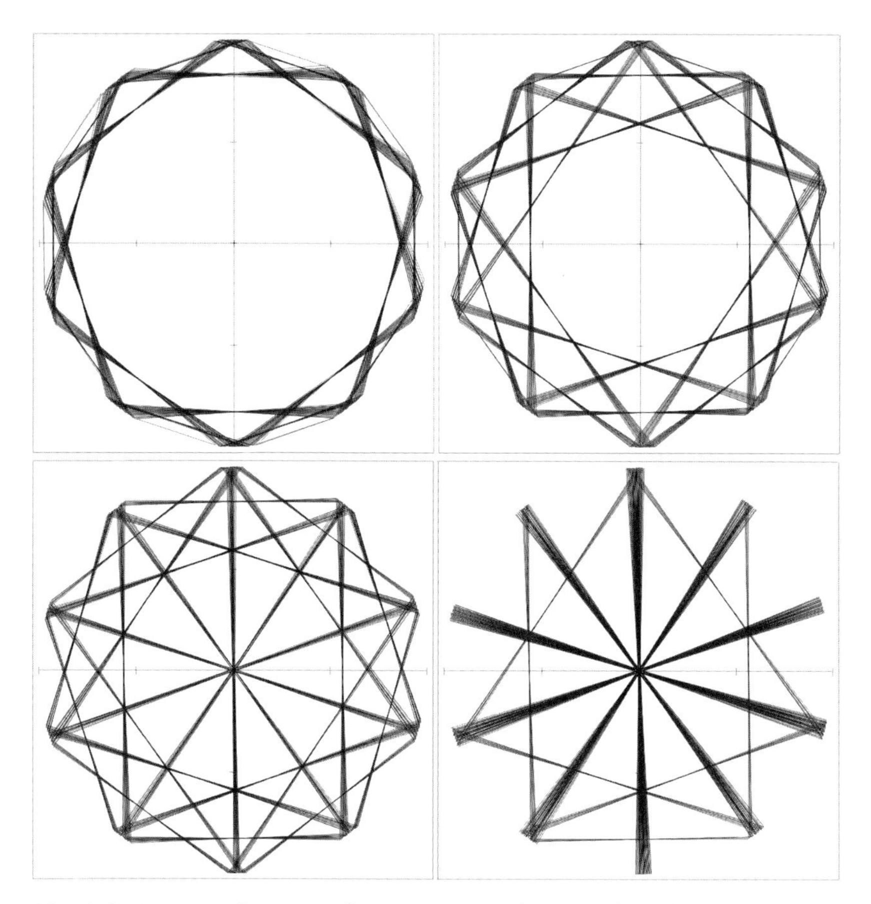

**Fig. 9.8** *Sequence of rotations of Sun, Mercury and Venus when all three axes are parallel, 400 times in each case, beginning when the axes are oriented to the vernal point (at the top in this case). Top left: Accuracy <15°, period 312.25 years. Top right: accuracy <10°, period 458.02 years. Bottom left: accuracy <7.5°, period 606.12 years. Bottom right: accuracy <5°, period 911.51 years*

whenever all three axes are oriented in the same direction, i.e. are parallel with one another in space. (Fig. 9.9 shows in sequence how this type of figure develops.) As in the case of threefold conjunctions of orbiting planets there is, here too, no one hundred per cent correspondence (see explanation preceding Fig. 7.4). This would only occur where the ratios are quotients of exact whole numbers. So the figures obtained are calculated on the basis of varying exactitudes.

The pictures are valid for all three celestial bodies. The actual depictions are of the solar rotation, but the figures arising in the case of Mercury and Venus are only minimally different (as correspondence is never quite 100%; if it were exact, the figures would be identical). The higher the degree of accuracy that is required, the longer it takes on average before all three axes are (approximately) parallel. An accuracy of 0° would signify absolute parallelism; one of, for example, 15° would mean that all three axes lie within this angle at a specific moment in time. Smaller than 15° also denotes that the time or the position is plotted in the depiction when the three axes are closest; this could be, for example, 3.5° or 7.8°.

Each of the figures thus arising consists of three different types of line. This corresponds to three different intervals of time (with very minor fluctuations) which are repeated in a fairly but not absolutely regular rhythm. In the top left figure, for example, these are 45.96, 712.40 and 758.34 days; and in the bottom left picture 45.96, 804.30 and 850.26 days. The interval of 850.26 days was that of the 37-19-18 ratio of the three rotation synods. Here the Sun rotates 33.5 times, Mercury 14.5 times and Venus 3.5 times, which leads to the lines running through the centre as occurs for the first time in the third rotation diagram. This picture involving the 10-rayed star-figure, the 10-pointed star-figure and two pentagons—i.e. all the star-figures possible within the decagon except for the double pentagram—is the one which we could call the virtual depiction of the mutual rotation figure of Sun, Mercury and Venus. With this as an example, Fig. 9.9 uses the first six lines to show how the figure takes shape. Only Venus and Mercury are shown, since the Sun has the same sequence as Mercury, the only difference being that there are some more revolutions between the points.

The strong 10-rayed star-figure (Fig. 9.8, bottom right) with its accuracy of <5°, also appears if a deviation of 3° or 1° is set, except that the lines at the edge then occur in different places. It is not meaningful to set yet greater degrees of accuracy, since the correspondingly longer periods of time lead to accumulation of possible errors caused by the inaccuracy in determining the rotation of Venus (see footnote on p. 145). The calculations on which the figures here are based were also carried out

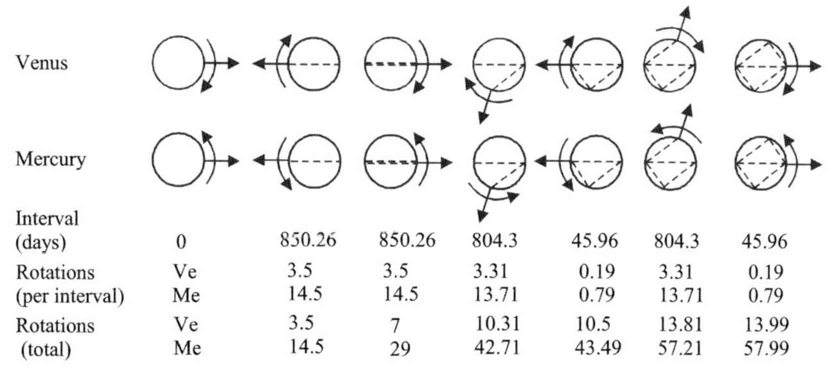


| | | | | | | | |
|---|---|---|---|---|---|---|---|
| Interval (days) | 0 | 850.26 | 850.26 | 804.3 | 45.96 | 804.3 | 45.96 |
| Rotations (per interval) | Ve | 3.5 | 3.5 | 3.31 | 0.19 | 3.31 | 0.19 |
| | Me | 14.5 | 14.5 | 13.71 | 0.79 | 13.71 | 0.79 |
| Rotations (total) | Ve | 3.5 | 7 | 10.31 | 10.5 | 13.81 | 13.99 |
| | Me | 14.5 | 29 | 42.71 | 43.49 | 57.21 | 57.99 |

**Fig. 9.9** *Development of the common rotation figure of Sun, Mercury and Venus with an accuracy of <7.5°. The dotted lines inside the circles show the figure beginning to form (vernal point 0° on the right)*

with a value for the Venus rotation of 243.019 days, with checks made using 243.025 and also 243.005 days. Even with the latter numbers the same figures arose although they were a little less resonant, i.e. the lines do not lie quite so close together. The well-established[*] geometrical resonance of the rotations of Sun, Mercury and Venus is thus a unique, or indeed an almost incredible phenomenon.[†] Other rotational relationships investigated in the same way (using the periods of other planets and also fictitious values) likewise in some cases give regular formations, though very rarely as clearly, and certainly not of a kind that, independently of the accuracy applied, would always cause the lines of a single n-sided polygon to appear.

'This, therefore, is what the Pythagoric hymn says about number: *That it proceeds from the secret recesses of the monad, until it arrives at the divine tetrad. And this generates the decad, which is the mother of all things.*'[9]

We have pointed to an inner connection between the rotations of the Sun and the two planets closest to it. Later we shall also investigate how the Moon, which has a similar period of rotation, fits into the picture we have uncovered. From Earth outwards the planets have much shorter

---

[*] To speak of total certainty would be untrue. If the calculations were to be carried out using the value at the lowest end of the measurement accuracy for the Mercury rotation (see footnote on p. 174), the 10-pointed star-figure would no longer appear. But equally there would also no longer be a genuine 3:2 resonance between the orbital period and the rotation which all astronomers now take as their point of departure.

[†] For this reason the Table in Appendix 6.8 gives some of the data that lie behind the figures in Fig. 9.8.

periods of rotation amounting to about one day or less, so we shall not be dealing with them here. Instead let us consider what is now known about the origin of planetary rotation.

Most of the planets have two features in common: apart from Venus and Mercury they rotate very rapidly, and apart from Venus, Uranus and Pluto the angle of their axes is less than 30°. Regarding the latter point it has been calculated that the probability of the observed distribution of the axial tilts having come about at random is less than 1:100,000. Since retrograde rotation such as that of Venus corresponds to an obliquity of more than 90°—with Venus it is 178°—the main question which arises concerning the calculated improbability asks why the rotation of almost all the planets is prograde, i.e. turns in the same direction as their orbit. Based on the theory discussed in Chapter 5 according to which the planets were created out of small chunks, the opposite ought to be the case. Imagine two planetesimals, one of which is already larger and is seeking to crash its way to becoming a proper planet. Both are travelling around the Sun along Kepler orbits, i.e. the inner one moves faster. If the smaller one is also the one closer to the Sun it is likely to ram the larger one on the side facing the Sun thus giving it an impulse which sends it spinning counter to the direction of orbit. In the opposite case it is the larger chunk that does the overtaking and is thus more likely to collide with the smaller one with the side that faces away from the Sun. The result is the same: it receives a bump which causes it to rotate backwards.

However, we can only be as definite as this if the orbits are circular. Eccentricities make the situation more complicated because when a planetesimal on an outer orbit is nearly at perihelion it can for a time be travelling more rapidly than one on an inner orbit. Taking this into account, a number of researchers have calculated, or found out by means of computer simulations, that the effects will add up differently depending on which eccentricities dominate among the multitude of planetesimals. If most of the eccentricities were relatively large, this would lead to prograde rotations, albeit very slow ones. But if on the whole they were rather small the result could yield both types of rotation, though again rather leisurely ones. In the case of circular orbits, as we have just seen, fast retrograde rotation arises. Only if a specific mean level of eccentricity dominates will the consequence be prograde rotation. But since there does not appear to be any obvious reason why this mean level in particular should have been dominant in the earliest times of the planetary system two theories were developed to solve this problem. The one states that the distribution of axial tilts and thus also of the general direction of rotation, although very improbable, was nevertheless a random phenomenon. The other maintains that a growing planet will at some point have swept clean and devoured everything in the circular region in which it orbits. Once this has happened it can only have collisions with chunks that are more eccentric, some of which will deliver tangential blows that will favour a prograde rotation. How realistic this scenario is appears to be unclear. At any rate this is what I gather from the concluding words of an article written by the originator of this model, who ends by saying that much remains to be solved in this field by future generations of astronomers.[10]

This excursion into the history of the rotations will be completed by our considerations concerning the origin of Venus' unique rotation. People attribute its extreme slowness to the tidal forces of the Sun. (Originally its speed is thought to have been similar to those of the other planets.) Current scientific knowledge also states that in a similar way the Moon is gradually slowing down the rotation of Earth, although to a very much lesser extent. One theory regarding the retrograde rotation of Venus assumes that it has probably always been the way it is. Another is based on calculations which show that the axes of rotation can be subject to chaotic fluctuations, as already mentioned in Chapter 5. According to this the axis of Venus was at some point tilted to almost 90° after which the solar tidal forces as it were dealt it the final blow.[11] Whatever the case may be, all the theories are alike in that they see the exceptional rotation of Venus—and also the measure of the rotations of the other celestial bodies except those that are not linked by gravity—as products of chance.

But perhaps a force quite different from those we imagine today had a

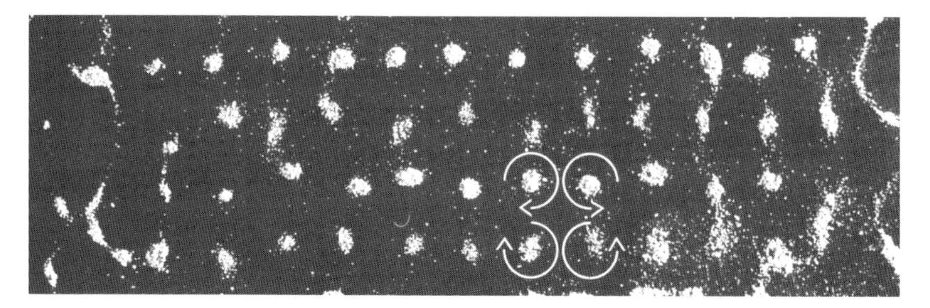

**Fig. 9.10** *Rotational sonorous figure according to Hans Jenny. 'The powder congregates in small circular areas which continue to rotate regularly as long as the note is sounded. This rotational effect is not merely adventitious, but appears systematically throughout the vibrational field. The picture shows a plate with a number of rotating areas in which the direction of rotation of each area is contrary to that of its neighbors. The arrows show the direction of rotation. The circulation continues steadily and its course can be followed by marking with colored grains.'*[12]

part to play in the origin of the rotations. Let us recall the theory we discussed in Chapter 5 according to which waves—shock waves—may have triggered or promoted the conglomeration of the planets. We followed this by showing sonorous figures which prove that musical vibrations can bring order into amorphous matter. Research carried out by Hans Jenny also showed that suitable notes can bring about a rotational effect. Once again, sand or powder is scattered onto a plate which is then caused to vibrate as described briefly in Chapter 5. Prograde and retrograde rotations ensue very close to one another. So we might imagine that a music of creation could be capable of causing various harmonious rotations tuned to one another with rotational directions fitting a superior harmony.

## *Sun, Moon and planets*

Many experts are content to assign to chance the credit for quite a variety of developments and interplays, and it is no exaggeration to regard chance as one of the two divinities of the modern scientific view of the world, the other being the law of nature. A good many scholarly works claim that it has merely been chance which, with the distances between Earth, Moon and Sun being what they are, has caused the Moon and the Sun to appear the same size from our earthly point of view. It is solely because of this that we are able to experience an eclipse of the Sun in the way that we do here on Earth. Just as with the planets in relation to the Sun, the distance of the Moon from Earth is linked to its orbital period by the law of gravity. And the orbital period, in its turn, is identical with the period of rotation. The two are linked by gravity in a 1:1 resonance. So we shall now see whether chance has done a good job also in the matter of the Moon's movements in relation to other celestial bodies, especially the Sun.

The Moon's period of rotation of 27.32166 days can be expressed to within fractions of a second. The rotation is perfectly uniform or, if there are small deviations, these are so minimal as to be negligible here. (The same goes for the rotations of the other celestial bodies; in the case of Earth some very subtle fluctuations have meanwhile been detected.) The Moon's orbit around the Earth, on the other hand, is the most complicated of all in the whole solar system,* so there are also correspondingly

* This is the case at least by comparison with the planets. I do not know whether there are other moons whose orbits are equally, or even more, difficult to calculate.

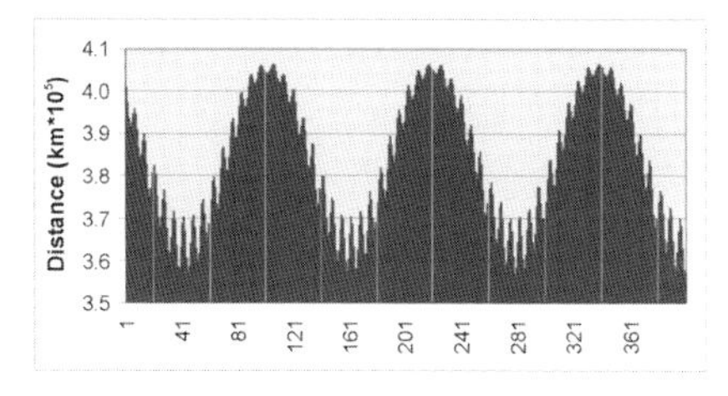


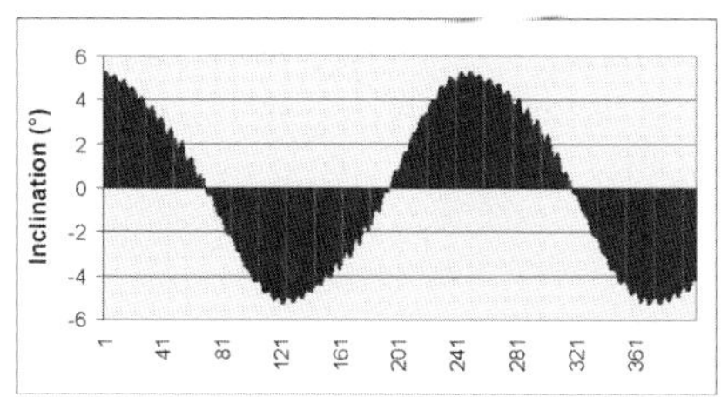


**Fig. 9.11** *Distance of Moon from Earth and inclination of its orbit to the ecliptic; stepping interval 27.32166 days, 400 times, period 29.92 years, beginning 1 January 2000*

many rhythms superimposed on one another in the movements of the Moon. The reasons for this are the various effects of gravity from Sun, Earth and other planets and also the very rapid precession of the lines of apsides and nodes. What covers tens of thousands to hundreds of thousands of years in the case of the planets takes the Moon *c.* 8.85 and, respectively, 18.61 years. Examples of the rhythms are shown in the two diagrams in Fig. 9.11. The time involved is the period of rotation and thus also the sidereal period. Without all the many influences, two completely straight lines would arise. Instead, what we find are about $3\frac{1}{2}$ turns of the apsides and $1\frac{1}{2}$ turns of the nodes as well as other overlaid and never fully identical fluctuations.

In order, as it were, to put us in the right frame of mind, let us now have a look at a picture of the uniformly rotating Moon. It comes about in the same way as, for example, the solar pentagram in Fig. 9.2. The positions of a specific point on the surface of the turning Moon are consecutively plotted, and sequential points are joined up with one another. In this case this is done every time Earth is in conjunction with Jupiter (see Fig. 9.12).

A flower-like shape forms in the centre based on the number seven. But nothing of this shows in the lunar orbit (right). Although there is some degree of structure involving the number nine, there is no way in which this could be described as embodying any kind of geometrical order in the movement of the celestial bodies. By the same method a few more figures, some of even greater regularity, can be traced in accordance with the rotating Moon. What remains of all these when the positions of the Moon in its orbit rather than its regular rotation are plotted is either nothing at all or, in very few exceptions, a scant, barely discernible remnant. As this example demonstrates, the complexity of the Moon's movement transforms the flower- and star-shapes into a rather chaotic tangle of lines.

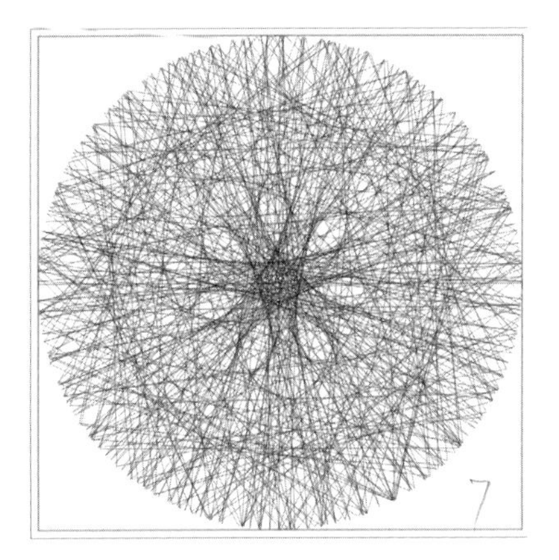


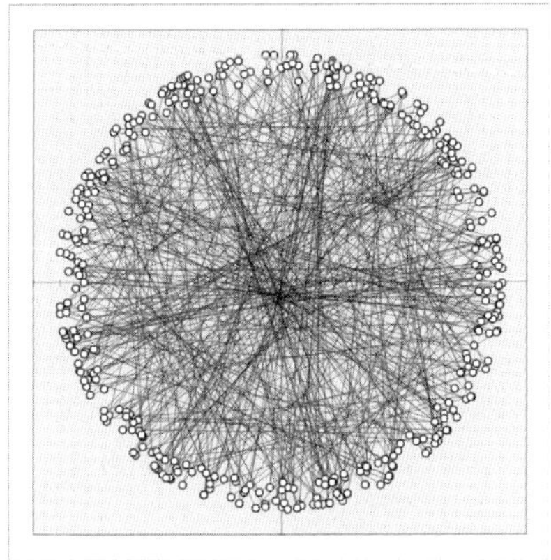

**Fig. 9.12** *The Moon at Earth/Jupiter conjunctions, 400 times, beginning 28 November 2000, period 436.83 years. Left: rotation. Right: the Moon in its orbit*

But it also tells us something else. The points showing the position of the Moon in the right-hand picture all lie within a ring formed by the precession of the line of apsides. This is the orbital sphere of a rotating ellipse which appears in Fig. 14.2 in Appendix 1.1. In the next illustration this sphere is shown more clearly by two added circles, and its geometrical scale is emphasized by means of a hexagon.

We might imagine the Earth/Moon system to be a single unit in which the centres of the two components are linked in a gravitational coupling as

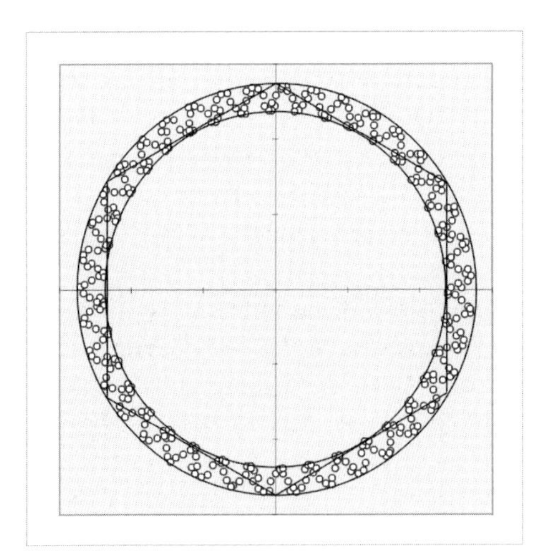

Greatest distance: 406.740 km
Smallest distance: 356.410 km
Mean distance: 384.403 km

Largest/smallest distance ($=r_c/r_i$)
1.141

Area of outer/inner circle ($=1.141^2$)
1.302

Area of outer/inner circle of the hexagon 1.333

**Fig. 9.13** *The Moon in its orbit around Earth at new moon positions, 400 times, beginning 6 January 2000, period 32.34 years. (The subscript c and i stand for circumcircle and inner circle respectively.)*

though by a rigid bar (whereby Earth would still be able to rotate freely). The orbiting Moon would then represent the tip of this bar or of the horizontal axis of the whole system which rotates. At new moon the axis always points straight at the Sun. So the new-moon position corresponds to a view of the Earth/Moon unit towards the Sun. The period is the same as that of full moon and can be calculated as the synodic period of Moon and Earth amounting to 29.53059 days. The degree of fluctuation here amounts to a little over half a day. This Earth/Moon-Sun-View gives us a first shimmer of geometrical order. A figure takes shape which is structured in accordance with the number thirty-eight, a number we have just recently encountered as an intermediary. It also shows that the lunar orbit possesses its own law, since taking the numbers alone would have brought us to a 37-pointed star-figure, because 37 synodic moons correspond to 40 sidereal moons (39.991 in 2.991 years).

The Earth/Moon team, then, turns its gaze towards the Sun with some degree of regularity. So obviously we must now look to see how the Moon reacts when the Sun (with its defined axis) returns the compliment and looks towards us. The distance is great enough for it to see Earth with its orbiting Moon in one glance, i.e. with regard to the interval of time it would make no relevant difference which of the two cosmic companions were receiving the Sun's attention. So we might as well speak of a Sun-Earth-View (SEV). Its period is of course identical to the Sun's synodic rotation period of 27.2752 days, an interval of time which here on Earth expresses itself in magnetic storms and short-wave fade-outs.* And we cannot fail to notice that this period is very close to the sidereal rotation and orbital period of the Moon. In other words, the Moon always looks towards the stars when it is being viewed by the Sun. More accurately we have to say that there are 588 synodic solar rotations to 587 sidereal lunar rotations. Even more accurately there are 587.0008, but 587 is accurate enough, as we shall see when we explain the next picture.

And sure enough, chance has done its bit. It has indeed, even if there appears to be a considerable danger of overdoing the superlatives in the solar system. This discovery of the Sun-Moon hexagon certainly calls for some celebratory music—perhaps the final movement of Mozart's Jupiter Symphony. On a personal note I have to say that the Moon has not deceived me with its shining light, for it did indeed once seem to be dropping me a hint that there was something special about it.

---

* The period of magnetic disturbances is not known exactly, but it is at least very close to the synodic period of the Sun's rotation.

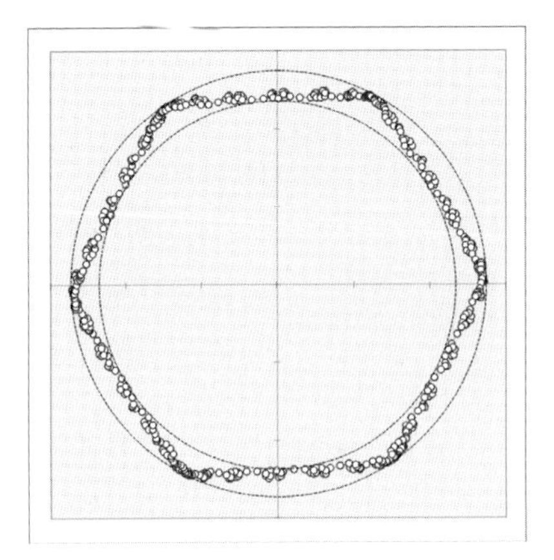
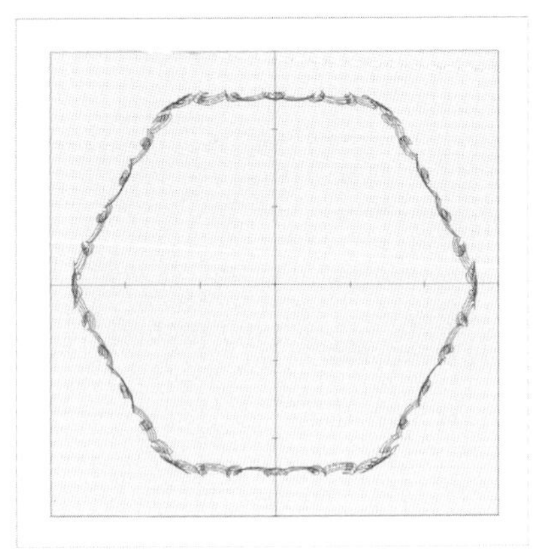

**Fig. 9.14** *The Moon at Sun-Earth-Views. Left: 588 times, period 43.91 years. Right: 2352 times (4 * 588); in this case only the connecting lines are shown*

This hexagon takes shape step by step as the Moon keeps returning almost to the same spot. Not until 588 Sun-Earth-Views have been achieved does it reach completion, just as the calculated ratio said it would. But the Moon does dally somewhat along its hexagonal path. Especially in the right-hand picture we see small bends and bows caused by the way the intervals in the SEV are also subjected to certain fluctuations of at most 0.15 of a day. In combination with the Moon's special rhythms, this suffices to cause the plotting of the next point sometimes a little in front and sometimes a little behind the previous one. The reason for this regular hexagon, or rather we should say the arithmetical reason for it, is to be found in the way all three of the periods involved fit together. In addition to the two periods of rotation, the third measure of time is not the orbital period of the Earth but the velocity with which the precession of the apsides takes place. (In this sense there are actually four participating measures of time.) This velocity can also be indicated according to how long it takes the Moon to pass from one perihelion position to the next. It needs slightly longer for this than for its sidereal period, namely 27.55455 days; the perihelion goes on ahead of it, so to speak. This period is termed the anomalistic month.

So 588 synodic solar rotations = 587 sidereal months = 582 (582.04) anomalistic months. It is perhaps surprising to realize that the hexagon takes shape simply because 588 minus 582 equals 6. However, all great things, just like the music of Mozart, are also simple in a way. For instance we could produce a hypothetical heptagon or pentagon by performing the same calculation with periods in ratios of 711:710:704 or, respectively,

468:467:463. To do this we would merely have to change the synodic period of the Sun, to 27.283 or 27.263 days. The polygons arising in this way do, however, have dented or bent sides so as to fit inside the orbital sphere of the Moon or, in the case of the heptagon, still touch the two circles. It's just that the one we have shown is the optimum figure. When we change the periods by as little as 1 to 2 thousandths the lines no longer coincide even after completing only one round of the circle. But on the other hand, as the resonant periods are so very close together it is easy to work out that the formation of a regular polygon is not so very unlikely. But of course this applies only to the very close areas in which the synodic rotation of the Sun (or of any other interval) and the sidereal month are almost identical, or small whole number multiples.

However, the actual periods with which it is still possible to connect the Moon's orbit do not fulfil these conditions or any others that might lead to the formation of a genuine figure; or else any likely looking beginnings are soon made unrecognizable by the fluctuations. The Moon is very sparing in the dispensation of its favours in this respect. But in the rare circumstances when it does grant them it does so in all their fullness with phenomenal results such as the Sun-Moon hexagon. There is only one other instance in which this capricious individual is willing to oblige, and this completes the picture of the hidden order in the interplay of these four rotating celestial bodies.

So in order to uncover the mystery, let us return to the mutual rotation of Sun and Mercury with its period of 44.7433 days. As with the Venus rotation, almost all the lines possible in a decagon were covered. And the axis parallelism of Moon and Venus also leads to a decagon since its rotation synod is 24.5604 days which is about one tenth of a Venus rotation. But nevertheless this is no more than a fairly approximate relationship. The cycle-resonance is relatively high at 3.83°. Together with the constantly changing lunar orbit the decagon becomes virtually indistinguishable after only 4 formations, corresponding to 40 parallel positionings of the axes. So it will be better to look and see how Venus, whose rotation is perfectly linked to those of Sun and Mercury, moves in its orbit when the axis orientations of the two inmost planets correspond (see Fig. 9.15).

The queen of heaven appears to enjoy embellishing her pentagrams with corresponding pentagons. The Sun/Mercury period amounts fairly closely to one fifth of a Venus orbit or, vice versa, five rotation conjunctions amount to one (0.996) Venus year. With a cycle-resonance of −1.55° this causes the pentagon shown to be formed. Furthermore, two periods of 44.7433 days amount to one Mercury/Jupiter synodic period

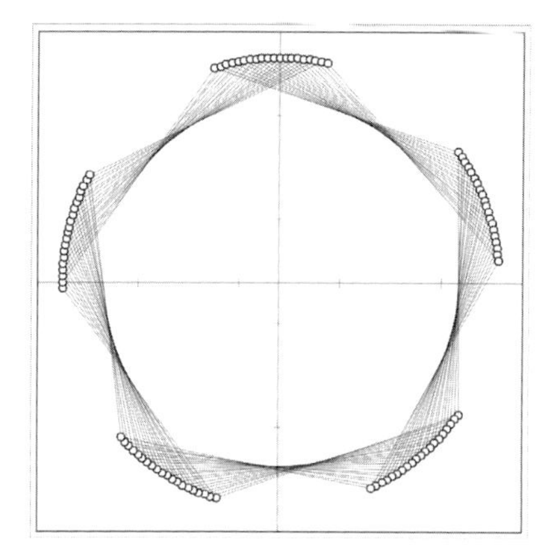

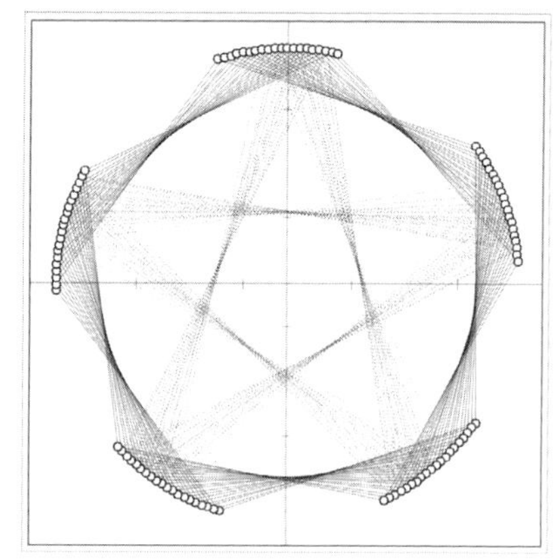

**Fig. 9.15** *Venus at Sun/Mercury rotation synods, 100 times, period 12.25 years. On the right the dotted lines show Venus' movement at Mercury/Jupiter conjunctions, here at minimum distances (cf. Fig. 9.4)*

which means that one Minor Pentagram and two pentagons are drawn in an almost identical span of time. So the 20 pentagrams in the picture take twice as long as the pentagons to be formed. Since the Mercury/Jupiter conjunction is also in an octaval ratio to the Venus-Mars-View, the Sun/Mercury period is also the lower double octave of the time interval of the Venus-Mars-View. The three periods are in an exact ratio of 4:1.993:0.997. So one can say that the rotations of Sun and Mercury are linked, via the Minor Pentagram, in a perfect manner with the ratios of the hexagram in Fig. 9.5, and this means with the architecture of the inner planetary system as a whole. Since the mutual rotation of the Sun and 'its moon' Mercury is of such central significance, let us now see how Earth's own Moon behaves in this special constellation (Fig. 9.16).

Our capricious Moon thus completes that aspect of the 'forever indiscribable'* celestial order which is connected to the rotations. In doing so it is in harmony with the continuous 10-pointed star which it also traces: the Sun/Mercury/Venus 10-pointed star-figure with the double pentagon and the 10-rayed star-figure (Fig. 9.8). And to this it adds the still-awaited double pentagram. The polarity of the pentagram, sign of the human microcosm which the Moon envelops through its movements around humanity's home planet, is unravelled in an image that speaks to us like a parable. One of the mysteries of the solar system has been revealed; the golden vessels of the Egyptians which Johannes Kepler stole (see more later) can now be carried on into the third millennium. Audible

---

* See Goethe quotation in Fig. 9.17.

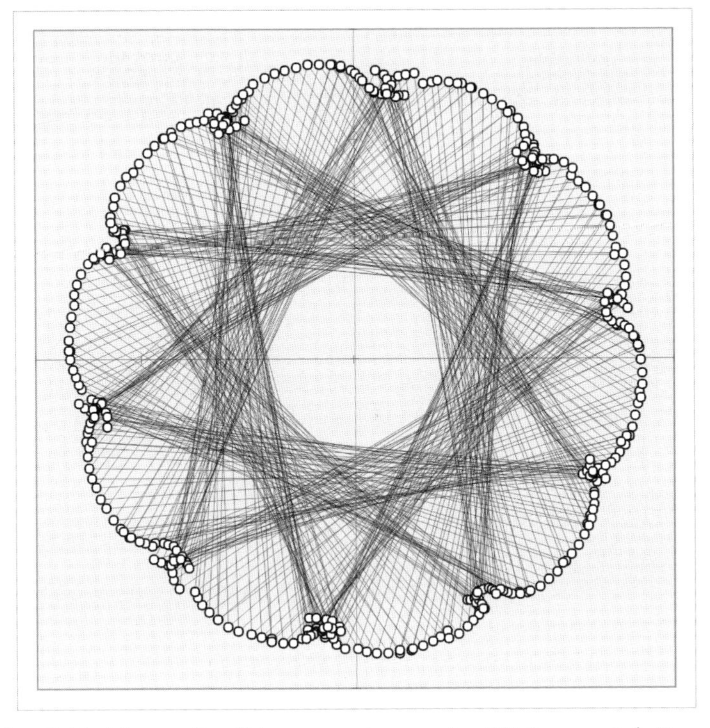

**Fig. 9.16** *Moon at Sun/Mercury rotation synods, 400 times, period 49 years*

solely to the ears of the spirit there now perhaps resounds, as a musical accompaniment, the as yet unwritten tenth symphony—by Beethoven, Bruckner, Dvorak and Mahler.

The mathematically comprehensible portion of the mystery lies in a fractional resonance of the three periods of time involved (as explained in connection with Fig. 7.13). Once again the measure of the precession of the perihelion relates to the two others in such a way that the regularity noted can come into being after a specific length of time. The figure begins to take shape after *c.* 80 to 100 constellations. The separate points unite to form an overall picture because 8 Sun/Mercury periods correspond to approximately 13 anomalistic months. During these 8 rotation conjunctions the Moon completes 13.1012 orbits (i.e. very precisely 13 and one tenth; the number after the decimal point is the relevant one). The next more exact correspondence is in the ratio of 101 Sun/Mercury periods to 164 anomalistic months. In 101 periods the Moon completes 165.4028 orbits, i.e. here, too, there are tenths after the decimal point, namely 4 tenths. The same goes for the following almost perfect correspondence of 210:341.

**Fig. 9.17** *Moon at Sun/Mercury rotation synods with added 10-pointed star and double pentagram*

# 10. A Symphony of Flowers and Stars

Another concept owed, like so much else, to ancient Greece is that of the symphony. The word originally meant 'consonance', 'harmony', a 'resounding together', which in those days involved chiefly the octave, the fifth and the fourth which were linked as a unity in the tetrachordal structure. In the Middle Ages it referred more to music in general, including singing, as a 'sinfonia'. Thereafter it underwent a long development before crystallizing out around 250 years ago into the musical form which we call by that name today. As a rule, a symphony is an orchestral work of four movements in which, when at its best, each theme is linked in some inner way with all the others. By this means a kind of world view is expressed in a symphony which is the complete opposite of that other view which merely places one phenomenon beside the next without relating them with one another.

The work we are about to investigate is the symphony that lies hidden in the harmonious interplay of the inner and the outer planetary system. The individual melodies or themes are in each case made to resound by three planets. In detail this means that we shall be looking at the longer-term geometrical figures that arise when the positions of two planets are plotted at moments when one of them is in conjunction with a third. We have already seen, in Chapter 7, a series of such movement pictures drawn by the inner planets Venus, Earth and Mars. We also saw that in addition these figures made it possible to demonstrate the overall gravitational interplay between the planets involved. In addition it became apparent that the essence of a melody is independent of the instrument upon which it is performed (certainly in most cases). For example a 10-pointed star-figure is traced by Venus and Mars (Fig. 7.12) both when their positions are linked together from the planet-centred point of view and also when they are shown from the heliocentric point of view by means of the linklines. We shall soon be hearing the melody performed by another instrument as well.

But a few more preliminary remarks are needed before proceeding any further. I can tell you in advance that we will be seeing a variety of very astonishing geometrical figures.* Questions will arise concerning the

---

* Depicted from the planet-centred viewpoint, these figures correspond to various forms of cycloid. Normally one distinguishes between prolate and curtate epicycloids (a circle rolled round the outside of another circle) and hypocycloids (a circle rolled inside another) in these variations. Spirals of Archimedes (with two branches) also occur. Mathematical dictionaries will give more detailed explanations of these figures.

probability of such movement figures being formed. In view of the abundance of possibilities to be taken into account, an exact mathematical approach to this question would no doubt form the basis for a degree in applied statistics. However, it is possible to simulate the formation of such pictures with fictitious planets by using a computer. This shows that it is not all that unusual for a well-ordered figure to arise when three random orbital periods are linked together as mentioned above. A rough estimate arising from a series of such tests shows that at a little below every tenth time a very clear figure of small whole numbers emerges, and that a further 25% of the random constellations lead to fairly recognizable images based on numbers up to the number twelve. This would correspond more or less to the conditions in our solar system. In the same way regular figures can arise in the actual planetary system, for example when one determines the interval of time through the sidereal periods instead of the conjunction intervals. However, actually quantifying the probability of the figures being formed is difficult for two reasons. Firstly, one repeatedly arrives at images which lie somewhere between copperplate perfection and a chaotic tangle of lines, so that the subjectivity of the observer plays a not entirely negligible role. And secondly, to arrive at a statistically more or less safe assertion one would have to investigate a number of artificial planetary systems and ascertain whether the different figures are subject to an overarching order in the way they interact. It is easy to see what a time-consuming exercise this would be, even for a single system of virtual planets. So in this case I have no statistical estimate to offer, since I have preferred to use the time available to me to investigate what is going on in our own, real planetary world. Whether the symphony we are about to hear is, as it were, the product of a cosmic wind playing on an Aeolian harp or whether some other creative force was involved in its creation can in any case be neither proved nor disproved by means of numbers. But we have shown sufficiently clearly in Chapter 4 that the musicality existing in our solar system is highly improbable.

So let us now turn our attention to the celestial orchestra. The work we are about to hear is, as is customary, performed in four movements. Since the symphony is an optical one I cannot of course recommend that you close your eyes in order to avoid being distracted while you follow the stream of melodies and themes. Instead, to help you tune in, we shall kindle a planetary light, a spiritual flame as it were. This will be made to shine out by the new instrument already announced above. How this works will be explained later by means of another example.

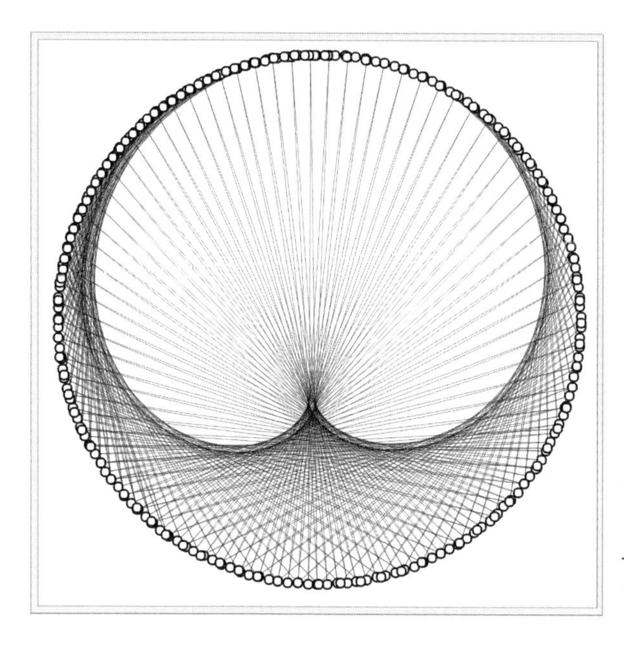

**Fig. 10.1** *Earth in its orbit at Earth/Uranus and Earth/Neptune conjunctions, 400 times, beginning 27 July 2000 (E/N), period c. 201.23 years (rotated around 145°)*

## *The first movement*

When the musical symphony first arose it involved the hearing of melodious sound brought into being by the sonorous intervals of octave, fifth and fourth. Their ratio in numbers is 2:1, 3:2 and 4:3. So let us begin here, too, with the first numbers. The geometrical figure that corresponds to the number one is the circle. Every star-figure resulting from a depiction of the conjunctions of two planets leads after a certain amount of time to a circle or to an almost circular ellipse. The only situation in which this would not be the case would be in the event of the resonance being perfect. The pentagram of Venus/Earth encounters, for example, turns once upon itself in about 150 occurrences. When one of the planets is plotted at every conjunction as it orbits around the Sun, the archetypal geometrical form is the only one that can arise after a longish time span. This is also the case in the planet-centred method of depiction. However, movement pictures involving the interplay of three planets normally do not lead to the formation of a circle.*

* Several concentric circles would theoretically be possible if the synodic period of the two planets plotted in the planet-centred viewpoint and the relevant interval of time (e.g. the synodic period of one of the two planets with a third one) were in an exact ratio of small whole numbers. If the interval were an exact multiple of the synodic period, one circle would be the only result. But in fact neither of these eventualities occurs in the actual conjunction relationship.

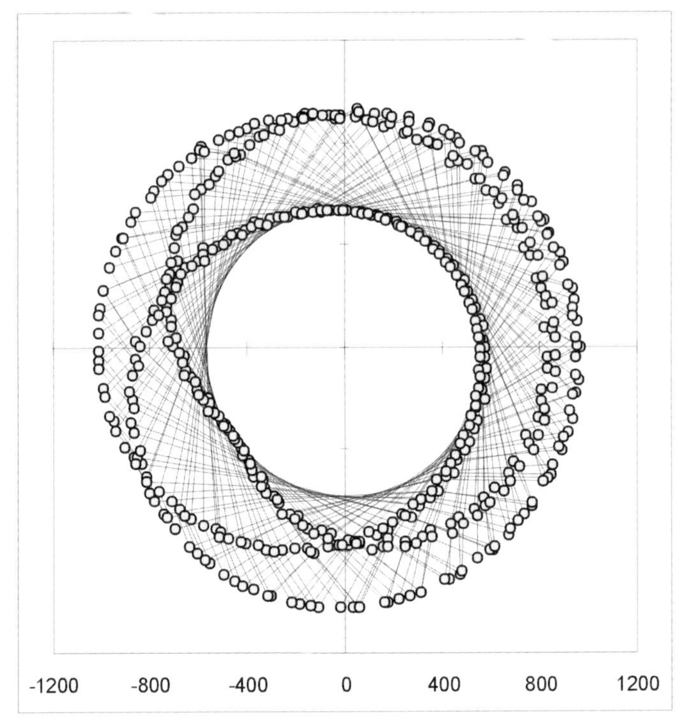


**Fig. 10.2** *Jupiter seen from Mars at Jupiter/Neptune conjunctions, 500 times, beginning 19 September 2009, period c. 6378.11 years*

So the smallest number that occurs geometrically in this case is the number two.

A double loop arises when Mars is related to the two outer planets. That this is really the case becomes clearer when the same figure is plotted using a constant mean value instead of the actual conjunctions which fluctuate by up to 64 days. A further diagram shows the first 50 steps in the creation of the figure (see Fig. 10.3).

The geometrical formation is not initially clear, but after about 100 positions the interrelation between the points begins to show. The double loop that then arises is one of the very few regular figures—and the only really clear one—which come about when an inner planet is related to the conjunction of two outer planets. In other cases the fluctuations of the intervals of time, which in some instances are longer than the orbital period of the inner planet, spoil all attempts. Thus the number two stands, as it were, as the only actual possibility at the boundary between the inner and the outer region. The figure itself is caused by a fractional resonance like the one we saw, for example, in the 10-pointed star-figure of the

*late 1: Venus-Earth linklines (stepping interval 3 days)*
*→ Fig. 6.2, page 102*

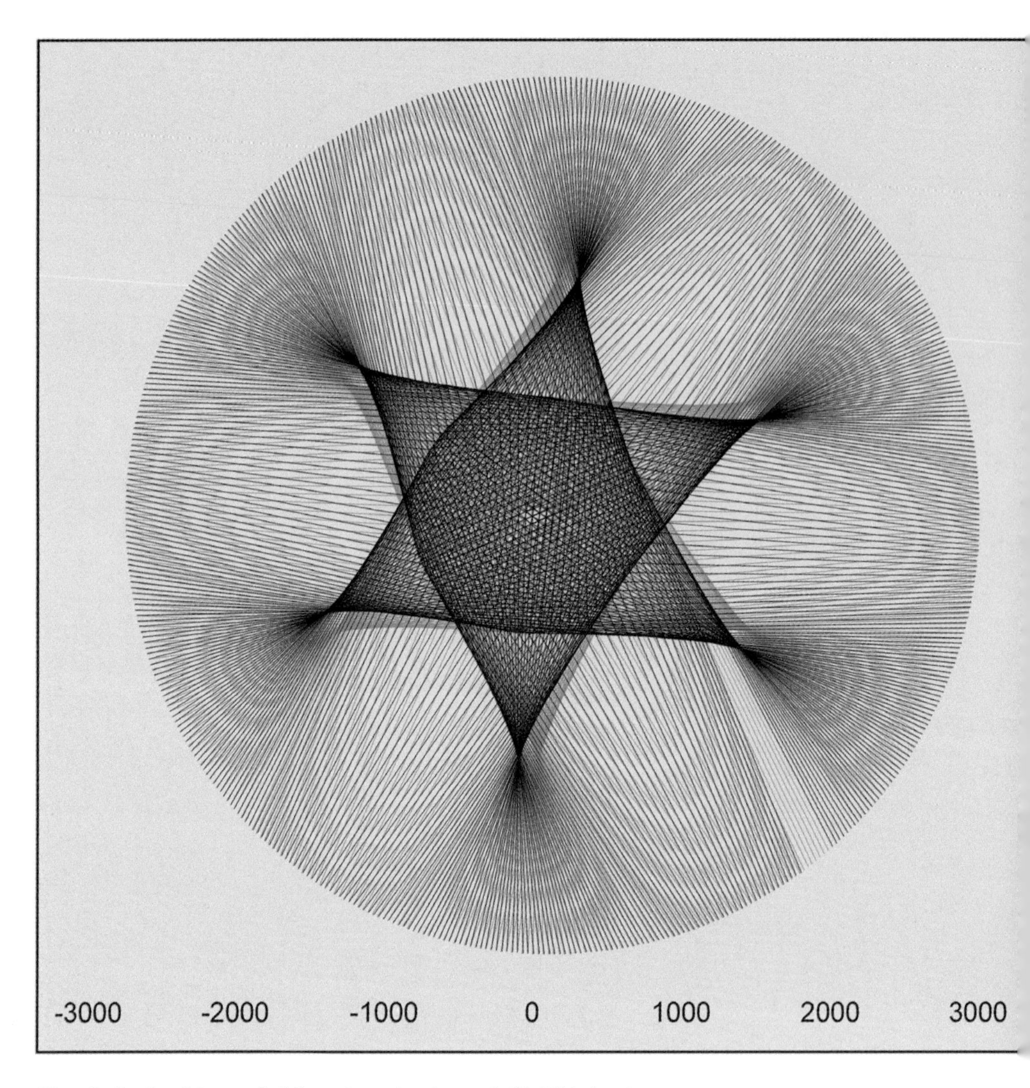


*Plate 2: Jupiter-Uranus linklines (stepping interval 60.781 days)*
*→ Fig. 6.9, page 108*

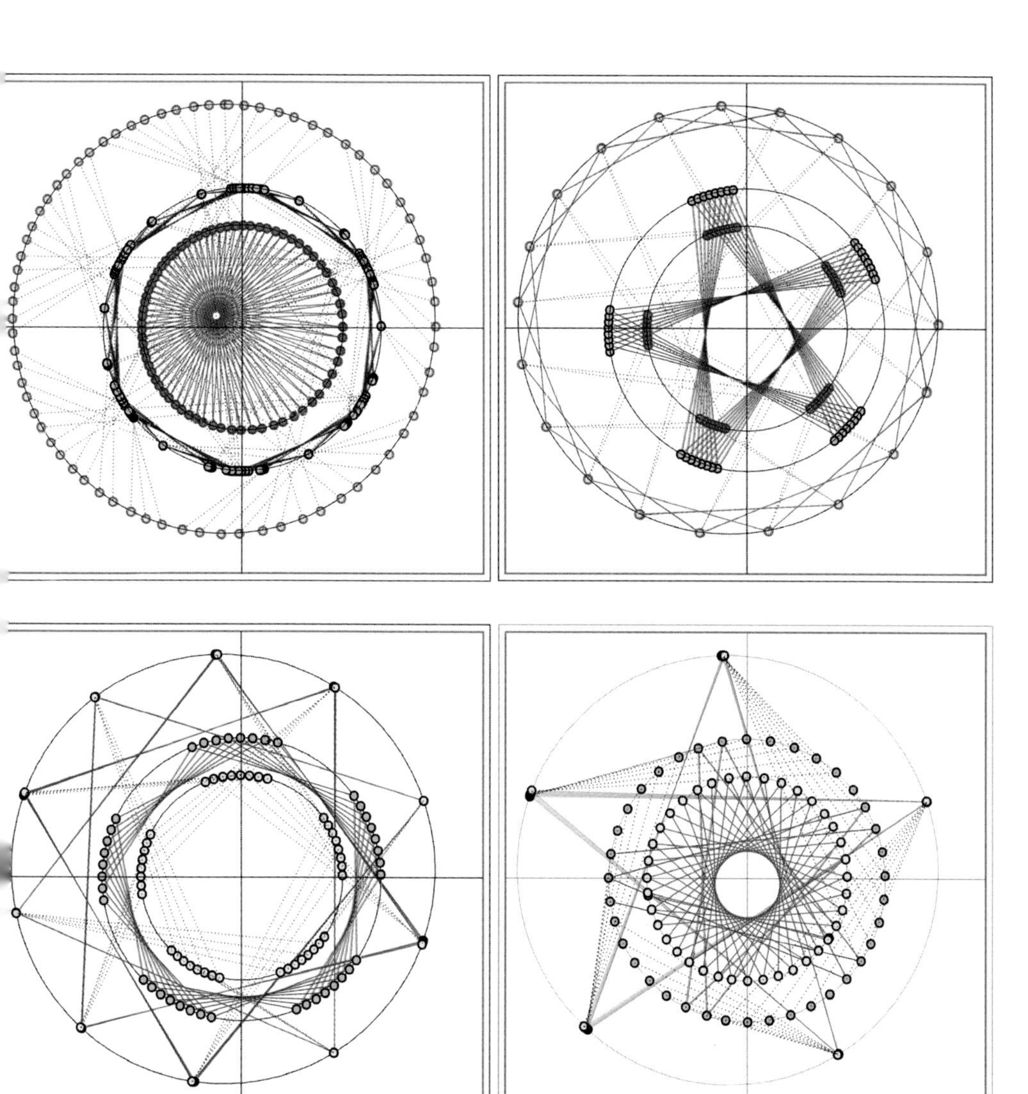

*[Pl]ate 3: Venus, Earth, Mars at Venus/Mars and Venus/Earth conjunctions*
*→ Figs 7.5/7.6, pages 130/131*

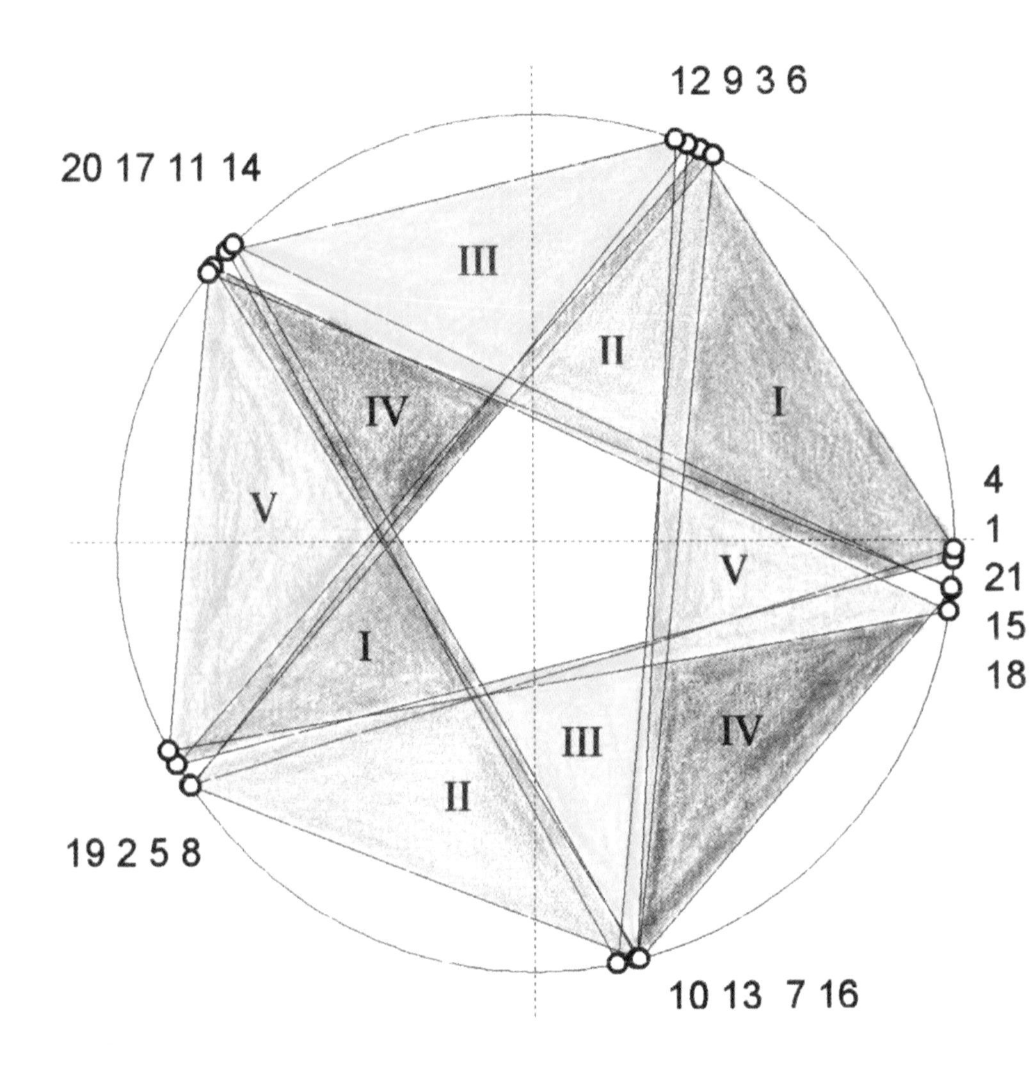


*Plate 4: Venus at Venus-Earth-Views*
*→ Fig. 8.4, page 147*

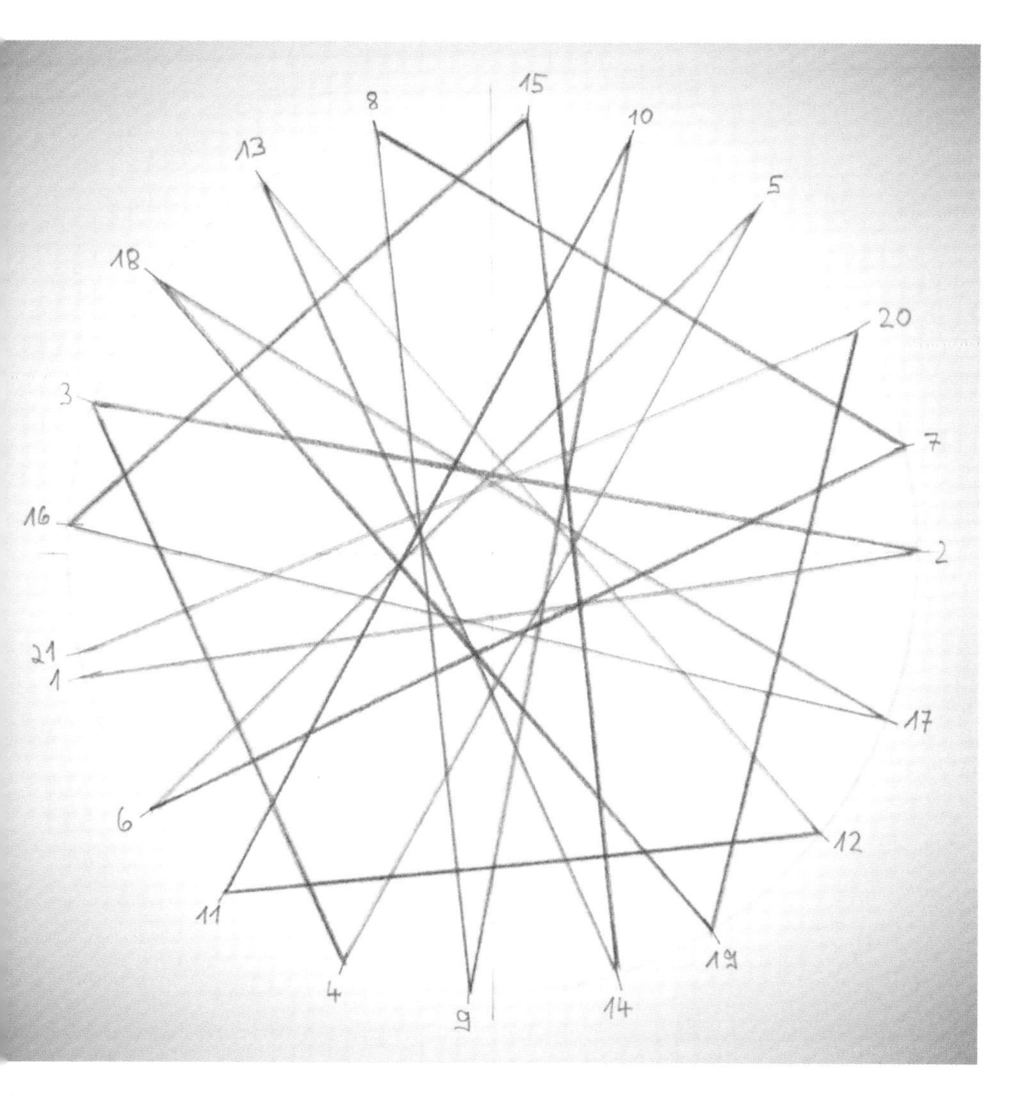


*Plate 5: Venus rotation at Venus-Earth-Views*
*→ Fig. 8.13, page 159*

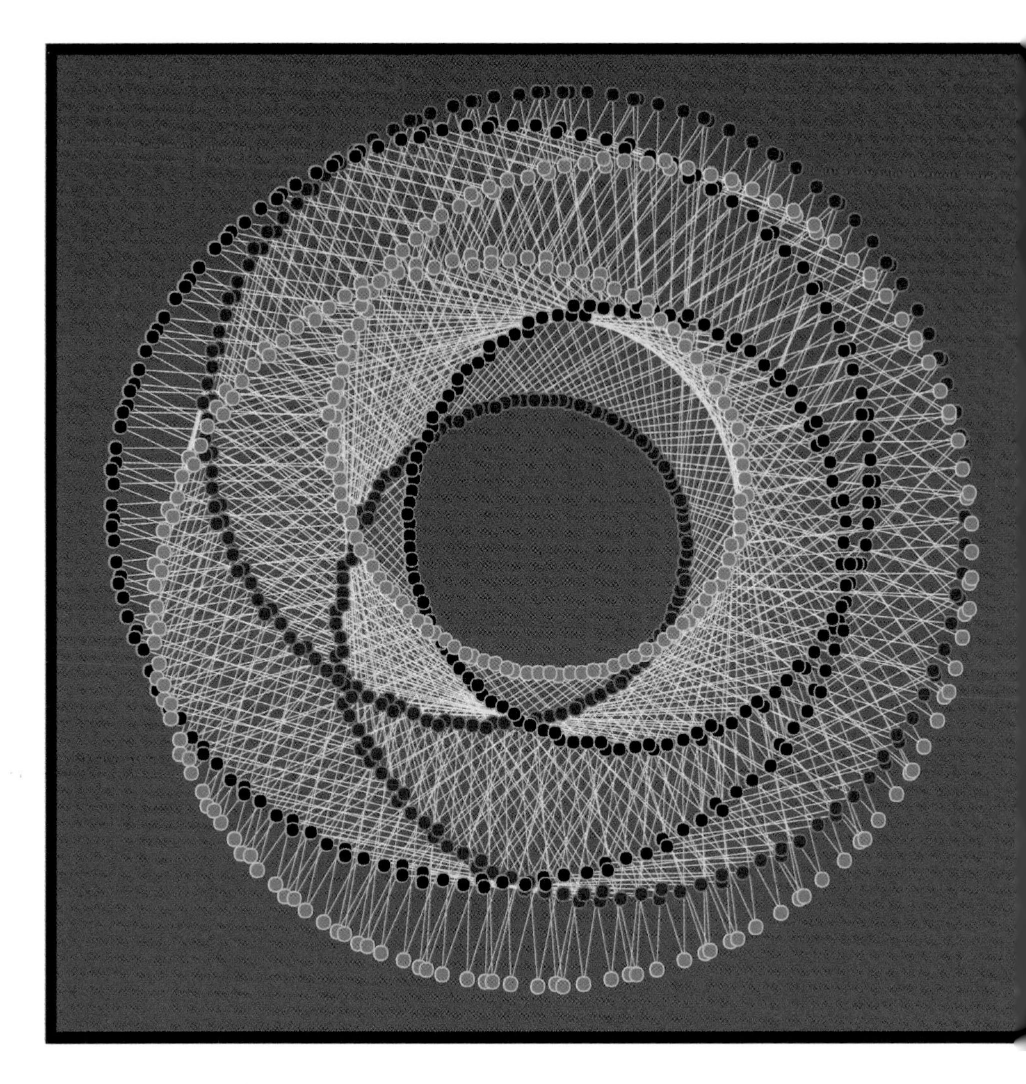

*Plate 6: Venus seen from Mercury at Venus/Jupiter conjunctions*
*→ Fig. 10.4, page 194*

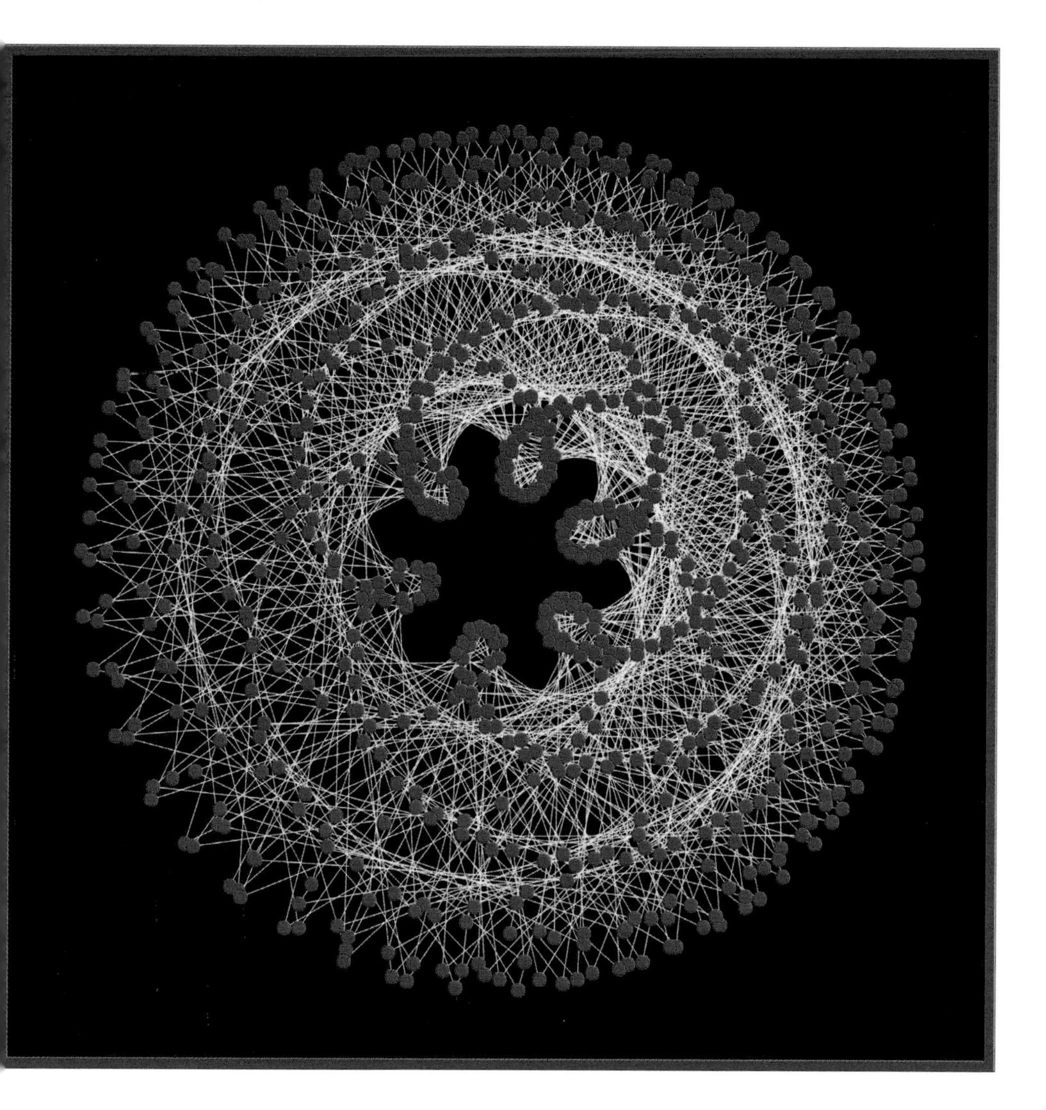

*Plate 7: Earth seen from Venus at Earth/Pluto conjunctions*
→ *Fig. 10.15, page 204*

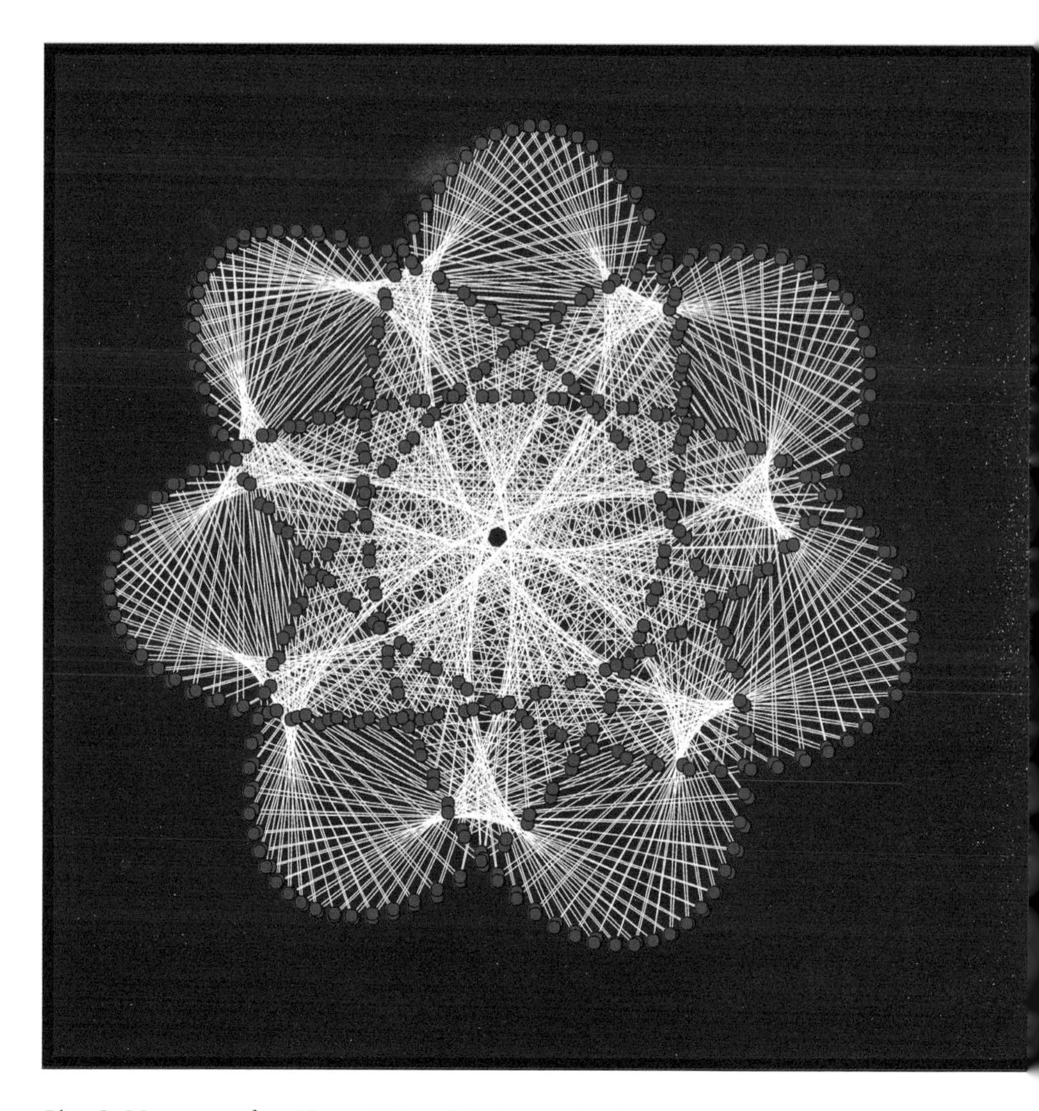

*Plate 8: Mars as seen from Venus at Venus/Neptune conjunctions*
*→ Fig. 10.24, page 213*

*'late 9: Earth at Venus/Earth conjunctions and Venus-Sun-View*
*→ Fig. 11.7, page 237*

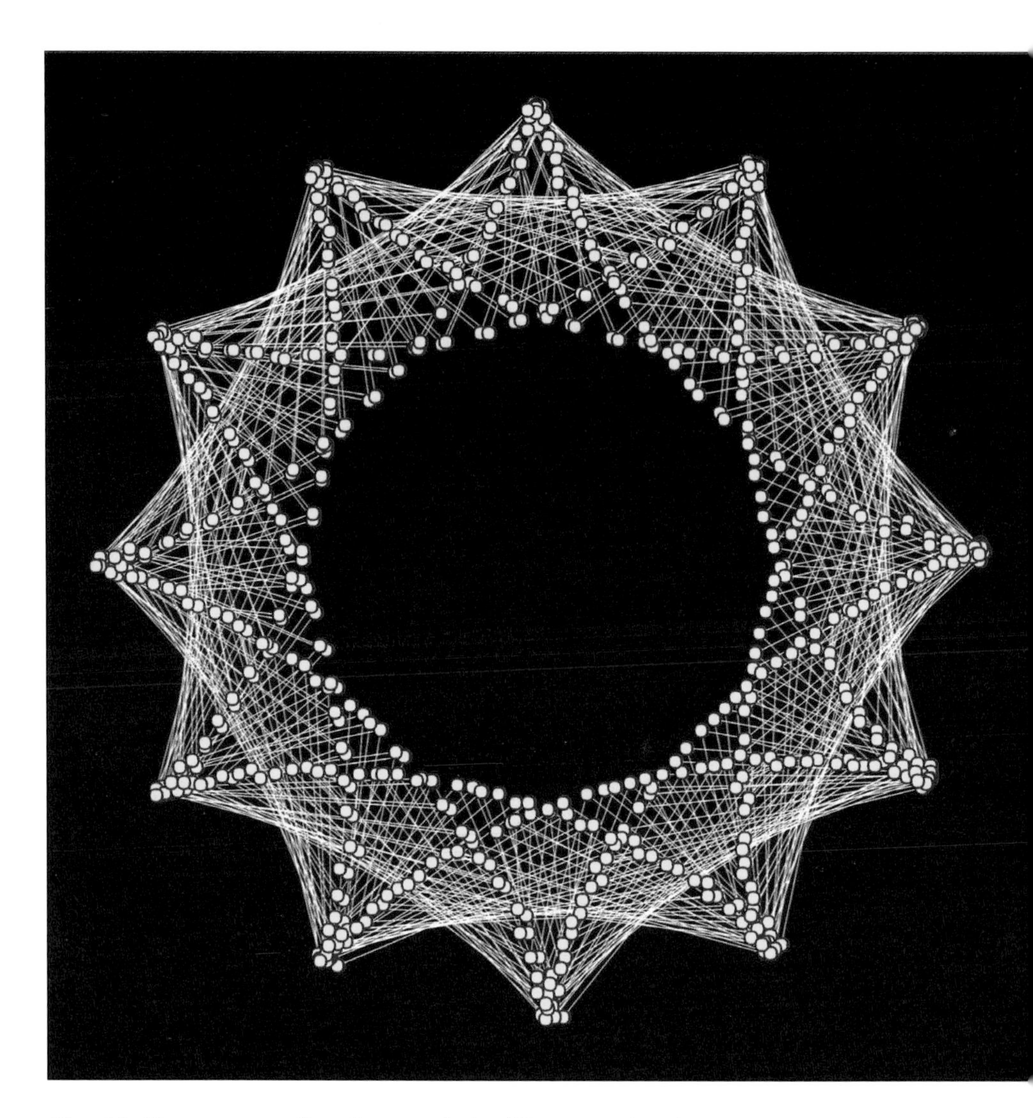

*Plate 10: Neptune as seen from Saturn at Jupiter/Neptune conjunctions*
*→ Fig. 12.4, page 244*

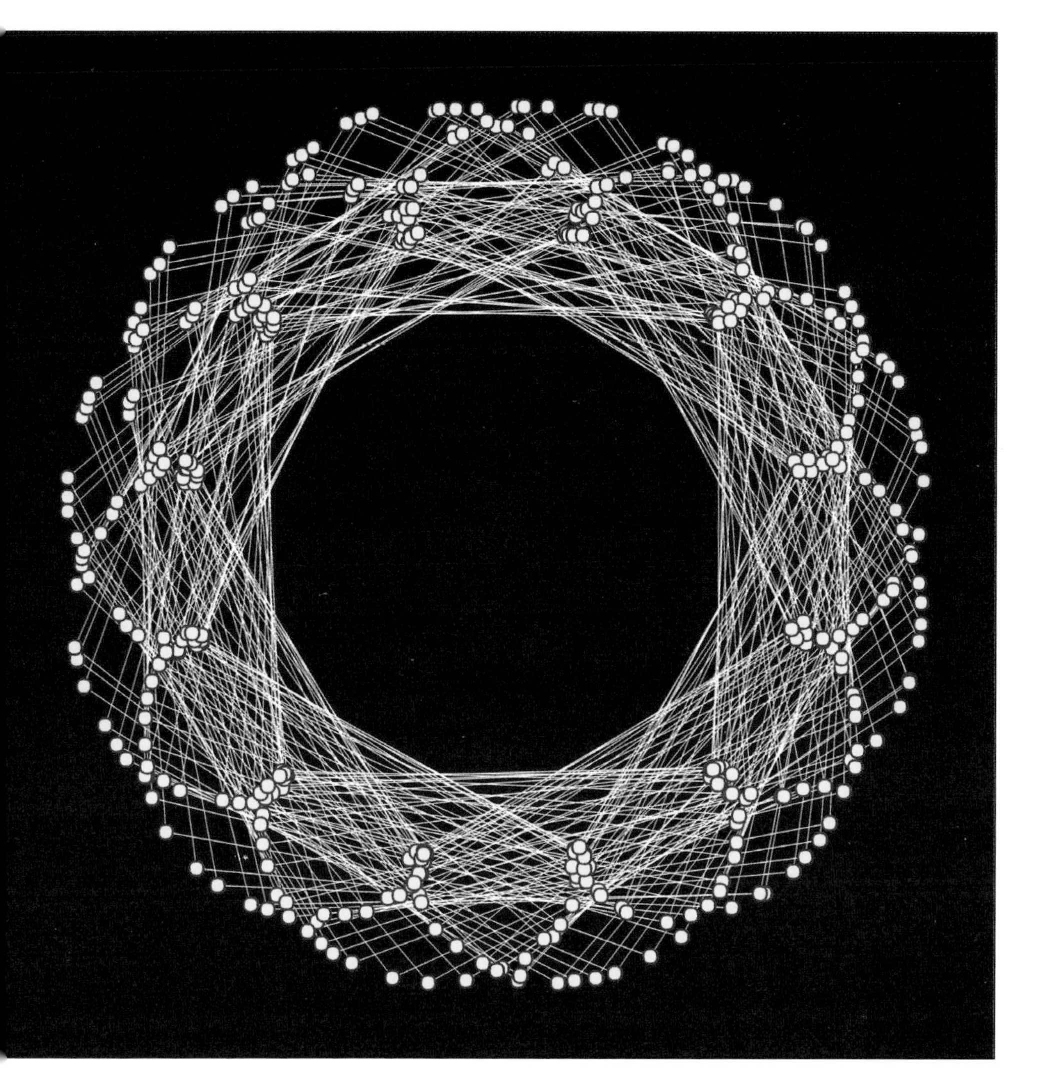

*Plate 11: Neptune as seen from Jupiter at Saturn/Neptune conjunctions*
*→ Fig. 12.4, page 244*

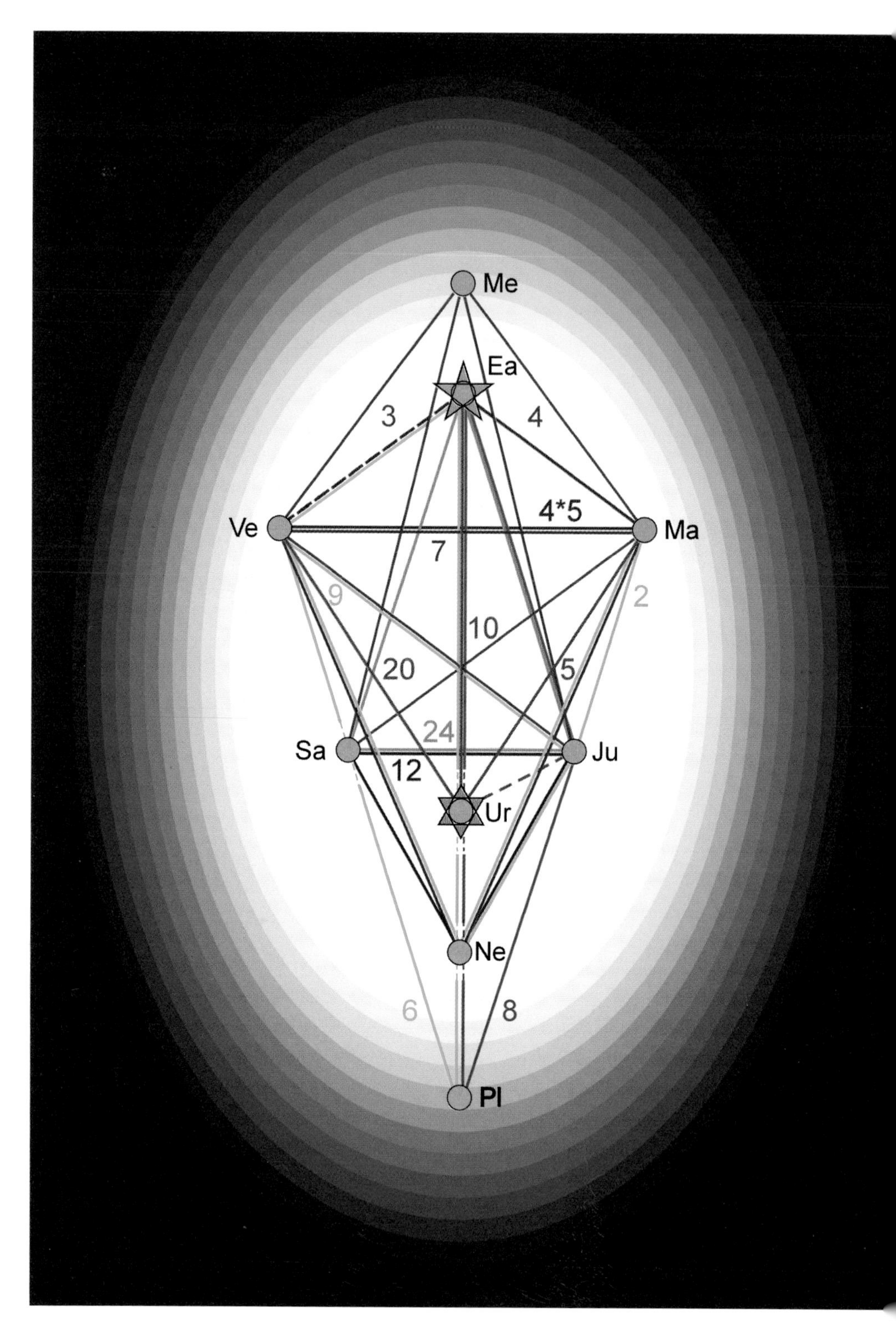


*Plate 12: Conjunction relationships between, in each case, 3 planets depicted by means of triangles and numbe[rs] of the same colour*
*→ Fig. 12.19, page 265*

*Plate 13: Jupiter-Neptune linklines as seen from Saturn at Jupiter/Neptune conjunctions*
→ *Fig. 12.22, page 269*

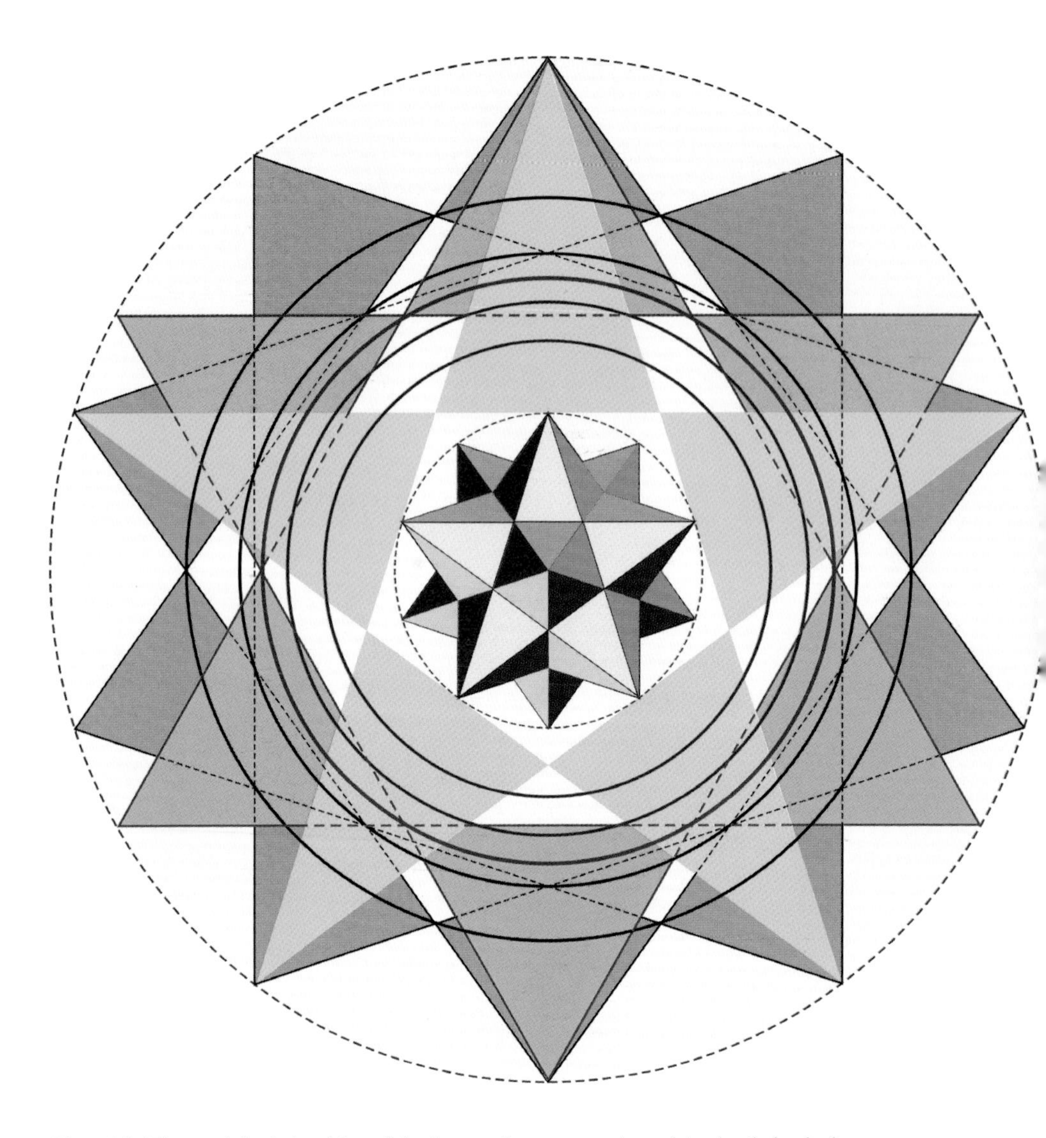

*Plate 14: The spatial relationships of the inner solar system mirrored in the dodecahedron-star*
*→ Fig. 13.9, page 283*

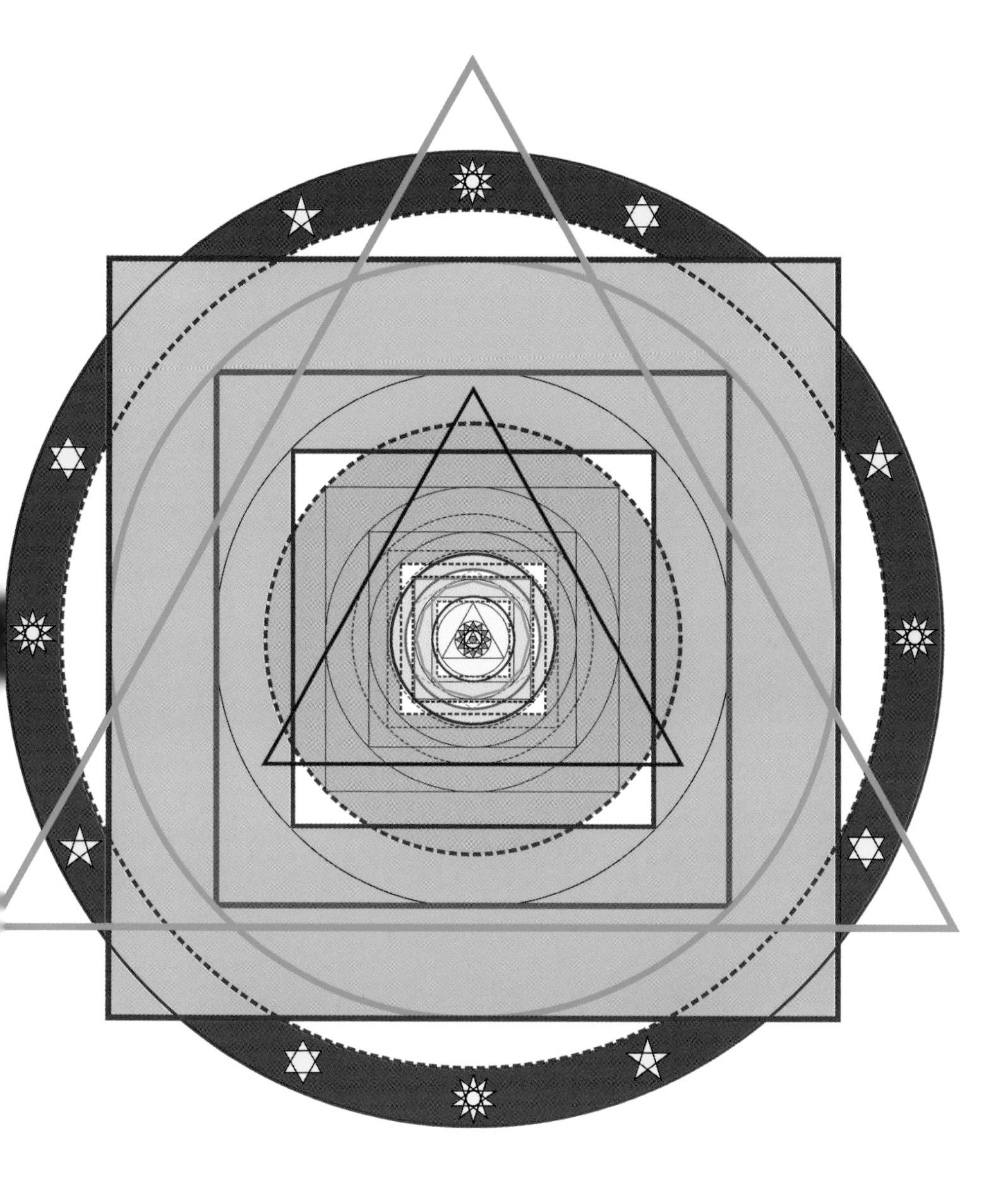

*ate 15: Ordering of the semi-minor axes and the Sun's diameter*
*→ Fig. 13.14, page 291*

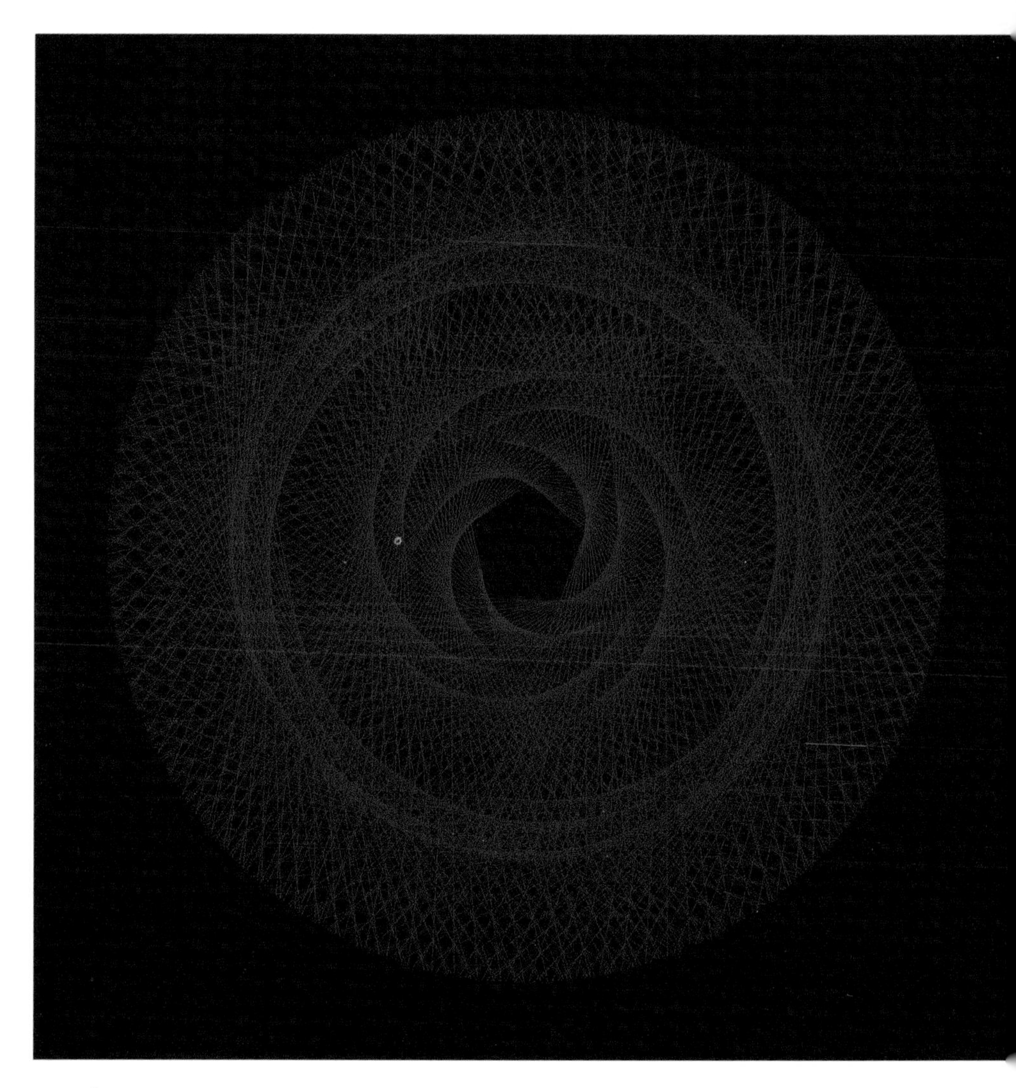

*Plate 16: Venus-Earth linklines at Venus-Sun-Views*
*→ Fig. 8.7, page 152*

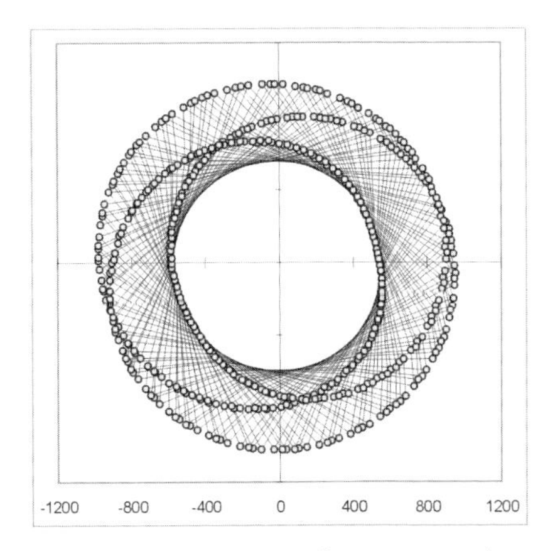


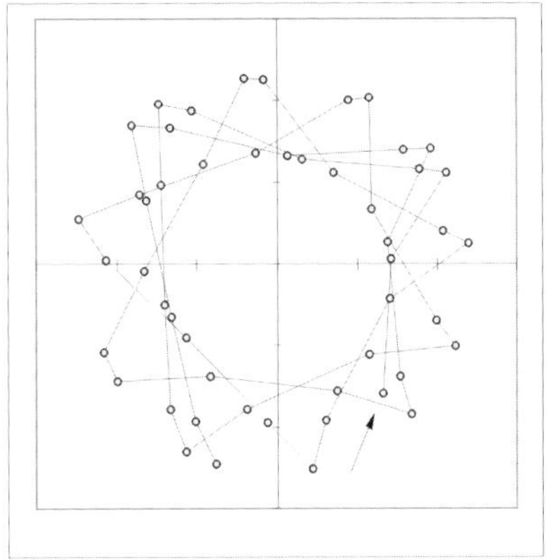

**Fig. 10.3** *Jupiter seen from Mars at the mean value of Jupiter/Neptune conjunctions, 500 times; right, at actual conjunctions, 50 times*

Moon in Fig. 9.16. In the longer-term resonances of Jupiter/Neptune and Mars/Jupiter synodic periods (183:32, 406:71 and 995:174), both Mars and Jupiter cover a fairly to very accurate x plus one half of an orbit. In this way, over a longer span of time, the points form a figure that is ordered according to the number two. The king of the planets is also present when the number three appears on the scene.

Fig. 10.4 manifests not only a structure, somewhat dented by Mercury's eccentricity, which is ordered according to the number three, but also in addition a fivefold underlying structure resulting in 15 partial areas. As we have seen, in a different configuration the relationship between the three planets involved can also lead to the number five of the Minor Pentagram (Fig. 9.4). But the main aspect of the figure shown here is that, of all the possible interactions between three planets, it gives us the clearest embodiment of the number three. And this comes about in a surprising way. When you follow the curves made by the adjacent planetary positions you discover that there is not one continuous line but rather three separate ones, each with an inward-turning loop. The mathematical reason for the overall formation of this figure* lies in the fact that Mercury and Venus, in the periods of the fractional resonances, complete a specific number of orbits plus one third of an orbit. This conforms fairly accurately to the ratio of *c.* 41:25 of the Venus/Jupiter period to the Mercury/Venus period, and with almost one hundred per cent accuracy to the ratio of

* However, I cannot give a reason for the creation of the individual curves which appear to be following a long-term plan.

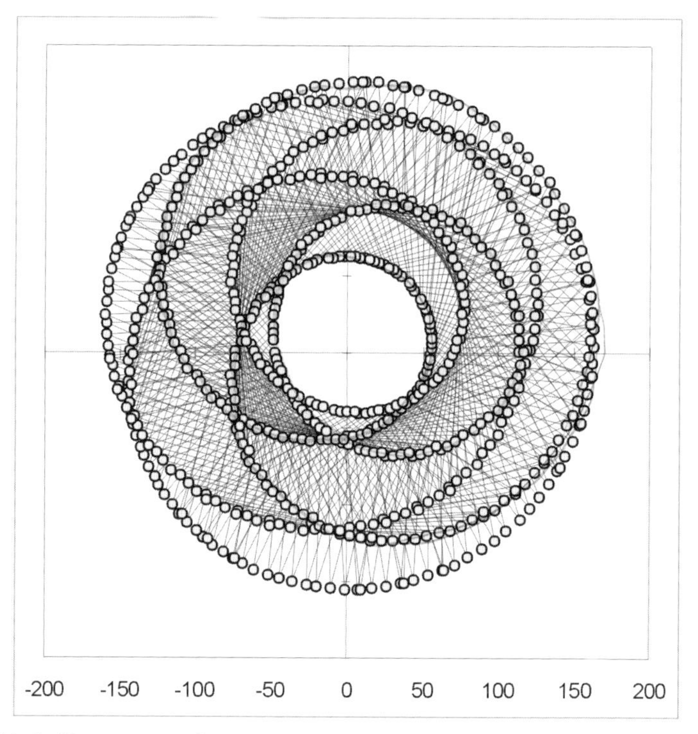


**Fig. 10.4** *Venus as seen from Mercury at Venus/Jupiter conjunctions, 700 times, beginning 22 May 2000, period c. 454.19 years* → Plate 6

100:61. So let us now turn to the number four. Mercury, the messenger of the gods, is once again involved. Perhaps we should have expected it, since we have already once before seen him bearing a message involving this number.

Here we see that in some of the six constellations possible between three planets an order according to the number four emerges (see opposite). While this is possible, it is not always the case. In some instances all six of the resulting figures point to a structure involving the same number. In the ratio of widely separated planets such as Mercury and Saturn this structure often only arises in the centre of the picture drawn by the linklines, i.e. in the inner region of the orbit of the planet closer to the Sun. The right-hand picture shows a corresponding enlargement which makes it possible to discern what is revealed there. This is even clearer in the lines traced in the heavens by Mercury and Mars when the former encounters Saturn, the god of sowing and reaping, the ruler of time and symbol of transitoriness. The picture has been rotated by −65 degrees in relation to the vernal point of 1 January 2000.

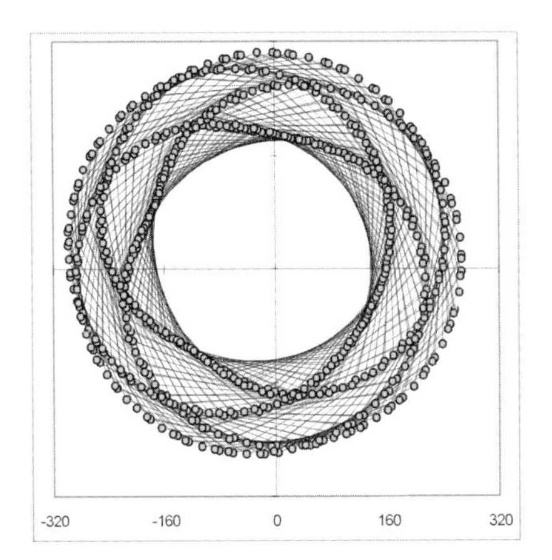


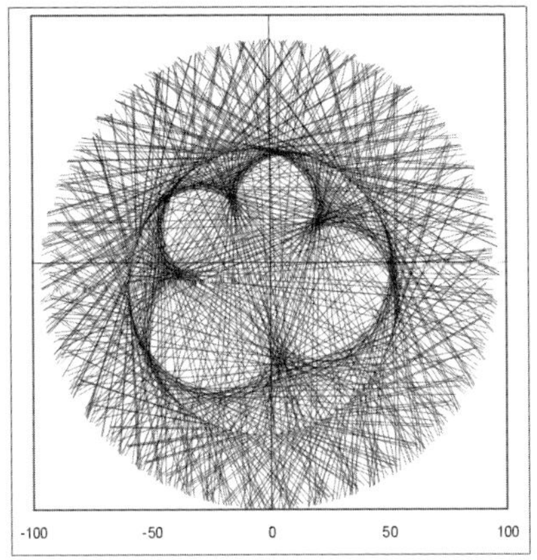


**Fig. 10.5** *Left: Mars as seen from Mercury at Mercury/Saturn conjunctions, 600 times, beginning 10 February 2000, period 145.7 years. Right: Mercury-Saturn linklines at Mercury/Mars conjunctions, 750 times, beginning 6 February 2000, period 207.16 years (detailed enlargement 15:1)*

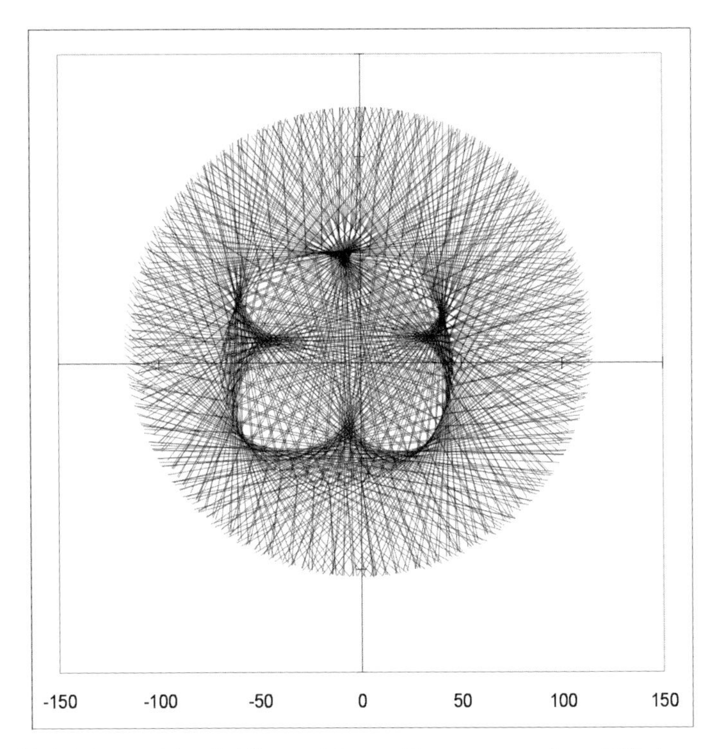


**Fig. 10.6** *Mercury-Mars linklines at Mercury/Saturn conjunctions, 750 times, beginning 10 February 2000, period 182.12 years (detail enlargement 2:1, rotated by − 65 degrees)*

Rays shine out from the cross, especially from its pinnacle. The light bursts forth there as do the beams of the hidden Sun among trees on the summit of distant mountains. It resembles a promise, an ongoing assurance given by the harmony of the planets that Saturn's might is not all-powerful, that beyond the all-destroying cross of time an inextinguishable flame is burning. And this parable-like image comes about precisely because of the participation of the ruler of transitoriness. Or one might also say: it is the cross which enables the vision of light to emerge. As the conductor Sergiu Celibidache might have said, the end of our first movement resembles the third and final movement of Anton Bruckner's ninth and last symphony, although of course this music is able to make an impression that is incomparably more powerful. With Bruckner, said Celibidache shortly before his death, time is that which begins after the end—in association with our inexhaustible hope that 'we shall once more be baptized in the light'.*

## *The second movement*

The theme for this movement will be set—in its interplay with Jupiter—by the planet Uranus (Greek *ouranos*: heaven) which was discovered in the eighteenth century at around the same time as the metal which shares its name. It was through the element of uranium that the possibility of transmuting the basic chemical substances came to be recognized. This amounted to a realization of the alchemists' ancient dream although in a different form since their efforts to transmute base substances into gold were actually only the external side of a search for a genuine inner conversion. 'Uranus, the Magician' is the title of the movement dedicated to this planet in Gustav Holst's orchestral suite *The Planets*. The seventh planet, which symbolically embodies the whole of the heavens, now opens our series of variations on the theme in its interplay with Jupiter and Mars.

Initially the hexagram of the Uranus/Jupiter conjunction with which we are already familiar is formed (see Fig. 6.9). The beginnings of this would even still show at a stepping interval of over 1000 days, so the interval of the Mars/Jupiter synodic period of 816.43 days lies within the scope which leads to this. But after a while, inside Jupiter's orbit, the magical art of Uranus distils from the lines of the 6-pointed star-figure a flower with five petals. This is shown in the case of the link-

* In the documentary film made by his son Serge Joan, *The Garden of Celibidache.*

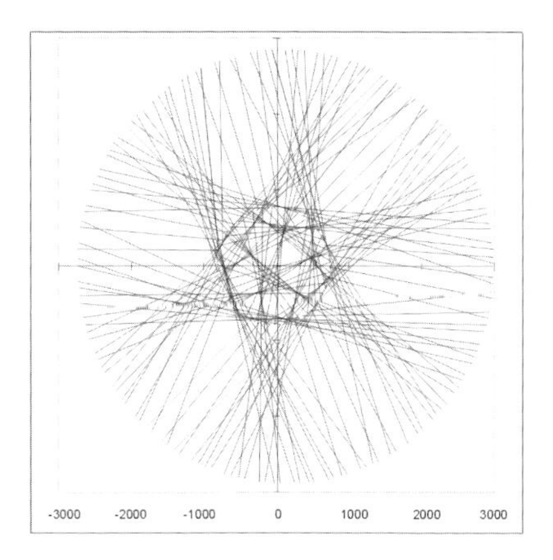


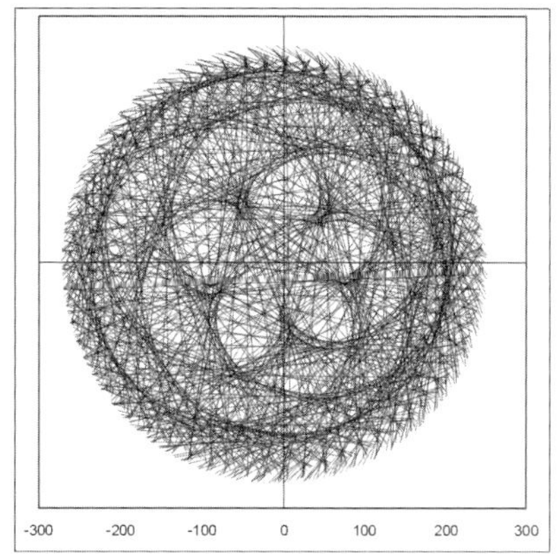


**Fig. 10.7** *Jupiter-Uranus linklines at Mars/Jupiter conjunctions, 150 times, beginning 13 March 2000, period 335.29 years. Right: Mars-Jupiter linklines at Mars/Uranus conjunctions, 1000 times, beginning 22 September 2001, period 1923.88 years (detail enlargement 3:1)*

lines of Mars and Jupiter at Mars/Uranus conjunctions. The configuration of these three planets is one of those in which principally the same figure arises in all the possible interrelationships (although with Jupiter/Uranus conjunctions as the interval of time the formations are only easily recognizable for shorter periods owing to the strong fluctuations). In the centre of the flower emerging so magically one then sees more or less clearly the quintessence of the transfiguration, now again by means of the left-hand example above—in its completion after just one-and-a-half millennia (see Fig. 10.8).

The only clear geometrical figure (of small whole numbers) of the conjunctions of two planets of the outer planetary system, of which it is therefore the proper sign, is recast as the corresponding symbol of the inner system which envelops the planet of humanity. As at the outset of our symphony, Mars, manly and active, stands at the crossover point of the two celestial spheres. The works of which the members of the planetary orchestra are capable when they play together are utterly amazing. And yet thus far we have only heard the prelude to this movement, which consists, as we have already hinted, of a theme and variations. The arithmetical basis of the figures shown is, once again, a fractional resonance. The ratio of the Jupiter/Uranus synodic period to that of Mars/Jupiter is 68:11 or, more accurately, 173:28. After 68 Mars/Jupiter conjunctions, Jupiter and Uranus will have completed 12.814 orbits or, respectively, 1.809, i.e. in each case approximately x plus four fifths orbits. And with 173 encounters we find the crucial correspondence

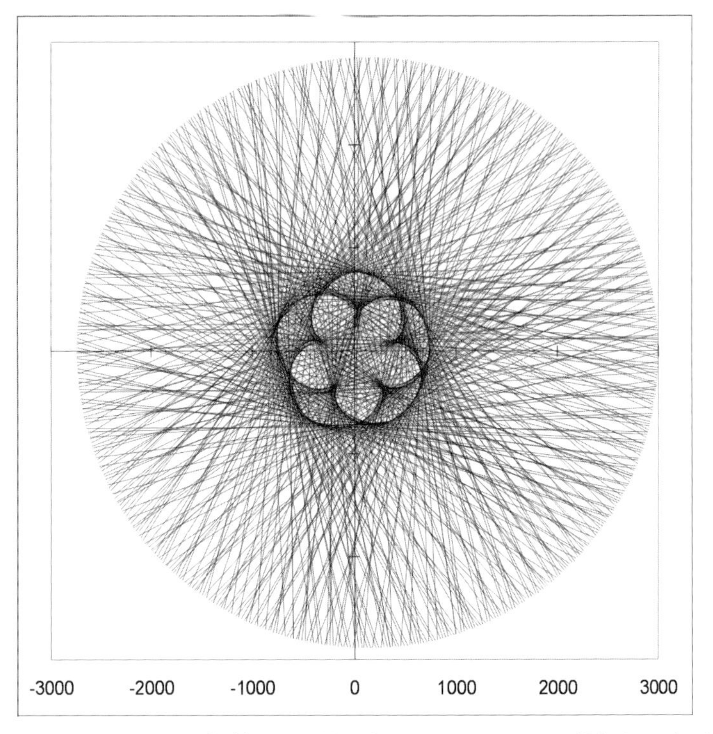


**Fig. 10.8** *Jupiter-Uranus linklines at Mars/Jupiter conjunctions, 650 times, beginning 13 March 2000, period 1452.90 years*

with 32.6002 or, respectively, 4.602 which equals very precisely x plus three fifths orbits. The pentagram-like formation is therefore clearly mathematically provable even when the actual form of the figure brought about by the bare numbers cannot be obtained.

But now as Mars lays down its musical instrument, Earth steps forward to take its place.

The art of the celestial magician can be discerned once again behind the 10-pointed star-figure which now emerges because, as we shall soon see quite clearly, the figure is again obtained from a transformation of the hexagram of the two large planets. How he manages yet again to bring about this highly amazing feat is most mysterious. It is true that here too there is a corresponding fractional resonance,* which leads to a geo-

* The ratio Jupiter/Uranus to Earth/Uranus comes close at 41:3 and even more accurately at 232:17. There are 41.494 Earth orbits and 3.498 Jupiter orbits to 41 Earth/Uranus conjunctions. With 232 there are 234.794 or, respectively, 19.794. By combining the values after the decimal point, approximately 5/10 and 8/10, one arrives at a figure arising according to the number ten.

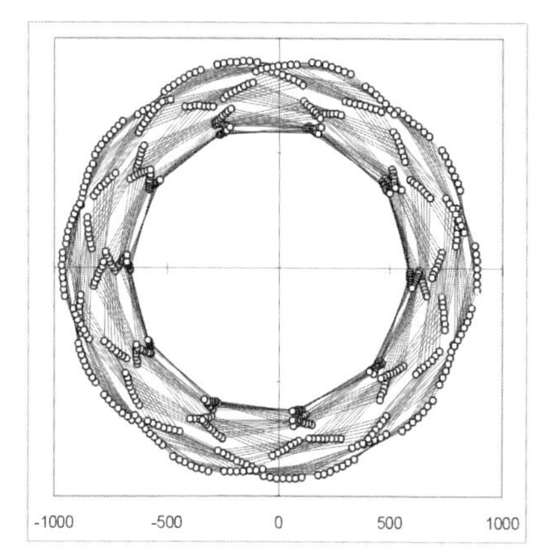


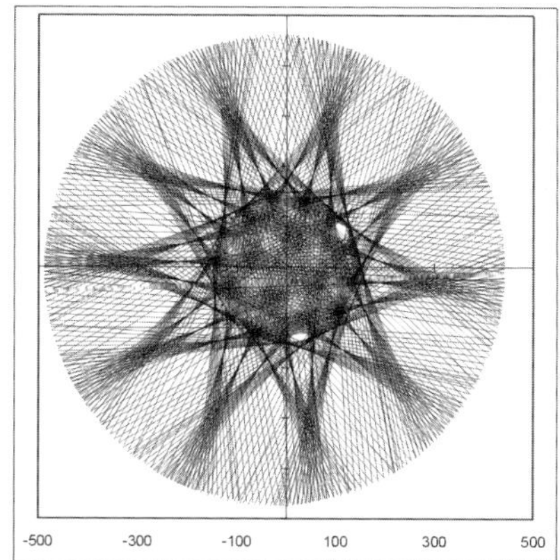


**Fig. 10.9** *Earth-Jupiter at Earth/Uranus conjunctions, beginning 11 August 2000. Left: in the planet-centred depiction 600 times, period 607.23 years. Right: linklines, 750 times, period 759.03 years (detail enlargement 10:6)*

metrical figure ordered in accordance with the number ten. But in spite of this one would never expect, on the basis of the ratios of the conjunction rhythms, to find in this constellation a doubling of the previous pentagram-like structure that was created together with Mars. Seen purely mathematically there is merely a replacement of one synodic period, that of Mars and Uranus, by that of the Earth/Uranus relationship. These two periods are fairly accurately in the ratio of 19:10 = 1.9 (and not two, so that in this ratio the number ten is unlikely to play a part where the geometric figure is doubled from five to ten). Whatever the case may be, celestial mathematics appears here to be pursuing a grand plan of which the arithmetical aspects cannot always, or not always easily, be unveiled. Perhaps a more profound analysis would go further. But our concern here is first and foremost with the wonderful geometry of planetary interrelationship.

This time (see Fig. 10.10) a flower with ten petals grows within the 6-pointed star-figure. It leads to an unusual geometrical drawing of which the significance will be revealed in the 'coda' at the end of this movement of our symphony. The outer star-figure with 83 points arises out of the 84:83 ratio of the Earth/Uranus synodic period and Earth's orbital period or, put differently, of the 84:1 ratio of the periods of both planets. And, by the way, if we create the same figure using the interval of the Earth orbit of 365.25636 days instead of the Earth/Uranus conjunctions, this mysterious planetary mathematics causes an 11-petalled flower to bloom. But we cannot now go into this more deeply because Venus has already

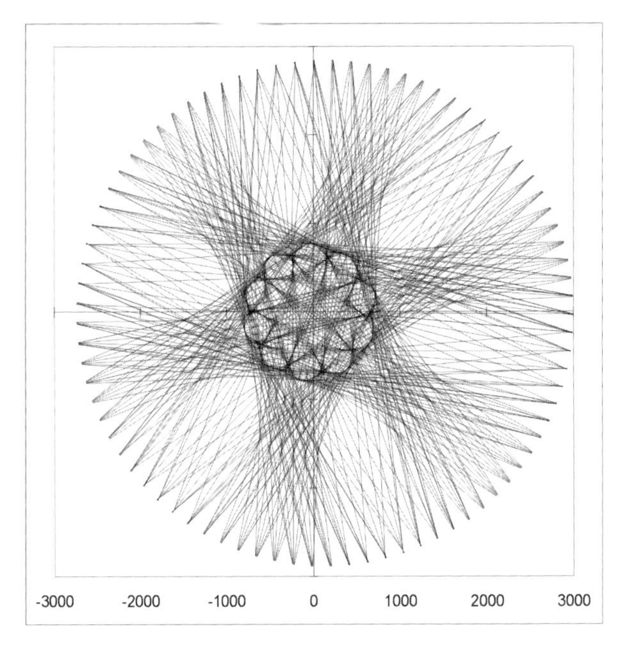


**Fig. 10.10** *Jupiter-Uranus linklines at Earth/Uranus conjunctions, 500 times, beginning 11 August 2000, period 506.02 years*

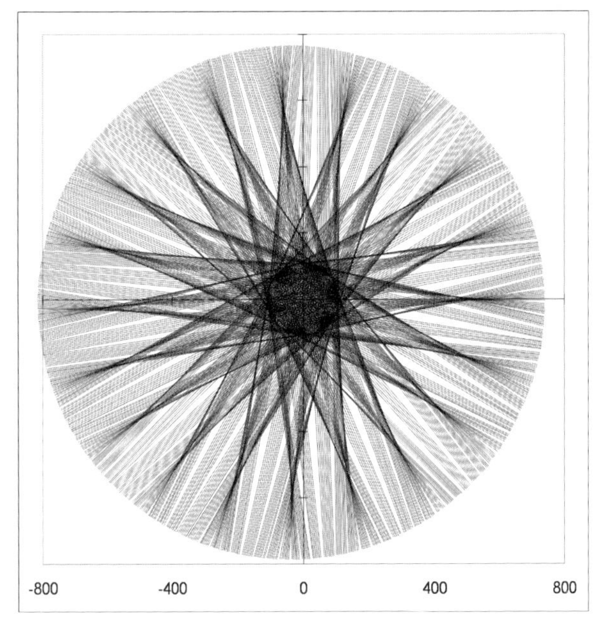


**Fig. 10.11** *Venus-Jupiter linklines at Venus/Uranus conjunctions, 1000 times, beginning 26 March 2000, period 619.72 years*

picked up its instrument in order for a while to take Earth's place in the next variation on the theme.

Here we are met with a 20-pointed star-figure of almost perfect resonance (see Fig. 10.11); it is hard to imagine a star more beautiful or more able to fit in with the structure as it has thus far emerged. We discover that the whole of the Venus, Earth and Mars interrelationship of 4★5 worked out in Chapters 7 and 8 is profoundly linked with the hexagram of Jupiter and Uranus. Once again the doubling of the figure cannot be derived from the ratios of the synodic periods of the planets (31★Ve/Ur ≅ 19★Ea/Ur ≅ 10★Ma/Ur). All we can do is take note anew of the fractional resonance.★ This geometrical structure of the planets' encounter under the influence of the hexagram even in a certain way includes Mercury, the planet closest to the Sun. If Mercury takes the place of the other inner planets in otherwise similar constellations, a 50-rayed figure arises, although this is not quite so clear. The arithmetical skills of the planetary bodies are, moreover, revealed in another very simple—and thus all the more astonishing—process of addition. These wandering stars appear to be quite aware of the fact that 10 + 20 = 30, and they give expression to this in a marvellous star-figure.

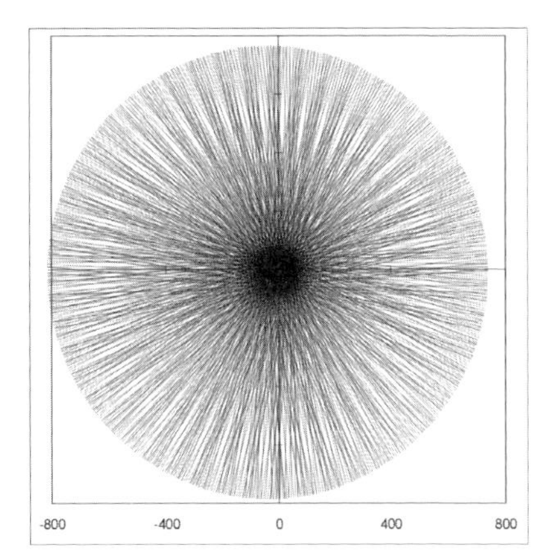


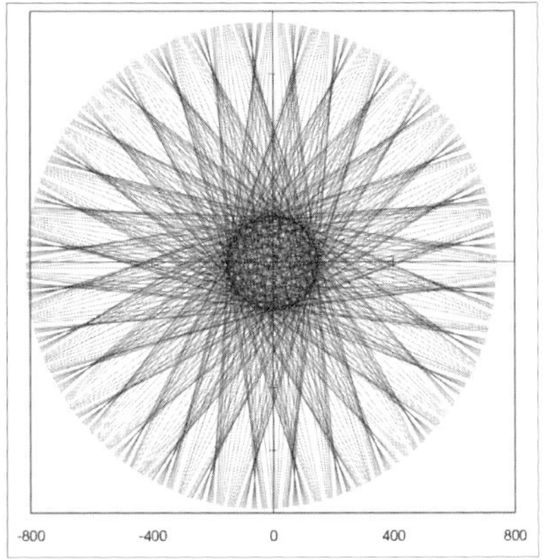


**Fig. 10.12** *Left: Mercury-Jupiter linklines at Mercury/Uranus conjunctions, 1000 times, beginning 22 January 2000, period 241.53 years. Right: Earth-Jupiter linklines at Venus/Uranus conjunctions, 750 times, beginning 26 March 2000, period 464.79 years*

★The essential ratio here is that 149 Venus/Jupiter synodic periods correspond almost exactly to 156 Venus/Uranus synodic periods. In the latter, Venus has completed 157.1507 and Jupiter 8.1503 cycles, in each case thus very precisely x plus 3/20.

The 30-pointed star-figure in the right-hand picture created by Venus, Earth, Jupiter and Uranus can, of course, also be seen as the synthesis of the number five belonging to the Venus/Earth pentagram and the number six of the Jupiter/Uranus hexagram. In the planetary orchestra, Earth is now preparing for the concluding part of this movement which will integrate and elevate everything that has thus far taken place. It will make use of the instrument which began to play before the symphony as such commenced. So let us first make its acquaintance. The simplest way to depict the conjunction relationship of two planets geometrically is to draw a star-figure. We plot the positions of an orbiting planet at the points where it encounters another planet. Or we can mark the places where it finds itself at the conjunction of two other planets—usually at different points in time—and link them chronologically with one another. In Fig. 10.1 these were the Earth/Uranus and the Earth/Neptune conjunctions.* In this way we gain an overall view of all the influences affecting one planet that are caused by two other ones. In the coming movements, too, we shall see that in principle this overall view corresponds to the other ways of depicting three planets. Each method has, however, its own advantages. Let us look at the beginning of the figure that emerges when Jupiter takes the place of Neptune:

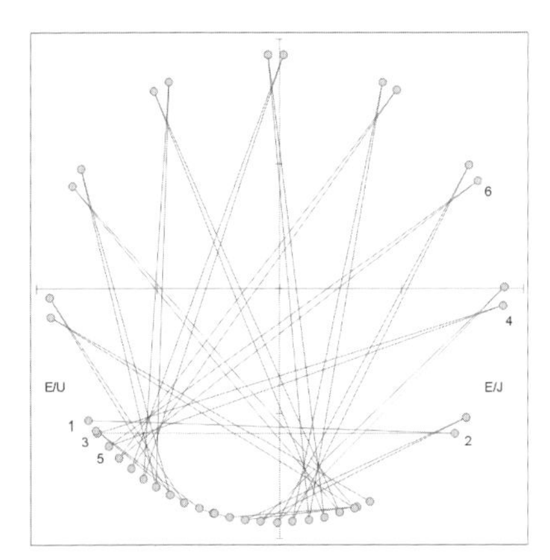


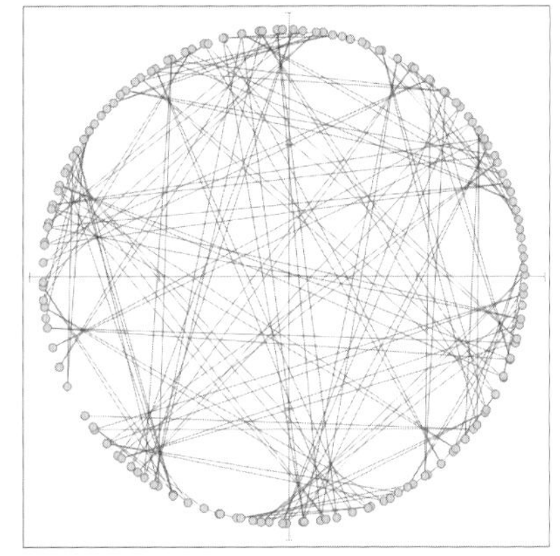

**Fig. 10.13** *Earth at Earth/Jupiter and Earth/Uranus conjunctions, beginning 11 August 2000 (E/U). Left: 40 times, period 20.22 years. Right: 150 times, period 77.93 years (both figures rotated by − 105°)*

*Just as in the variations of this movement the 7:1 ratio of the orbits of Uranus and Jupiter makes itself known as a hexagram or other six-numbered figure (sometimes in a hidden way), so will that spirit flame rotate 25 times before it completes its full and final form. In this way it conforms to the 51:26 resonance of the orbital periods of Neptune and Uranus.

At first the positions occupied by Earth at the two conjunctions skip back and forth. Two bundles of rays are formed, but after about 150 conjunctions one sees that in this method of depiction (let us call it the 'double conjunction', see Glossary), too, the number six of the Jupiter/ Uranus relationship governs what is going on. If one plots the 'double conjunction' of Mars with Jupiter and Uranus, after a while a five-petalled flower arises which approximately resembles the one shown on the right in Fig. 10.7. But with Venus or Mercury plotted in the same way, the structure is ordered according to the number six, just as is also temporarily the case with Earth. This one, however, appears to have the strength to outdo what has already been achieved. Or one could also say this depiction of movement is only seemingly carried out by Jupiter and Uranus, since the final composition to emerge also embraces the numbers

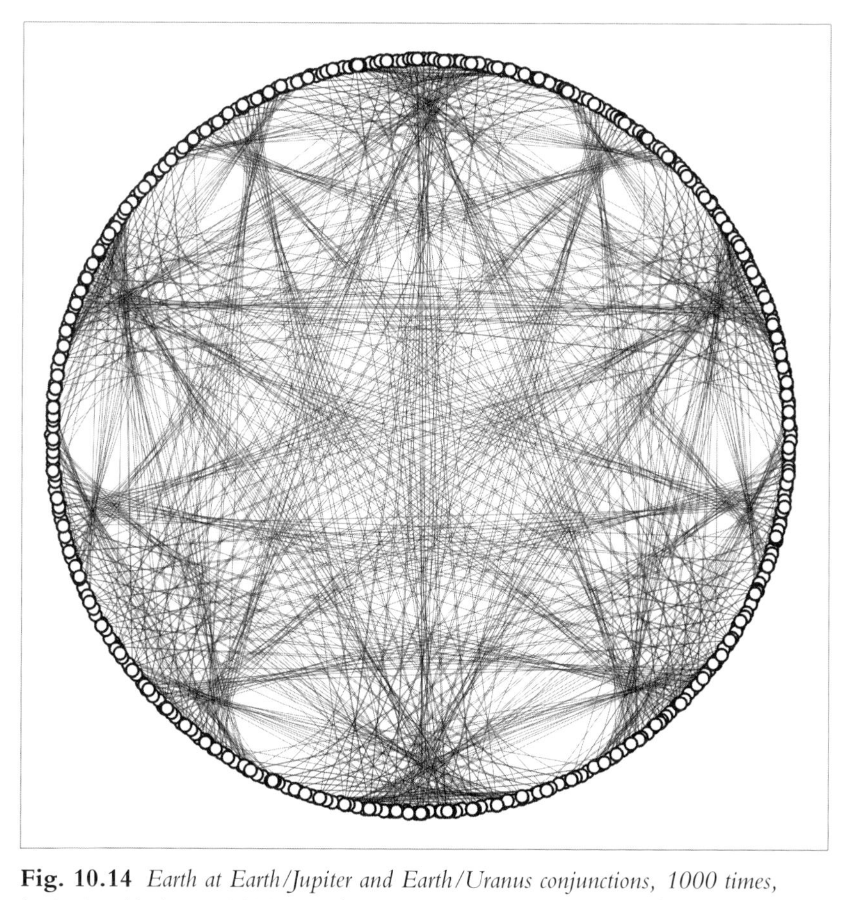

**Fig. 10.14** *Earth at Earth/Jupiter and Earth/Uranus conjunctions, 1000 times, beginning 11 August 2000, period 524.49 years (rotated by − 105°)*

three, four and five. And the geometrical figures these numbers bring about are composed in such a way that it is worth taking a somewhat longer look at them. Just as with the other figures, this one, too, is not unique, for it appears again with slight variations and slightly rotated during the next one thousand 'double conjunctions'.

## *The third movement*

The third part of a symphony is usually the scherzo which with Haydn and Beethoven took the place of what in earlier days had been the minuet. The scherzo is a relatively short movement in the A-B-A form with a marked rhythm, and it frequently has a contrasting more melodious middle section. In the overall harmony of the planets it is now the turn of Pluto, which has so far not featured often in these pages, to come to the fore.

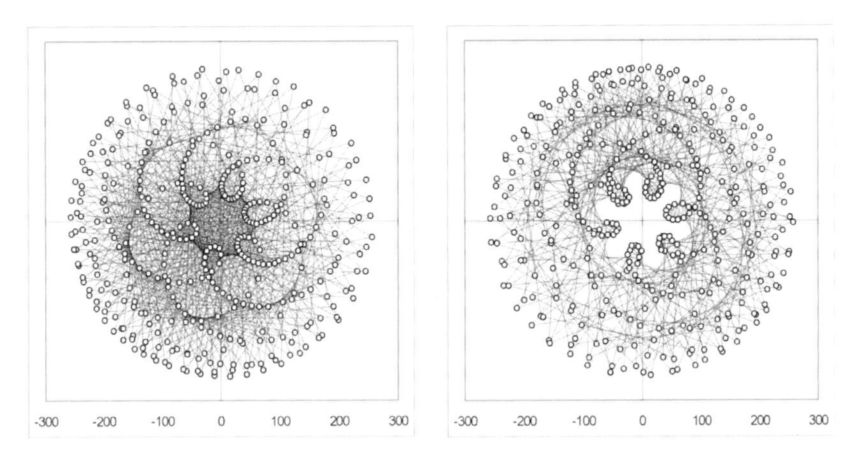


**Fig. 10.15** *Earth as seen from Venus in their interplay with Pluto, 400 times in each case. Left: at Venus/Pluto conjunctions, beginning 27 November 1882, period* c. *246.7 years. Right: at Earth/Pluto conjunctions, beginning 25 August 1799, period* c. *401.6 years* → Plate 7

The eccentricity of Pluto's orbit is clearly shown in the distortions of the pictures. However, in the inner region six obvious loops show up after one Pluto orbit. The time aspect is especially visible in the left-hand picture where the period is more or less equal to one Pluto year. So Pluto adds another loop to the five that appear in the pentagram of the Venus/ Earth relationship (see Fig. 6.1). In this way Pluto has, as it were, the opposite geometrical effect from that of Uranus which (in collaboration with Jupiter) transformed the sixfoldness of the hexagram into the number five and its multiples. In this way alone the reciprocal metamorphoses of the two star-figures, which characterize the inner and outer planetary system, demonstrate that a kind of symphony is being performed. Each theme is mysteriously linked with the whole in a way that it is most certainly difficult to presume could arise from the accidental collisions of chunks of matter in ancient times.

The beauty of the theme involving Venus, Earth and the planetary system's most distant member comes to expression even more clearly in the depiction using the linklines since the irregularities caused by the eccentricities are entirely evened out after a while.

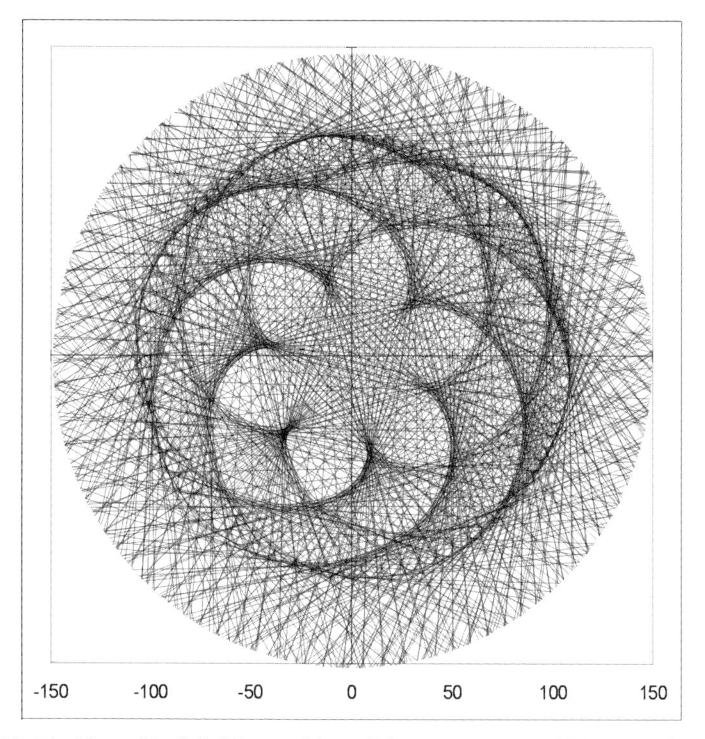


**Fig. 10.16** *Venus-Earth linklines at Venus/Pluto conjunctions, 1000 times, beginning c. 5 September 1691, period c. 616.7 years*

Let us now see the effect Jupiter has when it takes the place of Pluto. The Sun also makes an appearance now (although of course it is always invisibly present in conjunctions since it is with the Sun that the planets line up).

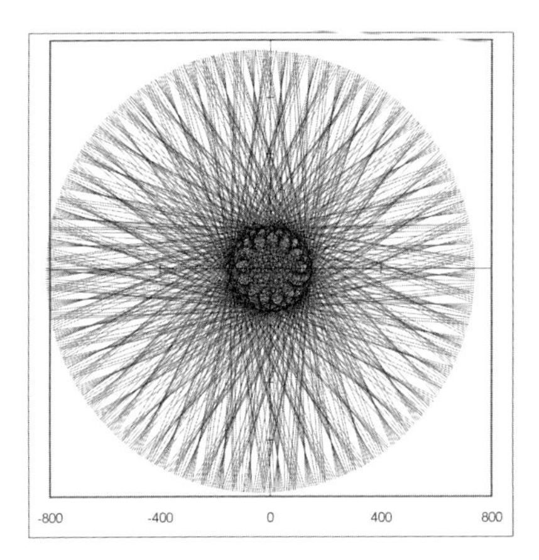


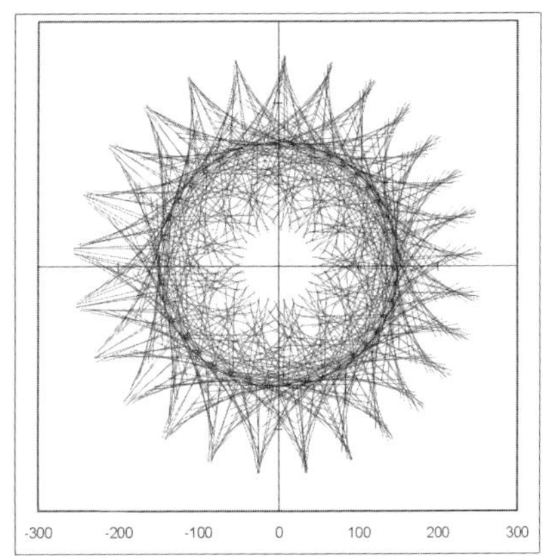


**Fig. 10.17** *Left: Earth-Jupiter linklines at Venus/Jupiter conjunctions, 750 times, beginning 22 May 2000, period 486.63 years. Right: Sun-Venus linklines from the geocentric point of view at Earth/Jupiter conjunctions, 1000 times, beginning 28 November 2000, period 1092.07 years*

There are three resonances in the relationship of Venus, Earth and Jupiter. In the outer part shown on the left a 40-pointed star-figure is formed, while within the Earth orbit a 13-petalled flower arises. Because of the density of the lines it is very much in the dark there, but it shows in greater detail in the right-hand picture. For this reason we have here drawn the linklines traced by the Sun and Venus—as seen from Earth—which reveal not only the flower but also a further very resonant star-figure with 27 points. This also shows in other depictions, for example the planet-centred view from Venus and Earth at the same conjunctions. But in combination with the Sun both this star-figure and the flower are at their most beautiful. The number thirteen, by the way, appears several times in the figures formed by three planets. In these cases the conjunction relationships of the large planets are mirrored in the configurations of Jupiter and Neptune or, respectively, Saturn and Uranus with, in each case, one inner planet. We were introduced to these in Fig. 6.7 with its 13-pointed star-figure and in Fig. 6.8 with its 13-petalled flower. Just as the hexagram of Jupiter and Uranus was expressed in the configurations of the previous movement of this symphony, so does the number thirteen appear in corresponding planetary interrelationships. There will be another example of this in some of the coming diagrams. But the flower in the above picture is independent of this, since Neptune is not involved in it.

Meanwhile, here we see Venus, Earth, Jupiter and Pluto, the four planets that take part in the music of this movement, preparing for the final section of the scherzo:

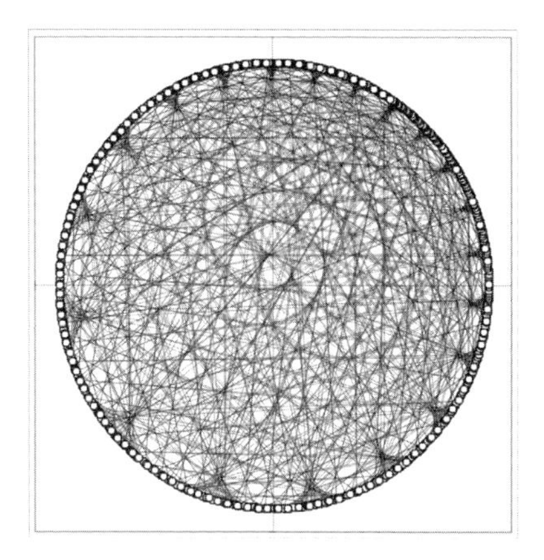
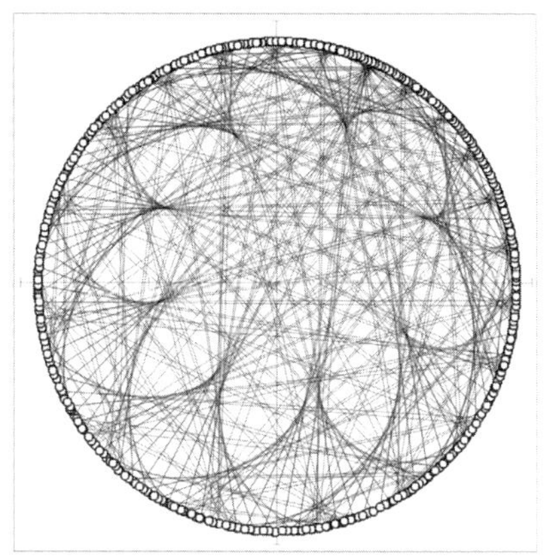

**Fig. 10.18** *'Double conjunctions'. Left: Venus at Venus/Jupiter and Venus/Pluto, 800 times, beginning 12 April 1874 (V/Pl), period c. 252.6 years. Right: Earth at Earth/Jupiter and Earth/Pluto, 500 times, beginning 22 November 1883 (E/Pl), period c. 261.0 years*

In both the pictures the 21:1 ratio of the Pluto and Jupiter orbits manifests in the 20 curves, or the concentrations of lines, showing at the edges of the figures. In the middle of these figures which arise from the 'double conjunctions'—and also in those involving the linklines not shown here—the flower-like structures that appear are rather distorted. Nevertheless, the eight-petalled flower in the right-hand depiction is clearly recognizable. That the strange figure on the left contains the remains of a four-petalled flower can only be guessed at. But it becomes obvious when we ignore the eccentricity of Pluto in the calculations that lead to this figure. So to sum up we can say that even when the outermost planet of our solar system is involved the celestial mathematics in the geometry shows itself from its more simple aspect in that the numbers six and eight and, in a restricted way, the number four all make their appearance.

So now, almost seamlessly, we can make the transition to the final movement which begins, you might say, with a drum-roll:

## *The fourth movement*

In the formation arising within the Venus orbit at its conjunctions with Saturn and Neptune over the course of about eight centuries the ratio between the two is mirrored in its entirety. During 23 of their conjunctions (in almost 825, i.e. 824.993 Earth years), Saturn circumnavigates the Sun

**Fig. 10.19** *'Double conjunctions': Venus at Venus/Saturn and Venus/Neptune, 2600 times, beginning 17 March 2000 (V/N), period 809.46 years*

28.007 times and Neptune 5.007 times. The Venus figure is dominated by the encounters of the two large planets which are expressed in the 23 intersecting pointed arches. In the middle region of the picture we can see, although much less clearly, a ring of 28 smaller arches, and in the centre there is just a hint of our old friend, the five-petalled flower. The 23-fold order also appears in the corresponding figures of Mercury and Earth, but only the goddess of love herself is capable of a devotion so pure that all the facets of the Saturn/Neptune constellation are able to appear.

But now Saturn withdraws, making way for Mercury to determine the next theme.

The number eleven, too, is represented in the celestial round-dance (although this figure is not quite accurate). The resonances which make this possible are the 64:39 and, more exactly, the 745:454 ratios of the

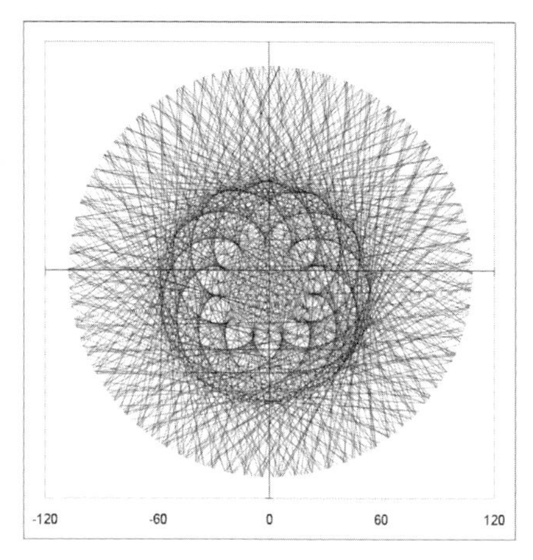


**Fig. 10.20** *Mercury-Venus linklines at Mercury/Neptune conjunctions, 750 times, beginning 18 January 2000, period 180.90 years*

*'As is well known, Pythagoras was the originator of this view of the celestial movements as rhythm and music. But equally well known is the fact that his ideas were little understood, and one may easily conclude that they have come down to us in a much-corrupted form . . . Pythagoras does not say that those movements* cause *the music but that they themselves* are *the music. This indwelling movement needed no external medium through which it might become music, for it was of itself music.'*

Friedrich W.J. von Schelling[1]

Mercury/Venus synodic period to the Mercury/Neptune synodic period. In the time-span of 64 or, respectively, 745 Mercury/Neptune conjunctions, Venus and Mercury (and of course also Neptune) complete—at first approximately and then very accurately—x plus $\frac{1}{11}$ of their journey around the Sun.

And now Jupiter is preparing to take over from Mercury, while Venus and Neptune still do not lay down their instruments.

And, sure enough! In the art of creating a rose window the skill of the wandering stars is to all intents and purposes no less marvellous than that of the medieval master builders (see Fig. 10.21). Furthermore, we notice in the outer region that the number thirteen's lack of sharp definition in the conjunction ratio of Jupiter and Neptune, which we already noticed in the corresponding right-hand picture in Fig. 6.7, evidently serves a very specific purpose: the lack of absolute exactitude is likely to be the very reason why the celestial rose window can here be traced in the firmament with such a high degree of precision. And even if this is not the only reason for the deviation from the resonance, it still gives us an inkling of how finely tuned the movements of the various planets have to be in order to cause the various themes of this symphony to resound. Music only ever reaches its mark if it is heard. It is only then that it acquires meaning, and only then that its meaning can be revealed.

In Fig. 10.22, the picture on the left shows the rose window again, using a different method of depiction. Here it is not so multilayered but it is still

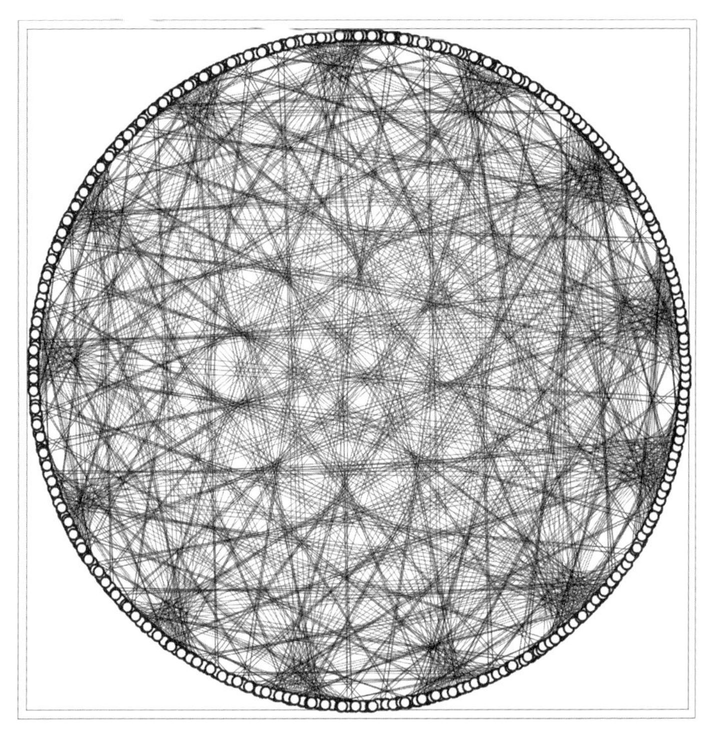

**Fig. 10.21** *'Double conjunctions'; Venus at Venus/Jupiter and Venus/Neptune, 1200 times, beginning 17 March 2000 (V/N), period c. 379.14 years*

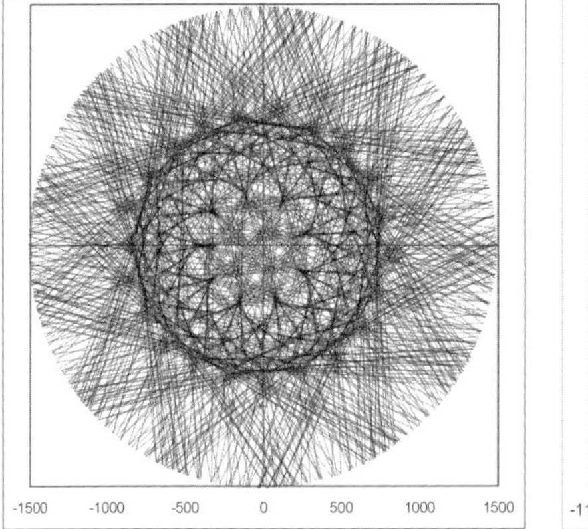


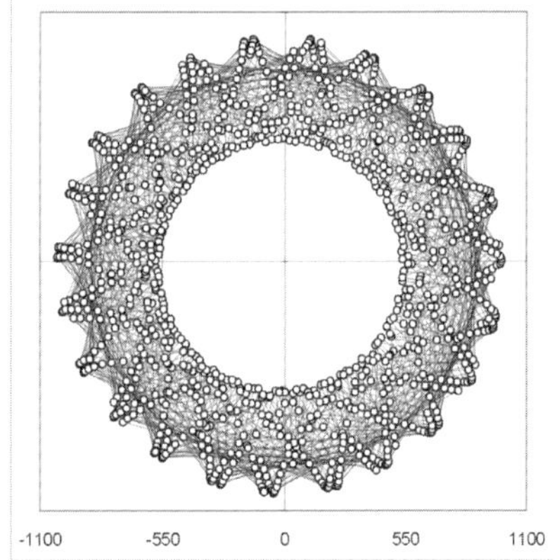


**Fig. 10.22** *Left: Jupiter-Neptune linklines at Venus/Neptune conjunctions, 1000 times, beginning 17 March 2000, period 617.49 years (detail enlargement 3:1). Right: Jupiter as seen from Mars at Venus/Jupiter conjunctions, 1100 times, beginning 22 May 2000, period 713.72 years*

readily visible in the midst of the Jupiter orbit and of the star-figure surrounding it. In our amazement we have almost failed to notice the recurrence of the theme heard at the beginning. Neptune now pauses to gather strength for the imminent finale. Meanwhile Mars has joined the interplay of Venus and Jupiter. The introductory melody of this movement now resounds in a different form, but it is unmistakeably the theme of the number twenty-three. It has detached itself from the Saturn/Neptune conjunction and now rises in wide curves almost up to the stars or, so to speak, up to the star-figures of not so small 'small whole numbers'.

But the task of reaching the skies is given to the theme of the number which we are still waiting to hear in the harmony of these planetary motions. To bring this to the ears or eyes of the beholder Neptune now takes up his trident once more and replaces Jupiter. Gradually the combined melody of Venus, Mars and Neptune approaches, then slowly withdraws a little in a somewhat different garment before showing itself once more in all its perfection (Fig. 10.23).

A not very precise 7-pointed star-figure appeared in the Venus orbit in the rhythm of the Mars/Jupiter conjunctions, but its points disappeared when instead of the geometrical conjunctions we took the minimum distances that are more important for the gravitational interaction (Fig. 6.19). That star-figure now reappears in almost perfect form in the interplay of three planets when Neptune takes the place of Jupiter. As with the other figures in this chapter, it does not matter which type of conjunction is used in the calculation. A whole series of 7-pointed star-figures arises because this configuration is one of the relatively rare ones in which the geometrical formation appears in accordance with only one number in all kinds of constellations of the three planets. Mathematically the figures can be explained by the following conformity of the synodic periods: 37 Mars/Neptune $\cong$ 77 Venus/Mars $\cong$ 114 Venus/Neptune (37.0003; 77; 113.9994). The decisive feature is that the resonance corresponds quite accurately to 3/7 Neptune, 37 3/7 Mars and 114 3/7 Venus orbits.

Meanwhile, the three planets are now preparing to play their theme for the last time on a different instrument. At the start they do so rather more slowly, perhaps so that one can follow the development of the melodies in greater detail (Fig. 10.24).

The final notes have died away and the symphony is at an end; but we have not quite finished this chapter. We still need to review what we have heard and look at the architecture of each movement. To our astonishment every number up to the number eleven appears once and only once in the interplay in each case of three planets. The exception is the second number four that appears in the eccentrically very distorted

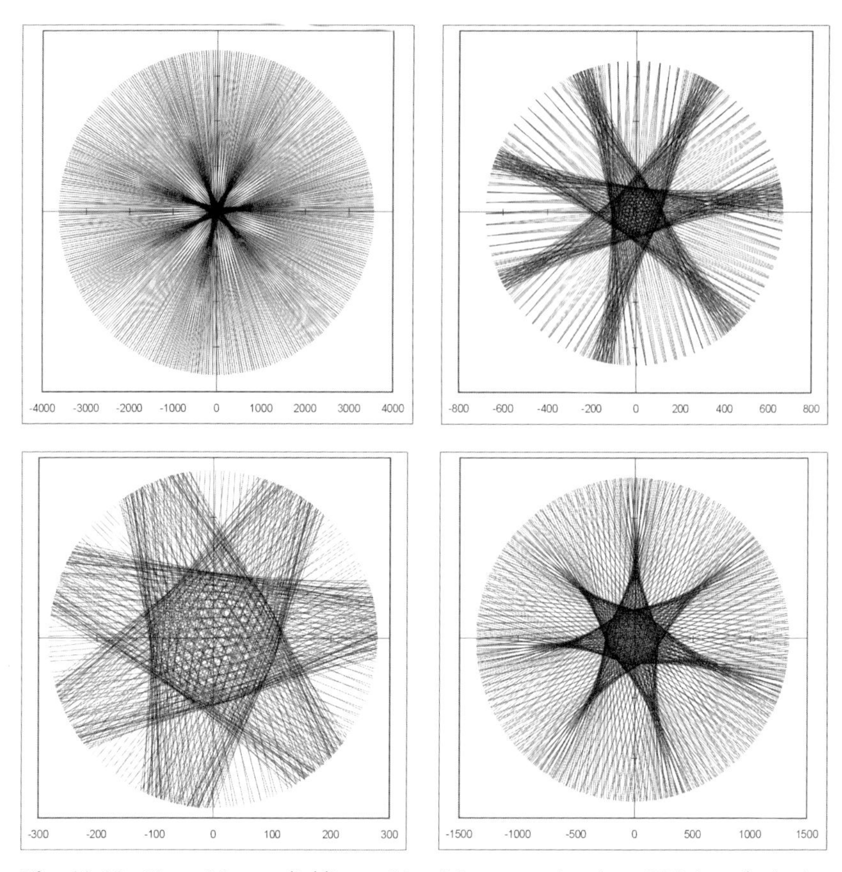


**Fig. 10.23** *Venus-Neptune linklines at Mars/Neptune conjunctions, 750 times, beginning 28 August 2001, period 1426.90 years. Detail enlargements, top left 5:4, top right 20:3, bottom left 16:1. Bottom right: Mars-Neptune linklines at Venus/Mars conjunctions, 750 times, beginning 19 June 2000, period 685.66 years, detail enlargement 10:3*

flower of the Venus/Jupiter/Pluto constellation. No doubt there are also some other constellations that lead to approximate figures of small whole numbers, but these are very much less resonant than those shown here.*

---

*We might mention an approximate 10-pointed star-figure of Venus/Earth/Saturn, an approximate 10-petalled flower of Mercury/Mars/Neptune, and two formations involving the numbers five and eight which can be seen in the inner part of the figures of Earth or, respectively, Venus in their conjunctions with Jupiter and Saturn. In the case of the last two, something occurs which cannot be found anywhere else in connection with constellations of three planets, namely that the gravitational perturbations have a very definite effect on the movement pictures which are very clear if one calculates them without including the perturbations. Perhaps there is some unknown reason why each small whole number makes only one appearance.

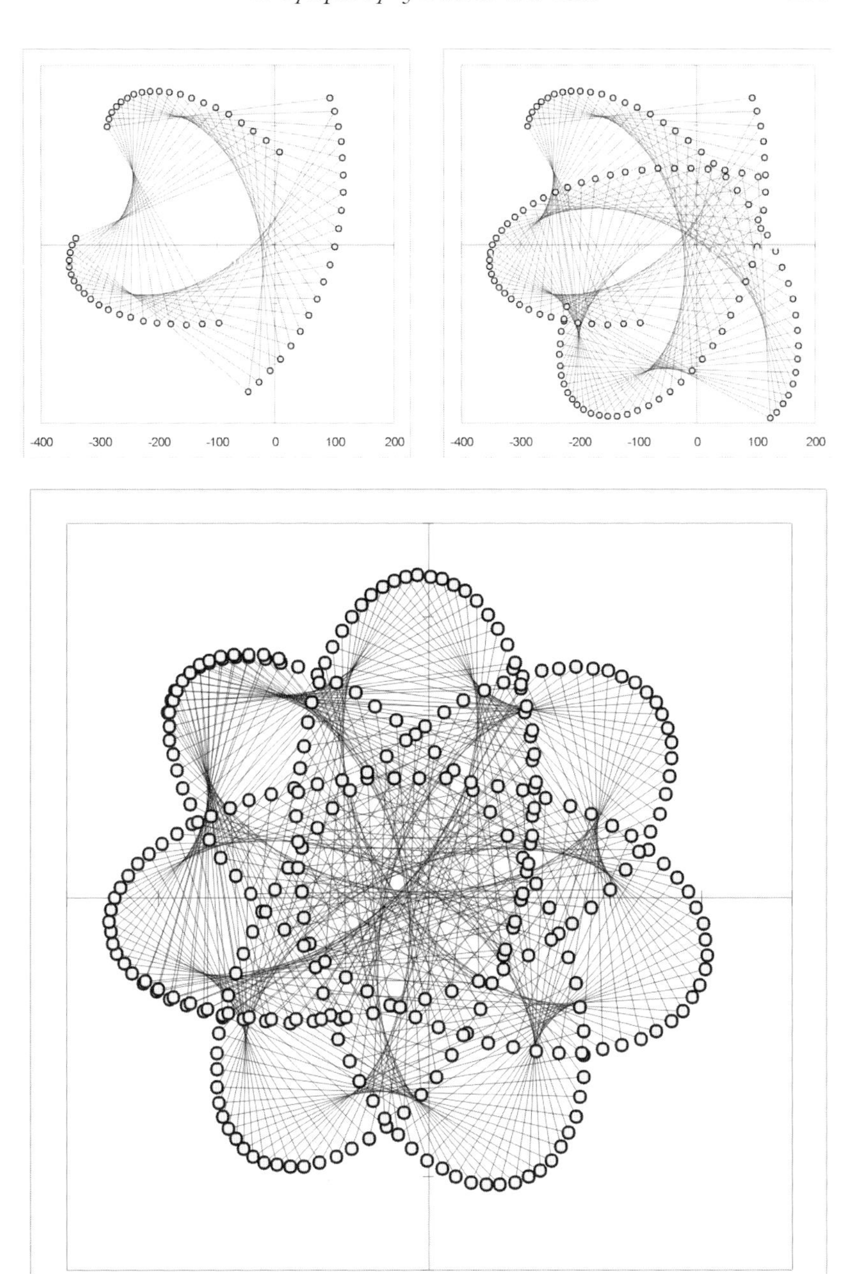


**Fig. 10.24** *Mars as seen from Venus at Venus/Neptune conjunctions, beginning 17 March 2000. Top left: 60 times, period 38.93 years. Top right: 120 times. Bottom: 300 times, period 194.65 years*

→ Plate 8

As mentioned, the numbers thirteen and twenty-three appear several times.

The main part in these events is played by Jupiter, which is involved in all the constellations of two outer planets. In this sense Jupiter is the fulcrum and pivot of the order in our solar system. An exception is the 23-fold Venus/Saturn/Neptune figure, but this is not an independent formation since it is caused solely by the resonance of the two large planets. (The same goes for a very similar 13-fold Venus/Saturn/Uranus constellation not shown here.) Among the inner planets, Venus is the one most often involved in the figures we have depicted. Moreover, the most exalted divinity and the goddess of love are the only ones to appear in all four movements of the symphony.

We hope the reader will agree that our division of the geometrical formations into four different 'movements' is not entirely arbitrary. The first and fourth are more closely linked because the same six planets appear in both. And it is not difficult to fashion a suitable structure for the separate pieces. The form presents itself very nicely out of the content of the central themes and is in harmony with them. Form and content in a great musical work always merge into an indivisible oneness, so why should this not also be the case in the symphony of the wandering stars?

In the following diagram, triangles within overall figures are used to show, in each case, the interplay of three planets. The first and fourth movements are combined and have been reduced to what are—perhaps—the bare essentials. It would be possible for the other formations that arise to be integrated into the whole depiction by means of suitable

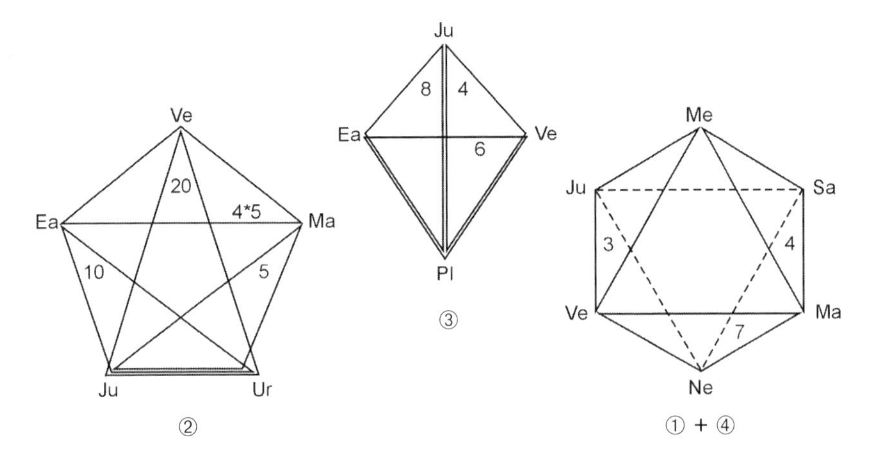


**Fig. 10.25** *Conjunction relationships of, in each case, three planets*

connecting lines. The three inner and the three outer planets have been united in the interpenetrating equilateral triangles of a hexagram.

In order to give some intimation of the bond that binds the architecture of the individual movements together in their depths it will be necessary to fill in the lines which are dotted in the diagram. But this will have to wait until later. We want to close this chapter with the words of a poet. Although he knew nothing about the symphonic harmony of the wandering stars in our solar system, phenomena in the earthly world led him to ask questions which also present themselves when one considers the manifestations that are apparent in the celestial spheres.

*And yet, though we strain*
*against the deadening grip*
*of daily necessity,*
*I sense there is this mystery:*

*All life is being lived.*

*Who is living it, then?*
*Is it the things themselves,*
*or something waiting inside them*
*like an unplayed melody in a flute?*

*Is it the winds blowing over the waters?*
*Is it the branches that signal to each other?*

*Is it flowers*
*interweaving their fragrances,*
*or streets, as they wind through time?*

*Is it the animals, warmly moving,*
*or the birds, that suddenly rise up?*

*Who lives it then? God, are you the one*
*who is living life?*

Rainer Maria Rilke[2]

# 11. Journeys into Macrocosm and Microcosm

The discoveries of modern astronomy surely belong among the factors that have most changed our view of the world. The progression begun by Copernicus, Galileo and Kepler has led in the twentieth century to knowledge which has given the word 'cosmos' an entirely new meaning. Philosophically speaking, i.e. with regard to the human being's place in the current view of the universe, we are utterly aghast at the vastness of scale opening up around us. Scientific theories, closely associated in the main with the 'big bang' catchword, are now being elaborated in an effort to find an overall concept capable of encompassing what telescopes and other instruments are revealing. But these theories are beginning to resemble a rather hasty retreat from the astonishment and awe engendered by a closer inspection of the mysteries that are coming to light.

On the other hand it is natural for us to want to find a supposed or actual context for our discoveries and to fathom the forces and laws that lie behind the phenomena. But surely it is important to distinguish clearly between a discovery and the conclusion we draw from it. The history of astronomy, however, shows that it is rather easy to overlook this line of demarcation. The 'observed fact' that the heavens rotate around the Earth was regarded for millennia as the physical reality. The movement of the wandering stars watched for in the skies followed more or less the epicycle model. To enable observations that differ from the model to be integrated into the system, Ptolemy thought up clever extensions of the original theory such as the equant—i.e. an eccentric point around which the outer circle (epicycle) was thought to move with constant angular velocity around the inner.

Let us consider the more important discoveries made in astronomy over the past century by looking not only through telescopes and at spectrographic images,* but also by peering into the realm of the very small. In recent decades, ideas about how the universe might have come into being have also involved experiments in particle physics which, among much else, have led to insights into the state of matter under extreme conditions such as those thought to have existed during the first moments of creation. Having travelled to the outermost edges of the universe and to its inmost regions we shall then touch on the middle

---

* Spectrograph: an instrument analysing light into its component wavelengths, one of the most important tools of modern astronomy.

realm—that of human beings and their relationship with one particular star in their closest environment.

Around the turn of the nineteenth to the twentieth century it became a certainty that the solar system was located in the plane of symmetry of a flat, very extended collection of stars which appears in the sky as the ribbon of the Milky Way. In 1918 the astronomer Harlow Shapley succeeded in proving that the Sun is situated not at the centre of this configuration but about 30,000 light years away from it.[1] Shortly after this Edwin P. Hubble (1889–1953) was able to clear up the controversy, current at the time, as to whether spiral nebulae, which had been known for about 300 years, were independent galaxies or belonged to the Milky Way. His investigations of certain variable stars in the Andromeda nebula showed that it must be about 700,000 light years away (today's calculations indicate about 2.5 million light years) and therefore without a doubt an independent galaxy.

As time has gone on we have become accustomed to hearing about distances measured in thousands or millions of light years, so that even the diameter of the whole (observable) universe, estimated today to measure 10–15 billion light years, has a ring of familiarity about it. But what is a light year? Light travels 9450 billion kilometres in 365 days. If a human being could journey once round the Earth with every breath drawn, he would have covered a mere 2 light years by the age of 60. The universe is also thought to be 10–15 billion years old. Our Milky Way, as are also other comparable stellar systems, is estimated to contain 100 billion stars. And the number of galaxies in the portion of the universe accessible to our telescopes is again set at 100 billion. So the first lesson to be learned from the universe is that creation is unimaginably vast and abundant. No one yet knows for certain whether it is finite or infinite. But measured on a human scale it is undoubtedly as good as infinite.

The next claim we can certainly make about the cosmos is that it has brought forth a form which is in no way second to that of a rose, the embodiment of beauty on our planet. A molecular biologist, though, would ascribe the spiralling twist of petals in a rose to the design plan anchored in its genes or DNA.* With the rose, gravity has at most only an insignificant part to play in the ordering of its petals. Similar spiralling eddies are also found in whirlpools or grand cloud formations where they arise through the interplay of gravity with the inner properties of water and air.

---

* Deoxyribonucleic acid, the biochemical substance the building blocks of which contain the code of hereditary dispositions to produce other substances (proteins, enzymes etc.).

**Fig. 11.1** *Top: Galaxy M 83 in the constellation of Hydra. Bottom: A red rose*

*'We can now claim that our account of the universe is complete. For our world has now received its full complement of living creatures, mortal and immortal; it is a visible living creature, it contains all creatures that are visible and is itself an image of the intelligible; and it has thus become a visible god, supreme in greatness and excellence, beauty and perfection, a single, uniquely created heaven.'*

Plato[2]

*'Cosmology has a reputation as a difficult science, but in many ways, explaining the whole universe is easier than understanding a single-celled animal. On the largest cosmic scales, where stars, galaxies and even galaxy clusters are mere flecks, matter is spread out evenly. And it is governed by only one force, gravity.'*

Two twentieth-century astronomers[3]

However, that spiral galaxies come into being solely under the influence of gravity must be seen for the moment as a hypothesis. Although astronomers on the whole assume gravity on a grand scale to be the all-dominating force in the cosmos, we must ask how we can be entirely certain about this. And even if it should turn out to be the case, we cannot presuppose that we know all there is to know either about gravity or about its effects. To this day no one really knows what gravity actually is. Isaac Newton saw it as a mysterious force of attraction with an inexplicable effect over long distances about the cause of which he declined to speculate. According to Albert Einstein's General Theory of Relativity, space and time are inextricably linked in a way that makes it possible to talk of a space-time continuum. In his view, gravity is an effect of the curvature of space-time on the material objects located within it. The concept of particle physics states that forces of every kind are mediated by exchange particles, for example photons which are held to be carriers of light or, more generally, of electro-magnetic effects. In this way gravity is said to be caused by gravitons, postulated elementary particles the existence of which, however, has not yet been proven.

Let us now return to Edwin Hubble, the man who—as the literature frequently states—discovered the expansion of the universe or even proved it. A much more likeable characterization would, I think, describe him as the person who first began to have inklings as to the actual vastness of the universe. He proceeded in his work by systematically carrying forward the spectroscopic investigation of the light of spiral nebulae already begun by others and bringing this into relation with the distances he had discovered. The first thing he noticed was that the light of most galaxies showed red-shifts. At the time a red-shift was known as the Doppler effect, after a nineteenth-century Austrian physicist. This states that the wavelengths of light are shifted towards the red end of the spectrum when emitted by a body that is travelling away from the observer or if the distance between the light source and the observer is increasing. One way of explaining this is to imagine the light source to be 300,000 km distant (= 1 light second), moving away from us at one tenth of the velocity of light and emitting one wave each second. The first light wave reaches us after 1 second and the next after 1.1 seconds because the source is now 330,000 km distant. So the arriving light has a lower frequency than the emitted light; or a longer wavelength. This is a shift towards the red because the wavelength of red light is longer than, for instance, that of blue light. If the light source is moving towards the observer, a corresponding blue-shift is detectable.

Hubble found that the light of most of the galaxies he investigated was red-shifted; and in the first instance the only interpretation permitted by the Doppler effect was that the spiral nebulae must therefore be moving away from us. His measurements then also showed that the red-shift increases with the growing distance of the nebulae, so that the velocity at which they are receding increases proportionally to the distance. We must, however, also take into account that distance measurement is one of the most difficult aspects of astronomy and that accuracy decreases with the increased distance of the object. To date it has not been possible to find an accurate, generally accepted measure for the receding movement of nebulae or, therefore, the corresponding assumed expansion of the universe. (I shall have more to say about the reasons for this assumption shortly.) Depending on the method used, determination of Hubble's constant, as this is termed, continues to vary between 55 and 85 km per second and megaparsec* (1 megaparsec = 3,262,000 light years).[4]

So when Hubble published his findings in 1929, astronomers found

* As at 2009 this has been further narrowed down to 63–77 km per second and megaparsec (see www.wikipedia.org).

themselves confronted with a universe in which the galaxies were systematically rushing away from them—not surprising, you might say, considering the way in which human beings frequently treat their fellows and all the other creatures in their care on their home planet. But of course for a modern astronomer it was anyway not acceptable to imagine Earth, the solar system or our home galaxy to be the centre of the universe, which would have to be deduced from a general tendency to increase the distance between us and the rest. Thirteen years earlier, fortunately, Albert Einstein had announced his General Theory of Relativity which provided an opportunity for solving this problem, among many others. As early on as 1922 the Russian mathematician Alexander Friedmann had availed himself of its assistance in working out that rather than being static, the universe must be in a state either of expansion or contraction—always assuming that the effects of gravity could be completely comprehended with the help of Einstein's equations.

But as ordinary human intelligence cannot visualize the curvature of space, or rather of space-time, deduced from the equations of the Theory of Relativity, the literature frequently favours the balloon model in matters of cosmology. Imagine a two-dimensional universe formed by the surface of a balloon. Dots painted on it represent the galaxies. When the balloon is blown up, all the points move further away from one another, whereby the distance between star systems that are anyway further apart increases to a greater extent. This holds good from the viewpoint of every galaxy, so that no particular one occupies the centre of that two-dimensional curved 'space'; or, you could say, every point in that world is equally entitled to be its centre. When seen in the context of our three-dimensional space, this means, so to speak, that there must be a fourth dimension present in our universe, or even a fifth, if the fourth—time—is included in the space-time continuum as is claimed by physicists.* In none of the relevant literature, however, have I found any mention of the question that asks: Who, then, is the one who inflates the balloon of our world?

So when people talk of an expanding universe they are not referring to galaxies that are moving further away from one another—unless the depiction is very outdated or else aimed solely at the popular understanding of science. The talk is of *space* that is expanding, with the bodies of the universe being, as it were, fastened to that space. This is a subtle

* According to the mathematics of non-Euclidian spaces developed by Georg Friedrich Bernard Riemann (1826–66), a three-dimensional space would even need to be 'embedded' in six dimensions, and four-dimensional space-time would require ten.

difference, but one that has important implications for cosmology. One has to ask, though, whether a putative observer looking at our universe from the outside would in any way be able to tell whether it is space that is expanding or the nebulae within that space. It is surely impossible for the inhabitants of our universe to distinguish between these two possibilities by means either of observation or of experiment. Moreover, a further question then arises, namely what the space is expanding into, or what was there before the arrival of the space which is obeying its urge to expand. Whatever the answer to these questions may be, treating the recession of the galaxies as the equivalent of an expansion of space introduces a new interpretation of the red-shift. One can now imagine a light-wave travelling through expanding space and thus itself becoming stretched. The measure of this change in the length of a wave, now termed the cosmological red-shift, no longer indicates the recession velocity but instead tells us by what factor the universe has increased in size since the ray of light was emitted.

The first person to follow the idea of the expansion of the universe to its logical conclusion, around 1931, was, it seems, the Belgian astronomer and priest Georges Lemaître. Following this idea to its logical conclusion entails, of course, going back to the beginning. If the galaxies are today consistently receding from one another or, respectively, if space is expanding, then by tracing this development backwards one comes to a point when they must have been very close together. Lemaître went back further still, right to a highly compressed state of all the components of the universe. He called this the primeval atom which broke asunder in an unimaginably violent explosion and then dispersed into ever smaller fragments right down to the size of an atom. While developing this idea further in 1948, George Gamow and two colleagues reached the realization based on physics that the very high initial density must also have involved an exceptionally high temperature. From this they deduced that the original explosion would have been followed by an era of radiation which should still be detectable today as a kind of afterglow.

The British astronomer Fred Hoyle coined the term 'big bang' in 1950. He was speaking with his tongue in his cheek, since he had developed his own model according to which the universe, despite its expansion, exists in a constant state (the steady-state theory). In order to resolve this seeming contradiction he assumed that new matter was constantly coming into being to compensate for the loss of density that accompanied the expansion. For him and others, at the time, the unsatisfactory aspect of the big bang theory lay in the fact that there was a discrepancy between the age of the universe as calculated retrospectively on the basis of expansion on the one hand and on the other hand the age of the stars or the solar system arrived at by other methods,[5] which would have indicated that some bodies in the universe must be older than the universe itself. This paradox reared its head a number of times during the history of the big bang theory, and it was always resolved by yet more refined measurements of distances and spans of time which made it possible to date the big bang even further back in time or else to shrink the age of the stars. But just as regularly, measurements that were even more accurate once again showed the stars to be older than the universe, and so on.

The big bang theory experienced its finest hour in 1965 when two American scientists working with a microwave aerial intended for the observation of artificial satellites noticed some microwave radiation which they could not explain. This radiation arrived from every direction and gave readings of approximately 3° Kelvin (3° above the absolute zero point of −273.15° Celsius). In consultation with astrophysicists they reached the conclusion that this microwave radiation must be the afterglow from the radiation era presumed by George Gamow to have been present when the universe was very young. However, although this 'cosmic background radiation' was initially registered with much enthusiasm by many scientists, it soon became evident that its existence, or rather the nature of its condition, also posed serious theoretical problems.

One of these has been named the 'light horizon problem' which concerns the radiation's almost entirely uniform distribution. From the point of view of physics this uniformity is only thinkable—unless one wants to presuppose inexplicable ideal initial conditions—if the photons had been thoroughly mixed during the radiation era. Another way of expressing this would be to say that there must have been a guaranteed flow of information able to spread out at a maximum of the velocity of light. The distance of a radiation source from a receiver at the moment of emission of the radiation can be calculated with the help of a certain formula.* This calculation puts the red-shift at z.[6] For the cosmic background radiation the literature gives the value of $z \cong 1000$, which arises out of the difference between the wavelength of microwave radiation arriving today and that of the photons originally emitted.

For the two extreme values of the Hubble constant—at present

* $r_E = \frac{2 * c}{H_0} * \left[\frac{1}{1+z} - \frac{1}{(1+z)^{3/2}}\right]$ where $r_E$ = distance at the moment of emission (megaparsec), c = velocity of light (km/sec), $H_0$ = Hubble constant (km/sec/megaparsec) at the moment of reception.

roughly 50 or, respectively, 100—the distances are thus 11.6 or, respectively, 5.8 megaparsecs. So the source emitting the microwave radiation into the cosmos must have been between 19 and 38 million light years away from us at the moment in question. But at that time the location at which we find ourselves today was also within the emitting region which would of course have had to include the whole of the cosmos as it then was. Therefore the diameter of the universe at that time must have measured at least 19 million light years. Other calculations in particle physics, which it would take too long to explain here, have led people to conclude that the background radiation was emitted *c.* 300,000 years after the big bang, i.e. after the zero point in time. So in other words the light horizon problem—which is often described in the literature in a rather nebulous or unnecessarily complicated manner—denotes that the universe at that time was very much larger than it should have been, in any case if one were to set the velocity of light as the upper limit for the expansion. Meanwhile, however, the conclusion has been reached that although this limit is valid for the objects in space it does not apply to the expansion of space itself.

Astrophysicists are also troubled by another consequence arising out of what is termed the standard big bang model, namely the 'flatness problem'. This has to do with the critical density of the universe. If the total mass in the cosmos or, respectively, the corresponding density, is greater than the critical quantity, then one must assume that at some distant time in the future, having become overwhelmed by gravity, the universe will cease expanding and begin to contract to the point where it implodes. Otherwise the expansion could go on forever or, if the actual density were to correspond exactly to the limit value, it could asymptotically come close to a standstill. The ratio of existing to critical density is called $\Omega$ (omega), and the borderline situation just mentioned corresponds to $\Omega = 1$. Of course it has only been possible so far to work out the measure of omega within certain limits, but one can speak with some degree of certainty of a range between 0.1 and 2. In the present context the exact dimension of omega is anyway irrelevant; it is sufficient to understand that shortly after the big bang its value exercised a decisive function. If omega had been very slightly larger than 1 at that moment, then, having just begun to expand, the universe would very quickly have switched over to contraction. And in the reverse case, if the value of omega had been just below 1, the expansion would have been so rapid that no galaxies or stars could have formed at all.

It has therefore been calculated that in the beginning, the precision of omega's closeness to 1 would have to be expressed by means of 60 places after the decimal point.[7] Other sources give slightly different figures—the smallest I found mentions 18 decimal places—but this is not really relevant. What matters is that the initial conditions are exceedingly narrowly defined. Expressed in a different way: the probability of our universe being able to come into being in its present form lies between $1{:}10^{18}$ and $1{:}10^{60}$, i.e. in both cases very close to zero. Assuming the big bang theory, which arrives at this outcome, to be basically correct, one cannot help but ask what this might mean.

Faced with this universe which so obviously appears to be made to measure, some scientists have developed the 'theory of parallel universes'. According to this there must almost or indeed actually be an infinite number of universes with equally many different initial beginnings. The basic idea or motivation for thinking up such a proposition seems quite clear to me. If there are $10^{18}$ or $10^{60}$ universes, then, according to the laws of probability, at some point one such is likely to arise which, over the course of billions of years, is capable of bringing forth life and the human being. By believing this I obviate the need to ask whether the world is ordered precisely in the way it is for the purpose of creating *me*—of course among many other things as well. But there is more to this question than many people can bear to contemplate. And anyway, the theory of parallel universes offers no more than an illusory solution to the fundamental question that lies behind it, the question of humanity. The validity of this statement will certainly hold until other life is discovered elsewhere in the universe. Humanity, then, is not capable as yet of working out the probability of developing life, feeling, consciousness and spirit out of what was previously inanimate.

Though much discussed, the theory of parallel universes is not scientifically attractive since there is no way of even beginning to check out what it propounds. So let us instead examine the second category of reactions to those initial conditions that have turned out to be so very ideal for us. These can be summarized in the assumption that we quite simply do not yet know what laws of nature were at work behind the happening which involved such perfect fine-tuning of the forces involved in what we believe to have been the origin of the universe. A step towards finding an explanation for those very select conditions was taken in 1980 when the concept of 'inflation' was brought to bear on the early times of the cosmos. By then the discoveries of particle physics had entered into the considerations of the cosmologists. According to this there are four basic forces: electromagnetism; the weak force that determines the decay of radioactive elements; the strong force that holds the nucleus of the atom together; and gravity. Physics today holds that under extreme conditions these forces gradually unify into a single force. Such conditions can be partially reproduced in particle accelerators. They are said to have been wholly present immediately after the big bang. The characteristics of matter in the earliest stage of its creation are derived from that theoretical unification of those forces.

In accordance with the inflation theory based on these ideas, it is thought that a moment after its birth, i.e. at time zero, the universe expanded at an inconceivable rate for an equally brief moment. For a span of time lasting from $10^{-34}$ to $10^{-32}$ seconds after the big bang its diameter grew exponentially by a factor of $10^{29}$. The size of the cosmos prior to inflation is said to have been $10^{-28}$ centimetres, and thereafter 10 centimetres.[8] Such phenomenally precise figures were made possible by the inclusion of particle physics in the calculations. These on the face of it seemingly rather innocuous numbers disguise the conclusion that the velocity of expansion in that fraction of a microsecond must have been billions of times greater (an understatement in itself!) than the velocity of light. Astrophysicists do not appear to have any problems with this since at least a majority of big bang theorists have accepted the inflation theory. For myself, though, I have to admit that it took my breath away when I first calculated this velocity on the basis of the time and distance data I found in the literature.*

In order to appreciate fully the scale and audacity of the ideas arising out of the union of astronomy and particle physics we must, in addition, take into account that the generation of matter (i.e. of its more massive components such as protons and neutrons) is stated to have been completed by about one ten-thousandth of a second after time zero. If, for the sake of simplification, we assume that—once the wildly amazing inflation had disappeared as quickly as it had come upon the infant universe—the expansion continued at the velocity of light, then it is easy to calculate from the above figures that the whole of matter must have fitted into a space with a diameter measuring about 60 kilometres. By 'the whole of matter' I mean all the building blocks of all the atoms from which subsequently 100 billion galaxies each consisting of a similar number of stars came into being. People of today who have faith in science are, it seems, prepared to believe a very great deal.

We now come to what it was that is said to have caused inflation, why it soon came to a halt again, and how it was able to solve the above mentioned light horizon and flatness problems. We shall only be able to

---

* The inflationary velocity of expansion arises from the above-mentioned data: $\frac{Dist.}{Time} = \frac{10 - 10^{-28}}{10^{-32} - 10^{-34}} \frac{cm}{sec} = 1.01 * 10^{33} \frac{cm}{sec} = 1.01 * 10^{28} \frac{km}{sec}$. The velocity of light is $c = 3 * 10^5$ km/sec, from which a $3.67 * 10^{22}$-fold velocity of expansion is calculated.

touch briefly here on the very complicated facts and ideas involved, so readers are requested to consult the relevant literature for a more in-depth discussion. As time has gone on, physicists have come to see the vacuum as a condition of space in which energies or energy fields lie hidden; this can be partially proven experimentally. It is said to be possible for pairs of particles to arise spontaneously out of the energies of the vacuum; in most cases these very rapidly disappear again into 'nothingness'. These particles are also termed 'virtual particles'. Over and above this, a 'false vacuum' has been postulated which is said to bear a very high density of energy that is linked with a negative pressure. To date 'no one has ever yet encountered' this false vacuum.[9] It is postulated on the basis of certain theories regarding the unification of the first three basic physical forces. When these special characteristics are included in the equations of the General Theory of Relativity, it is found that gravity is reversed and leads to an exponentially increased speeding up of cosmic expansion. Fortunately, this strange form of the vacuum is said to be fairly unstable but at the same time also very generous, so that as it vanishes it releases its high energy in the form of heat which can then be used to generate particles or, rather, was used thus in the earliest days of the cosmos.

The answer to the flatness problem arises simply and elegantly from the fact that a combination of Einstein's equations with the false vacuum leads to solutions where omega = 1, thus always providing an exactly fitting density of matter regardless of the initial conditions.[10] The light horizon problem is regarded as having been dealt with by inflation because inflation has stretched the originally very small region (in which the possibility of information transfer with the velocity of light enabled there to be a causal link) way beyond the normally existing boundaries. Therefore the background radiation arriving uniformly here on Earth is said to stem from a region in which the original photons were able to arise homogenously, with their distribution being evened out by the explosive expansion.

So now let us take a look at the most recent development. This is based mainly on observations of distant supernovae of a particular type about which it is assumed that their maximum brightness varies only very slightly during their eruption. Astronomers are fairly certain about this because the critical mass leading to the eruption of this type of star is thought to be constant, so that the energy given off is also standardized. Building on this it is possible to determine the brightness to within 12% exactitude and thus calculate the distances with some degree of accuracy and then compare them with red-shifts measured in parallel.[11] Evaluation of a whole series of supernovae showed that the brightnesses were considerably below those expected from cosmological models.

The majority of researchers interpret this to mean that the universe initially expanded more slowly than had been thought or that it is now expanding more rapidly than model calculations might suggest. The only explanation for this is a mysterious cosmological constant or antigravity. This constant, also known as Λ (lambda), was originally introduced by Einstein into his General Theory of Relativity to make it possible to calculate a static state of the universe which, in 1917, everyone still thought existed. Its physical cause has since come to be seen as the energy of the vacuum, in this case the right one. Mathematical analyses conducted by a Russian astrophysicist as early on as 1967 showed that the energy of the virtual particles of empty space behaves in exactly the same way as that of the cosmological constant. However, the forces of all the thinkable ghostly particles are said to add up to a total energy which by far outstrips the energy density assumed by astronomers. The following conclusion is drawn from this discrepancy between the theory of the vacuum and cosmology: '*Some feat of fine-tuning must subtract virtual-particle energies to 123 decimal places but leave the 124th untouched—a precision seen nowhere else in nature.*'[12]

**Fig. 11.2** *The Sombrero Galaxy in the constellation of Virgo. Side-view of the spiral*

'*God therefore, wishing that all things should be good, and so far as possible nothing be imperfect, and finding the visible universe in a state not of rest but of inharmonious and disorderly motion, reduced it to order from disorder, as he judged that order was in every way better . . . In fashioning the universe he implanted reason in soul and soul in body, and so ensured that his work should be by nature highest and best. And so the most likely account must say that this world came to be in very truth, through God's providence, a living being with soul and intelligence.*'

Plato[13]

After this necessarily very abbreviated sketch of the history of the big bang theory I shall now also introduce an example of the arguments that speak against a cosmology of this kind while not forgetting that all these models of the universe have arisen on the basis of observations of red-shifts. The most important question to be asked is whether the wavelength shifts measured with our scientific instruments are indeed the yardstick by which we must measure the velocity of the receding galaxies or, expressed in another way, of the expansion of the space in which the celestial bodies exist. If the link between red-shift and expansion is indeed a law of nature, then it must be valid in every case. There is room for variation in the measurements because, apart from general expansion, there are also movements caused by gravity; but—speaking in pictures—it ought not to be possible for any apple to fall upwards.

There are indications, though, that must be taken seriously which show that the relation in question does not always apply. The American astronomer Halton C. Arp has come across a series of objects—galaxies and quasars*—which are very close together yet which display very varying red-shifts. Some of these bodies are thought to be linked by bridges of matter.[14] Conservative astronomers assume that the great closeness of the objects to one another and also those material links are merely coincidental, since because the dimension of depth cannot be observed the objects that appear close together might actually be very distant from one another. Regarding this controversy, Hans-Jörg Fahr writes that in the case of a number of objects with very varying red-shifts '*improved observation techniques*' are making it possible for the visible bridges of matter '*to be increasingly regarded as being adequately proven*'.[15] Well, as an opponent of the big bang theory perhaps he cannot be entirely impartial. But whatever the case may be, one must presume that Arp's findings concerning the general interpretation of the red-shifts still await more thorough clarification.

There are many other aspects that could be included in the list of difficulties and contradictions with which today's astronomers find, or ought to consider, themselves confronted. But one cannot expect them to declare publicly the extent of their lack of knowledge or their doubts. Astronomy is, if I may say so, one of humanity's most expensive hobbies. If supposed knowledge were to be put forward with less self-assurance one must assume that funds would be less readily available for the new astronomical satellites and the ever-more powerful telescopes which one needs in order to reach new insights into the grand scale of the cosmos. And, provocative though this might sound, to differentiate between discovery and knowledge is not the job of the scientist but of those for whom the scientist is working. By this I refer not to potential clients or

---

* Quasar is an abbreviation standing for quasi-stellar radio source. The general assumption today is that these are the core regions of very distant and therefore cosmologically young galaxies.

institutes but to those who wrestle intellectually with the findings of the researchers. In the case of music, the composer writes neither for himself nor for artistic specialists but for his audience. Indeed, Hermann Hesse considered music to be the only utterly indispensable form of art. So could astronomy perhaps be the only utterly indispensable form of science without which humanity cannot cope at all from the mental and spiritual point of view?

In rounding off our journey into the macrocosm we shall conclude by introducing some insights we have gained into the main subject of the present book, namely the structural order of the universe. Stars are often found in globular clusters. Both the clusters and the stars which stand alone are almost always in groups called galaxies (from the Greek word for the Milky Way). Several stellar systems, amounting sometimes to over a thousand, form a galaxy cluster. Our Milky Way is located in what is termed the Local Group which comprises at least 30 members, most of them dwarf galaxies. The next level of the hierarchy is that of the superclusters which usually comprise 2 to 6 galaxy clusters and measure several hundred million light years. On a yet larger scale the cosmos is interwoven by ribbon-like or filament-like structures along which the galaxy clusters are concentrated. Interspersed with these there are gigantic empty or almost empty regions termed voids. The largest such structure discovered thus far is formed out of elongated filaments *c.* 600 million light years apart. Whereas no overarching or consistent regularity has been discovered in the levels of order just mentioned, it appears that these ribbons—found when measuring a section of sky to a depth of two billion light years—are equidistant from one another. One of those who discovered it described this unexpected finding as follows: '*Although this modern endeavour may seem neither as pleasing nor as spiritual as those of the past, the concept of an isotropic universe wedded with an understanding of random fields now allows us once again to hear the music of the spheres.*'[16]

This, however, is the only 'note' that is thought to have been discovered. Other results are more like '*the gush of a waterfall than that of divine instruments*'. One should not forget, however, that investigation of galaxies in intergalactic space is based on red-shift measurements which are assigned a specific distance based on a supposed cosmological origin in accordance with a selected model of the universe with parameters not exactly determined. Therefore the discoveries to date must rightly be viewed as only the very first approach to the structure of the universe. But one may hope that we shall gradually gain clearer insights into the actual order or cosmic music of the spheres. And I am convinced that this will turn out to be even grander than anything 'the wise scholars of ancient Greece' or anyone else has been able or is able to dream up. But precise and clear measurements in every respect—as the experiences of astronomy have always shown—are the prerequisite for this. At present these are still some way off.

So let us now turn our attention to the microcosm, although not without first casting a final glance into the depths of that unfathomable living organism 'within which all other mortal and immortal organisms live and breathe and have their being'.

**Fig. 11.3** *Region of star formation in Galaxy Centaurus A. Picture by the Hubble Space Telescope*

The universe contains perhaps 100 billion galaxies with 100 billion stars in each, i.e. $10^{22}$ stars overall. The same unimaginably enormous sum is reached when trying to estimate the number of atoms in a thimbleful of water. In its numbers alone the world of the very small is just as astounding as the macrocosm. And furthermore, physicists in the twentieth century investigating the atom and its constituents have found themselves confronting characteristics that do not fit in with our normal view of the world. Under certain circumstances the building blocks of the subatomic world behave like discrete particles and under others like continuous waves. Niels Bohr called this phenomenon 'complementarity'. One can exactly determine the location of a concrete particle at two points in time but—in contrast with, for example, the orbits of planets—one cannot state the route it has travelled in the intervening moment. Either the smallest units of matter and light are in fact indeterminate or something called non-locality exists in their world—'spooky distance effects' as Einstein put it. The scientific world remains undecided as to the interpretation of these phenomena and their philosophical consequences. Fascinating though this subject is, we shall go no further into it here since there are plenty of books dealing with it.[17] The question we shall discuss here is whether something can be found at the level of the microcosm that resembles the 'harmony of the spheres' in the velocities of the planets put forward in Chapter 4. To do this we must first turn to the subatomic structure of our world.

An atom consists of negatively charged electrons, positive protons and electrically neutral neutrons. In addition neutrinos, without or almost without mass, which pervade everything were discovered in 1956. Initially we can say that our world consists of these particles although there are also quite a number of others which, however, are all extremely short-lived. These were found in particle accelerators and in the cosmic radiation which must not be confused with the background radiation. Some of these particles consist of various quarks which, though, can never occur alone but always only in combination, for example in a proton. All particles are said to have come, as it were, popping into the cosmos immediately after the creation of the world out of the abundance of energy in the vacuum, either with or without inflation.

Another fact, every bit as astonishing, is that the particles coming into being out of nothingness possessed the very characteristics needed for building a world. An electron, for example, has a specific mass, and a proton also has a fixed mass, and so on. If these values were not as they are it would not be possible for atoms to form. The masses of the most important elementary particles are fundamental constants which, at least as yet, cannot be further derived. The mass of the particles, which in most cases can now be very accurately measured, accords them a kind of individuality. The masses have fixed values, just like the pitch of a well-defined frequency in a musical system where various notes relate to one another by means of the intervals. So one obvious course of investigation is to see whether the elementary particles, too, show some harmonious order in the ratios of their masses or energies. According to Einstein's famous formula $E = mc^2$ (c being the velocity of light), mass and energy are linked, and in the physics of elementary particles the various values are given as the energy of the rest mass (measured in the unit of the electron-volt). In Chapter 4 and Appendix 4.1 and 4.2 we introduced the method for determining the harmony existing in a series of values or, respectively, the deviation of existing harmonic correlation from a random distribution. So let us now see what the situation is in this respect in the case of intervals which arise from the masses of elementary particles. We shall use the well-tried system of harmonically pure tuning with 12 possible half-tone steps. First of all, the 40 possible ratios of the listed 41 particles with all the others will be taken and then the means of the harmonic deviations will be calculated.

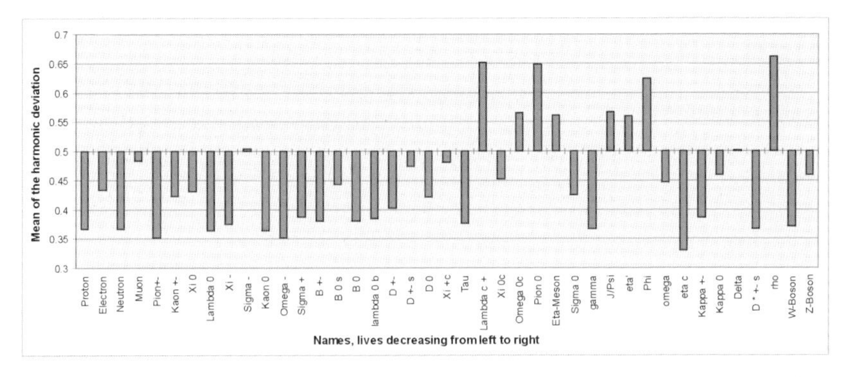


**Fig. 11.4** *Means of the harmonic deviations taken from the intervals of the masses of 41 elementary particles, related to each of the possibilities*

The particles have been arranged in order of their lives. The corresponding values are shown in the Table in Appendix 6.9. It was not possible to include the quarks because they cannot be isolated and only very approximate values for their masses are available. A considerable number of other particles exist but, as already mentioned, they are products of the ones shown here and have lives which would show at the right-hand edge of Fig. 11.4. The important point here is that there is an obvious relationship between the lives of the particles and their harmonic involvement in the totality of all the particles. Up to about one half, nearly all the values have a mean lower than 0.5, i.e. there is here a relatively good correspondence with harmonic intervals. Beyond a borderline of about $10^{-13}$ seconds, the values of the harmonic deviation from the statistical mean equalize. This is obvious even without the need for it to be calculated. Let us then look at those particles which do not disappear or, respectively, do not turn into others too quickly. But in order to gain an objective idea as to the extent to which the harmonies observed are not randomly distributed we shall also have to single out a few other particles as well. Protons and neutrons, for example, have almost identical masses. And the same goes for other related particles such as Kaon ± and Kaon 0; Sigma +, Sigma − and Sigma 0, etc. If both, or all three, partners were to be included in the calculation of harmonic probability, they would show almost identical intervals, and these would lead to unrealistically high deviations from a random distribution. So the selection will be made according to an overall criterion whereby in cases of very similar mass only the particle with the longer life will be taken into account. Out of the 22 particles with the longest lives 15 thus remain. Their harmonic relationships are shown in the next graphs:

*'So god, when he began to put together the body of the universe, made it of fire and earth. But it is not possible to combine two things properly without a third to act as a bond to hold them together. And the best bond is one that effects the closest unity between itself and the terms it is combining; and this is best done by a continued geometrical proportion.'*
Plato[18]

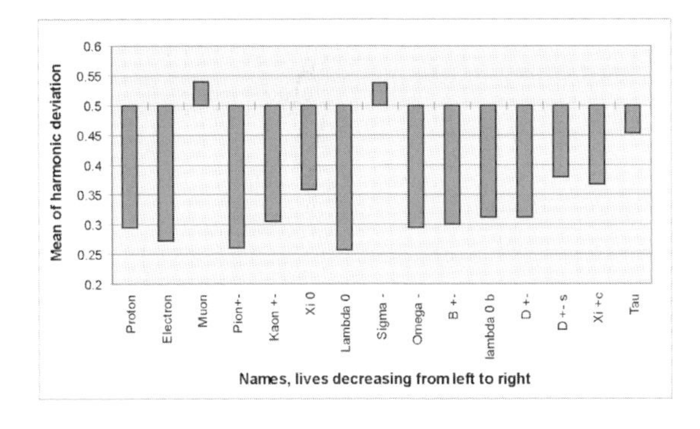


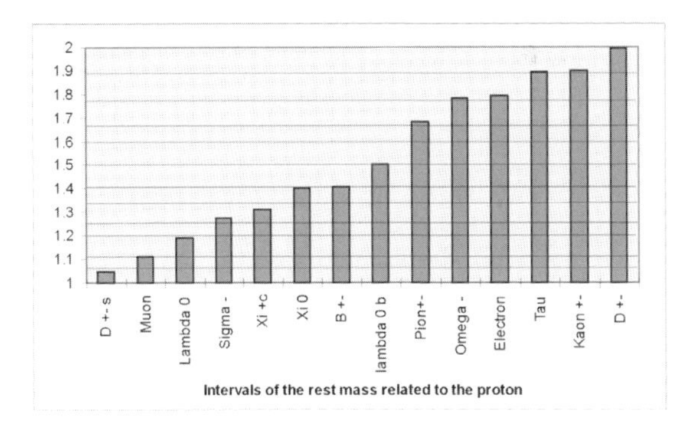


**Fig. 11.5** *Top: Means of harmonic deviations in the intervals of the masses of the 15 longest-lived elementary particles in relation to all possibilities. Bottom: 14 intervals (transposed to an octave), in each case related to the proton; musical intervals are shown by the horizontal lines*

What we find is that these particles, which once upon a time came popping out of nothingness, skipped into this world having been fine-tuned with one another in quite a harmonious manner. The probability of this structure being purely random is 1:2000. (For details of the calculation see Appendix 4.4.) So the basic building blocks of galaxies, suns, planets, plants, animals and man, together with all their somewhat more durable relatives, are linked by ratios which very closely resemble the musical order created by the human spirit. It is tempting to surmise that a music of creation, rather than the stupendous roar of a big bang, accompanied the first coming into being of the universe. Perhaps the intuition of the ancient mythologies and the mental powers of the Greek thinkers were after all superior in many ways to some of the theories arising today out of the findings of modern methods of research. '*A song lies sleeping in all things, yet they dream on and on. When you can find the magic word, the world will sing its song,*' wrote the poet Joseph von Eichendorff (1788–1857) in describing the connection he sensed between universe, man and music. Perhaps the next picture can give us an idea of the wonderful order at work in the sphere upon which our world is founded, a world in which there is no contradiction in being both a wave and a particle:

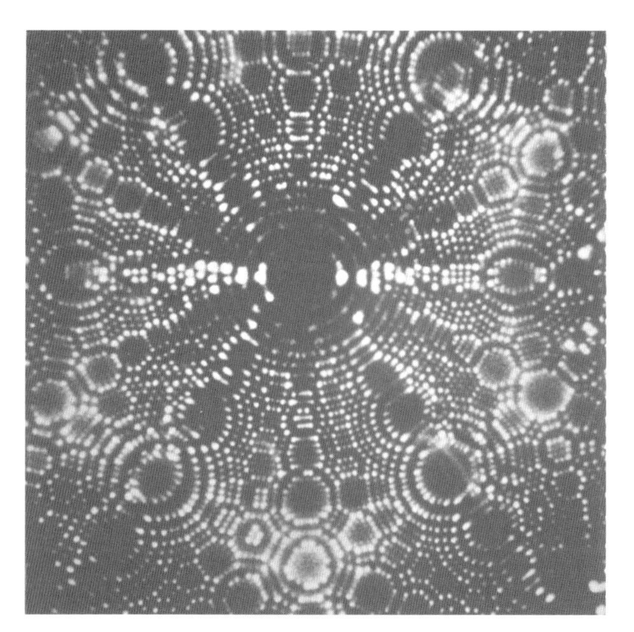

**Fig. 11.6** *Atoms magnified 1,200,000 times. The individual points of light show atoms of tungsten on a minute needle-point of crystal made visible by a field ion microscope developed in 1951 by E.W. Müller*

Human beings lead their allotted span of life amid the infinitudes of microcosm and macrocosm. One of the characteristics that marks them out is their search for the meaning of their existence and their quest for knowledge and truth. Some hope that the methods of science will help them solve the great riddles posed by life and nature or at least bring them closer to finding an answer. But certainly without thinking through the more important discoveries and findings of recent centuries they can do no more than feel their way in the dark. Yet their capacity for understanding appears to be limited. They imagine the elementary constituents of this world to be point masses. If these particles had dimension, they would be even further divisible. But seen as matter, a point is a nothingness which, nevertheless, in this case possesses bodily characteristics such as mass, charge and so on. Thus the seemingly so solid world appears to consist of a summation of nothingness.

In the same vein, it appears that the question as to whether the universe is finite or infinite cannot be answered by means of our thinking—even though there do appear to be some approaches to finding a solution. Infinity, at least, is unimaginable and if it existed it would come accompanied by many paradoxes. In an infinity with an infinite number of possibilities, for example, all things would have to occur an infinite number of times. But it is equally unsatisfying to think of the universe as possessing a boundary. Model universes with curved space-time, both limited and infinite like the surface of a sphere, might be an aid to our thinking. But on the one hand such universes are not yet proven, and on the other they merely relocate the boundary problem into another dimension.

According to our knowledge as it stands at present, the majority of astronomical observations indicate that the universe did have a beginning. If this hypothesis were untrue it would mean that the universe has always existed. A cosmos pulsating from one big bang to the next big bang via a collapse, as postulated by some astronomers and as it appears in Hindu mythology[19] in a somewhat different form, is equally eternal. Temporal infinity is just as unimaginable as spatial infinity. If there was a beginning, then the cosmos must have arisen, so to speak, out of nothingness. Some current ideas look on matter as possibly having emerged from the energy density of the vacuum. But this, too, is no real answer, for it supposes the pre-existence of energy. And if we knew, or were to develop a theory about, where this energy might have come from, then the next question would lead us automatically to ask where it, in its turn, had originated. However we may twist and turn, the only way we can think of a cosmos which has indeed entered into existence is that it must have been brought into being by some supranatural process, for otherwise it would merely be another variation of what had always existed. So our world cannot have come into being by chance; either it has always existed or it is a divine creation. Just as there are realms in which complementarity reigns, where two forms of existence which appear to us to be incompatible, such as particles and waves, can be contained simultaneously, so is there also a third possibility: the universe is eternal *and* the work of God.

Having journeyed into the macrocosm and the microcosm, let us now look to the stars once more, but only to those which are located in our immediate vicinity. Human beings appear to have a very special preference for one particular star-shape. Many countries show it in their flag, advertising frequently relies on its attractiveness, and all kinds of groups make use of its symbolism. In fact we come across it so frequently that we are surely not mistaken in our conclusion that the 5-pointed star as an archetype is deeply rooted in the human soul. As we have seen in earlier chapters, Venus and Earth and Sun collaborate in tracing the 'human star-figure' in the firmament with their movements. Let us now bring the present chapter to a close with a depiction which arises out of a combination of the two constellations that lead to the 5-pointed star-figures just mentioned (Fig. 11.7). There may be nothing special about the depiction from the purely astronomical point of view. Yet it, too, may succeed in showing us something of the cosmic order within which we have our place. And perhaps it will even cause something in our soul to ring out in concord with it.

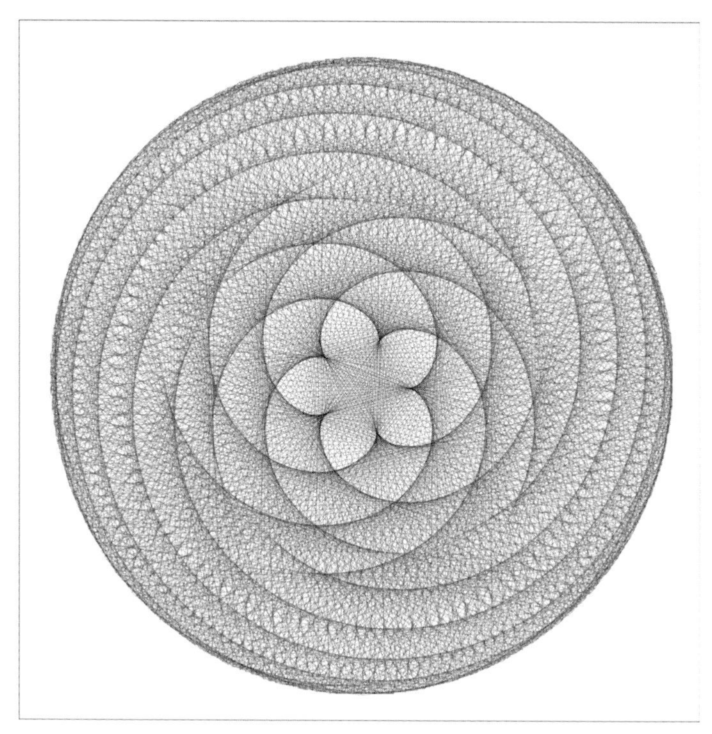

**Fig. 11.7** *The figure shown here arises from the sequential plotting of the positions of Earth in its orbit at two specific constellations and the connecting lines (as in the 'double conjunctions' shown and described in Chapter 10). These are firstly the Venus/Earth conjunctions and secondly the Venus-Sun-Views or, rather, every fifth one, i.e. whenever a pentagram as in Fig. 9.2 is completed (the horizontal 'Venus axis' is also defined here, as in Chapter 8). The mean values of the two periods are 583.92 and 583.75 days, so that displacements occur very slowly; 1700 positions of Earth were plotted, i.e. at 850 conjunctions and 850 Sun pentagrams in* c. *1358.5 years, beginning with the first position after 1 January 2000. In Plate 9 the first 700 lines are shown in red, the next 500 in blue, and the remaining 500 in green* → Plate 9

# 12. Ultimate Perfection

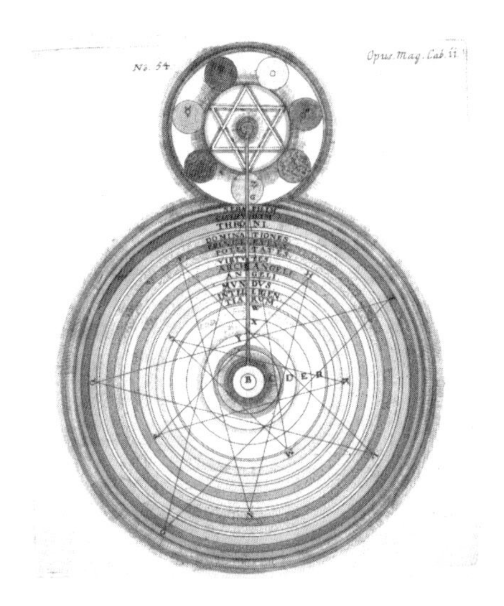

**Fig. 12.1** *Gregorius Anglus Sallwigt (alias von Welling),* Opus mago-cabalisticum, *Frankfurt 1719*

People in earlier ages knew of seven celestial bodies that moved independently of the fixed stars, and this alone was reason enough for them to attach special significance to the number seven. The order of the visible cosmos was also often symbolized by the hexagram, which expresses the union of opposites through its two interlocking triangles. And beyond the confines of the visible world some ancient ideas visualized 12 choirs of angels or angelic hosts as symbolized here in an eighteenth-century illustration. The distance from here to the 12 gods in the ancient Greek Pantheon, or similar conceptions, is not great, and the same goes for the apportionment of the fixed-star constellations to the 12 signs of the zodiac. The invisible heavens have always been linked with what people could see when they cast their eyes up to the skies. They found significance in what they saw, and from this grew both faith and superstition.

Modern astronomy has delivered us from the confines of the latter although in a most radical manner. It is not acceptable in any way to imagine spiritual forces being involved in a system of orbiting bodies formed by means of collisions between small rocks behaving in accordance with the laws of physics. That there is a fixed number of planets and that they move in specific ways does not amount to a special state of affairs in the eyes of a scientific view of the world. The *Pentagramma Veneris*, for example, is either entirely unknown or regarded as nothing more than an inevitable geometrical consequence of the more or less accidental distance between Venus and Earth. The previous chapters have shown, however, that the main structural principle of the inner planetary system involves the number five and its multiples and that they are also very conspicuous in the interaction between the outer and inner regions of the system. We saw furthermore that the pentagram, whether consciously or unconsciously, is considered to be important in connection with the human being.

It is often thought nowadays that religion arises out of a projection of human needs and fears into figures of divinity. No doubt this is so in many respects, but it is at best no more than half the truth. The sight of the starry firmament and—quite apart from whether we might be able to number it in billions of light years—the effect that this immeasurable expanse has on human beings should equally be regarded as a source, perhaps indeed the true fount, of ideas concerning a spiritual heaven or world. Boundless astonishment at the alien grandeur, combined with a sense of being intimately united with it, must surely have preceded any personification of cosmic phenomena in concrete personages. As their understanding grew, people became capable of comprehending that the celestial bodies moved in obedience to a regular and evidently meaningful order. Their knowing spirit was, at least in this respect, of the same order as that of the Unknown One who had set the celestial bodies in motion, and this in turn played a decisive part in bringing about their feeling of being bound up in it.

But as they then grew in their ability to penetrate the mathematically comprehensible laws of nature, they became less willing to allow for the presence of a creative spirit existing side by side with those laws. In their movements the planets obey gravity, and the initial circumstances—i.e. how many wandering stars arose where out of the original cloud of dust—are (why not?) purely coincidental. And the same must surely be valid for the orbits of the suns in the Milky Way, of the galaxies in larger numbers, and of those groups in superclusters, and so on. For several centuries these basic assumptions cast a spell of slumber over humanity's search for the true order in our solar system. Though there were, of course, exceptions, the discoveries we have been discussing here show quite clearly that this has been the case.

What now remains to be revealed is the way in which the outer planets interact in the region where distant gods follow their orbits. And we shall see that they do so in ways that are entirely appropriate for them and therefore full of significance for us.

We have already mentioned several times that the conjunction of Jupiter and Saturn, the two weightiest members of the planetary community, is of special importance, to the extent that on occasion it is termed the 'pulse of the solar system'. Within an organism the heart is closely interlinked with the other organs by its beat, so here, too, our main concern will be the relationship between Jupiter/Saturn encounters and those of other planets. First, however, let us take the pulse. Fig. 2.5 in Chapter 2 showed the Saturn/Jupiter trigon which arises out of the very approximate 5:2 ratio (2.483325:1) of the orbital periods of the two planets. We then saw that in the course of time a 43-pointed star-figure is formed which, though, does not make a particularly resonant impression. It is missed by *c.* 4°, the value of the cycle-resonance—i.e. the 44th encounter lags behind the first by this amount. And the 47th conjunction (corresponding to a 46-pointed star-figure), hurries ahead by about the same angle. As the following Table shows, only the sum of the two numbers, 43 and 46, leads to a ratio that is very accurate at 149:60 (2.483333:1) during the course of 89 conjunctions.

**Table 12.1** *Values and ratios of the Jupiter/Saturn relationship*

| | Ju/Sa synod | Jupiter orbit | Saturn orbit | |
|---|---|---|---|---|
| | 7253.4525 | 4332.5893 | 10759.2268 | Days |
| | 19.8585 | 11.8618 | 29.4566 | Years |
| n ⋆ Ju/Sa-synod | Years | n ⋆ Ju/Sa = x ⋆ Ju | n ⋆ Ju/Sa = y ⋆ Sa | Cycle-resonance(°) |
| 1 | 19.8585 | 1.6742 | 0.6742 | |
| 3 | 59.5756 | 5.0225 | 2.0225 | 8.094 |
| 43 | 853.9166 | 71.9889 | 28.9889 | −3.986 |
| 46 | 913.4921 | 77.0114 | 31.0114 | 4.108 |
| 89 | 1767.4087 | 149.0003 | 60.0003 | 0.123 |
| 2891 | 57410.9951 | 4839.9998 | 1948.9998 | −0.062 |

Even much higher values do not lead to much closer approximations. One can, by the way, also arrive at the number of Jupiter orbits, 149, by dividing the synodic period by the ratio of the golden section. The result is almost precisely $\frac{5}{12}$ Saturn years. Thus in 5 orbits Saturn (if we plot its positions after the interval in each case of the synodic period divided by 1.618..) would trace a precise, continuous 12-pointed star-figure. And if, after each completion of the star, i.e. in the rhythm of 5 Saturn years, we were to enter the positions of Jupiter, exactly the same figure would arise since on each occasion Jupiter completes $12\frac{5}{12}$ circuits. After 149 (148.9995) orbits, corresponding to 60 Saturn years, Jupiter would have completed its 12-pointed star-figure.

$$\frac{7253.45}{1.618..} = 4482.88; \quad \frac{4482.88}{10759.227} = 0.41665 \cong \frac{5}{12}; \quad \frac{5 * 10759.227}{4332.589}$$

$$= 12.41662 \cong 12\frac{5}{12} = \frac{149}{12}$$

However, for the time being these are purely fictional star-shapes arising out of the calculated combination of the time spans, the number twelve and the golden section. But perhaps this possibility, hidden in the numbers, points to a particular situation connected with the Jupiter/ Saturn relationship. So let us now embark on another short excursion into the inner planetary system which will take us to our good old Earth. Earth's orbit and its conjunctions with the two large planets are indeed very subtly and accurately linked to the golden section.[1]

During one synodic period with Jupiter or Saturn, Earth covers on average in each case one orbit plus 33.1438 or, respectively, 12.6508 degrees. The ratio of the two angles is 2.61989:1, i.e. the square of the golden section ratio ($g^2$) with a deviation of only 0.07%. Expressed in a different way, one can say that the angle which Earth traverses between two conjunctions with Jupiter and the difference between this angle and the angle created by the corresponding encounters with Saturn is almost in the ratio of 1.618..:1 (33.14°/(33.14°–12.65°) = 1.617..). In the long-term movement pictures of the three planets, this ratio results in a structure according to a specific number which is expressed in correspondingly varying ways in the different constellations and modes of depiction:

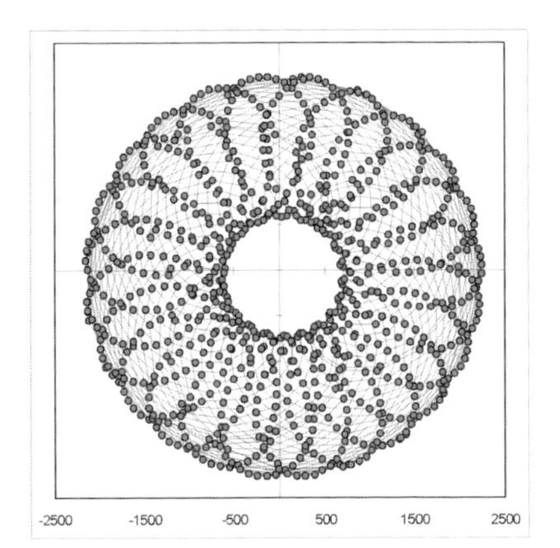


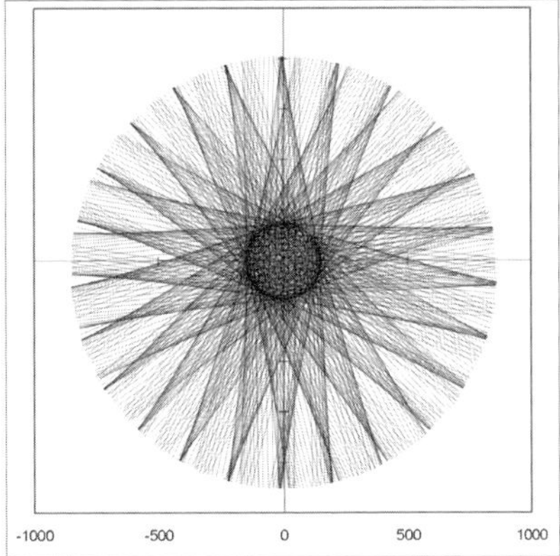


**Fig. 12.2** *Left: Jupiter-centred view of Saturn at Earth/Saturn conjunctions, 900 times, beginning 19 November 2000, period 931.63 years. Right: Earth-Saturn linklines at Earth/Jupiter conjunctions, 750 times, beginning 28 November 2000, period 819.05 years (detail enlargement 10:6); scale in millions of kilometres*

In the left-hand picture, some lines lying closer together show the three primary loops of the continuous movement of Saturn around Jupiter in the centre (here, too, of course, the mirror-reversal relationship is valid). During its path consisting initially of 43 loops which, in the period of time shown, are just about to be repeated (somewhat shifted), Saturn keeps arriving, during conjunctions with Earth, at points that lie on 24 loops oriented outwards in a different level of order. As in some of the examples shown in Chapter 10, the same number also occurs in the geometrical formations of other configurations that are possible between the three planets. The star-figure on the right (shown for clarity in a slightly smaller than double detail enlargement) arises from an outer 27-pointed star-figure (not visible here) which derives from the fact that almost 27 Earth/Jupiter conjunctions take place during one Saturn year.

Thus the double golden section in combination with the next but one inner planet (as seen from Jupiter) causes the number twenty-four to occur. Perhaps it is not so very important that certain relationships are ordered in accordance with this specific ratio, for we have been shown a number of regular figures which arise without its presence. So we can say with some certainty that the golden section is only one of the points of view able to throw a little light on the marvellous interplay of the planets. A brighter light can be hoped for if we manage to clarify which numbers in which position and in which geometrical manner are made manifest by the celestial bodies. So let us now turn to the second planet further out from Saturn. First we see the figures woven in our cosmic home by Jupiter and Neptune and, respectively, by Saturn and Neptune in the rhythm of the heartbeat of the solar system.

The Number of Perfection appears—and it appears in perfect form. In the outer planetary system, in the region bordering on the stars with their signs of the zodiac, the symbolic number appears before our eyes and before our minds which, since time immemorial and in varying cultures, has been associated with the heavens in the one sense as well as in the other. And in some ways this comes about in a manner that best realizes that ancient allocation. In the period which can be investigated by means of the astronomical algorithms available here (see Appendix 3.7) we observed only the smallest turning of the figures arising, which was in keeping with a minimal deviation from complete resonance. The number twelve is attained in all the possible constellations arising from the encounters of the planets Jupiter, Saturn and Neptune. For one thing these pictures of Jupiter/Saturn conjunctions show the continuous 12-pointed star-figure which can be constructed by joining every fifth corner point of a dodecagon. Thus, in collaboration with Neptune, real form is given to what otherwise, in combination with the golden section, announced itself as a purely mathematical possibility in the ratio of the Jupiter and Saturn orbital periods. In view of the overarching connections that are now appearing we shall have to remove any more profound

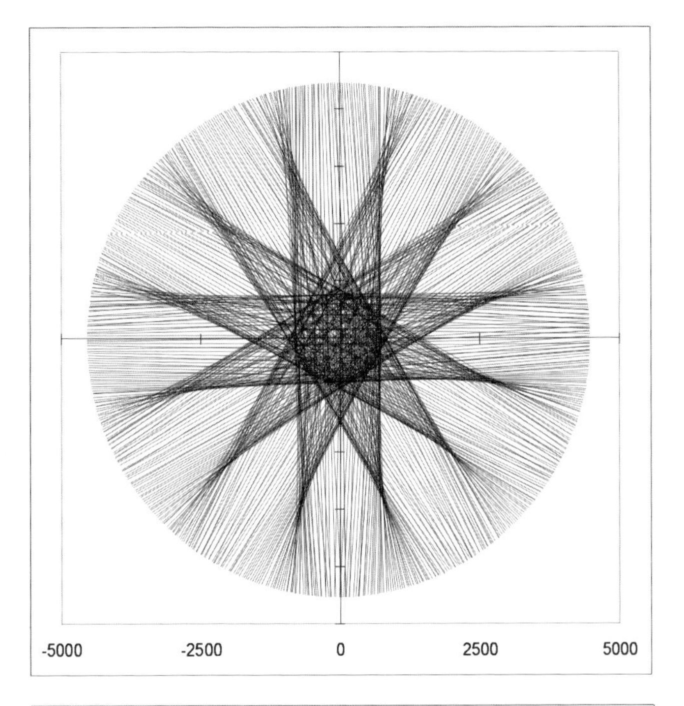


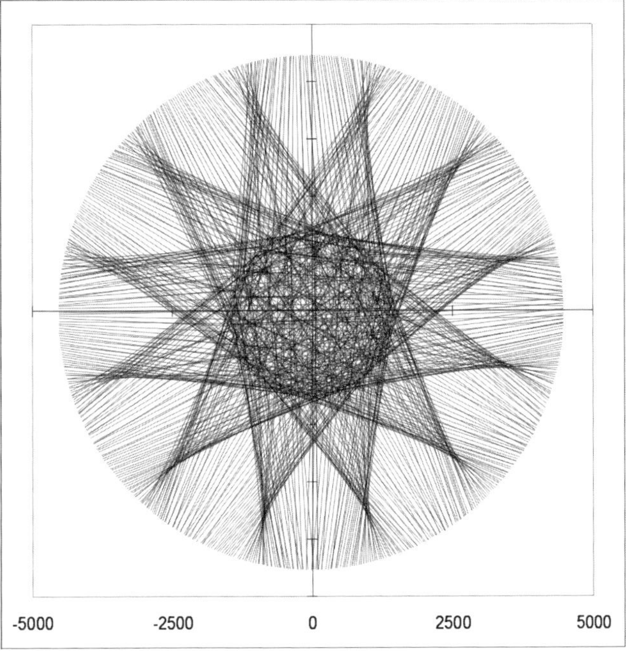


**Fig. 12.3** *Linklines at Jupiter/Saturn conjunctions, beginning* c. *5 April 4950* BC, *750 times, period 14,893.9 years. Top: Jupiter-Neptune; bottom: Saturn-Neptune*

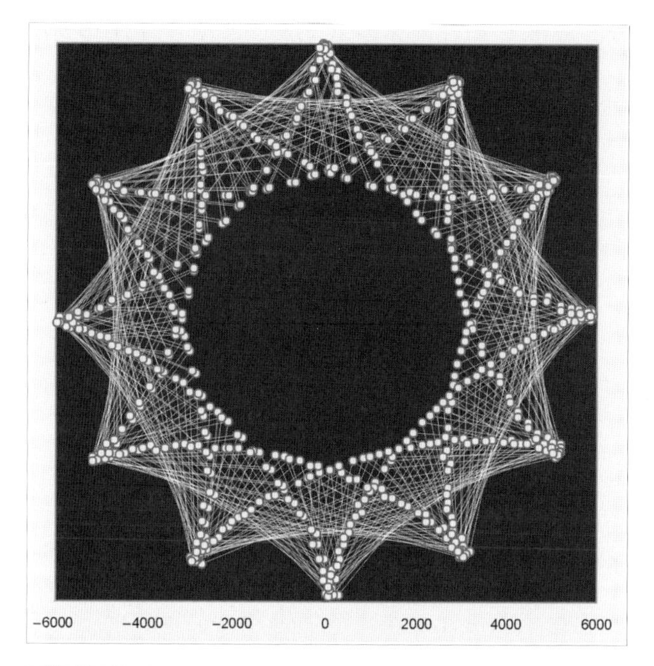


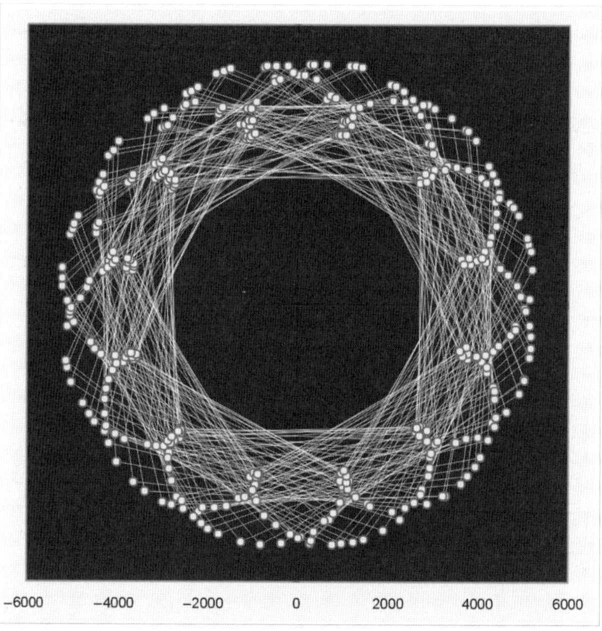


**Fig. 12.4** *Movement figures at conjunctions with Neptune. Top: Neptune as seen from Saturn at Jupiter/Neptune conjunctions, 700 times, beginning 4 March 2477* BC, *period 8947.37 years. Bottom: Neptune as seen from Jupiter at Saturn/Neptune conjunctions, 400 times, beginning* c. *15 June 5150* BC, *period 14,347.7 years.* → Plates 10 and 11

significance from the frequently used term '5:2 resonance', even when we see it expressed in the beautiful threefold curves shown, for example, in Fig. 5.5. Neither does the period of 43 or 46 have much of a part to play. The ratio of the orbital periods is 149:60, which means a deviation of exactly 1/149 from the ratio of 5:2. We shall soon see that on this basis a consideration of the gravitational interplay of the three planets also shows that things are as they are and cannot be different.

In addition to the continuous 12-pointed star-figure, the rhythm of the central conjunction of our solar system also forms the 12-pointed star-figure consisting of four triangles, as shown in the lower diagram in Fig. 12.3.

Let us now have a look at the figures traced by the light of the planets in the cosmic night when the two large planets have encounters with the third member of their alliance (see Fig. 12.4).

The time has come to turn once more to the man who always remained deeply convinced that our solar system is ordered in accordance with geometrically harmonious principles and whose ideas triggered my own attempts to clarify the secrets involved. We also now understand what he meant by those mystifying golden vessels of Egypt which he purloined: '*Now, eighteen months after the first light, three months after the true day, but a very few days after the pure Sun of that most wonderful study began to shine, nothing restrains me; it is my pleasure to yield to the inspired frenzy, it is my pleasure to taunt mortal men with the candid acknowledgment that I am stealing the golden vessels of the Egyptians to build a tabernacle to my god from them, far, far away from the boundaries of Egypt.*'[2]

**Fig. 12.5** *Johannes Kepler*

Johannes Kepler wrote these words shortly after his discovery of what later came to be called his Third Planetary Law. One must assume that they arose out of that discovery which he saw as the final, missing building block he needed in order to construct his edifice of the harmony of the planetary movements and their link with the model of the intervals structured by the Platonic solids. The golden vessels*

* Historically this concept goes back to Exodus in the Old Testament which reports that the Israelites carried artefacts of gold with them on their departure from Egypt.

are the ideas deeply anchored in humanity which say that the cosmos is ordered harmoniously in a manner that can only have originated in the work of a divine force. Older ideas about the movements of the planets were summarized by the Egyptian* Ptolemy (second century AD) in his *Almagest* and were regarded as valid right up to the time of Copernicus or in some instances even later. He also published writings concerning the presumed harmony of the spheres, so that Kepler, who saw himself as having placed the old ideas onto a new, scientific footing in his *Harmony of the World*, regarded him as a like-minded thinker. It was in this sense that Kepler wrote of 'stealing the golden vessels' which he rescued for the modern age. And now, in the present work, the majority of the most important discoveries we have been discussing provide the ancient idea of a harmony of the spheres with a footing that can, for the first time, be calculated and checked by anyone who wishes to do so. The marvels brought to light have surpassed all my expectations. Those golden vessels can now be passed on in a new guise, and it was this to which I was referring at the end of Chapter 9, rather than the incomparable revelation of the relationship $T^2 = a^3$.

So now let us look once more at the star-figures traced by Jupiter, Saturn and Neptune. The ratios of their distances from one another and thus also the ratios of their movements—well regulated in Quadratic Time—are ordered in such a way that, through the conjunctions of the large planets with Neptune, the skies can also be adorned with the other possible star-figures in the dodecagon. Fig. 12.4 shows 12-pointed starlike formations made up of three squares and two hexagons. Both are present simultaneously in each picture but with the emphasis reversed. In the first the lines linking the planetary positions form the two hexagons but the sequence of positions as such appears in three four-cornered starlike formations. The single figure is termed astroid and is a special case of the hypocycloid. Three astroids interweave to form a 12-pointed star-figure and in combination with the figures traced by the lines the result is a geometrical expression of perfection capable of touching the very core of our being almost as music does. In the lower diagram the planetary positions form two sixfold looped figures, and in the middle the connecting lines trace a 12-pointed star-figure made of three squares. Once again the formation appears to be guided by an invisible hand in accordance with plans devised over long ages. An example of the chronological sequence of the positions will be shown later in another figure.

---

* Ptolemy may have been a Greek living in Alexandria, Egypt. Indications as to his ancestry are unclear.

As to the numerical foundation of these geometrical revelations, the first approach to a concordance of the three conjunctions of Jupiter/Neptune, Jupiter/Saturn and Saturn/Neptune is:

14 Ju/Ne ≅ 9 Ju/Sa ≅ 5 Sa/Ne i.e. also: 7 Ju/Ne ≅ 4.5 Ju/Sa ≅ 2.5 Sa/Ne

Here, firstly, is once again the cycle of just under 179 years which plays the most important part in the Sun's movement round the barycentre of the planetary system (see Chapter 9). And then, to our amazement, we also notice that our old acquaintance, the 7:4:3 ratio, slightly changed, of the synodic periods of Venus, Earth and Mars is here too. In the second term of the equation we have four conjunctions plus one opposition, and in the third, three conjunctions minus one opposition, in comparison with the inner planets. However, the mathematical correspondence with the actual values is not quite as exact as in that Pythagorean constellation. But instead, the precision of the structure appears in a different way which is once more highly surprising. When listening to certain musical compositions a moment is sometimes reached when we feel that the highest pinnacle of beauty has been attained. But then a new variation follows, and the composer leads us into even more astonishing realms of our soul's musical landscape. Perhaps a comparable experience will ensue when we observe the order that lies hidden in the way the orbital periods of Jupiter, Saturn and Neptune are attuned to one another.

Even if the orbital periods are of no direct importance for the gravitational interplay of the planets, we cannot help but raise our hat to what

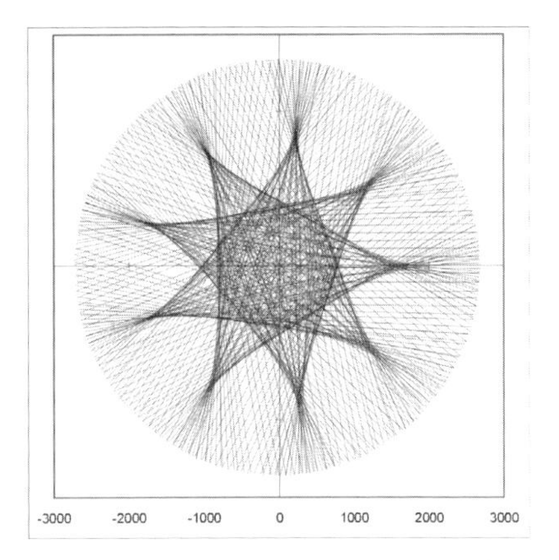


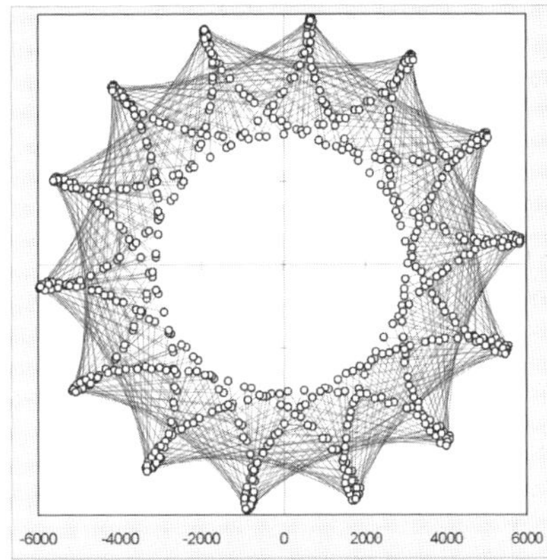


**Fig. 12.6** *Left: Jupiter-Neptune linklines in the interval of the Saturn year, 500 times, period 14,728.32 years (detail enlargement 10:6). Right: Neptune as seen from Saturn in the interval of the Jupiter year, 700 times, period 8303.24 years*

is revealed here. As if by magic there appears before us in the rhythm of the Saturn years a perfect 9-pointed star-figure, and in that of Jupiter's orbital periods an equally resonant 14-pointed star-figure, here in its planet-centred view. What we are being shown is the fact that the numbers according to which the ratios of the synodic periods are ordered express themselves, through the corresponding orbital periods, in the geometrical formations. In all my investigations I have not come across this phenomenon anywhere else (or at any rate not with such clarity, for there are signs of it in the Venus/Earth/Mars relationship). The number five of the series is the only one which does not appear and, as we shall soon see, it is indeed 'not permitted' to appear in the outer planetary system. The following Table summarizes the ratios of the three synodic periods and the closest numbers which show better approximations:

**Table 12.2** *Equalizing rhythms of the Jupiter, Saturn and Neptune conjunctions*

| Ju/Ne (Days) | Ju/Sa (Days) | Sa/Ne (Days) | | | | | |
|---|---|---|---|---|---|---|---|
| 4668.6936 | 7253.4525 | 13101.473 | Mean value | Orbits completed | | | |
| x*Ju/Ne | = y*Ju/Sa | = z*Sa/Ne | (Years) | Jupiter | Saturn | Neptune | ≅ n+.. |
| 14 | 9.0111 | 4.9889 | 179.0068 | 15.0911 | 6.0770 | 1.0864 | 1/12 |
| 87 | 55.9977 | 31.0023 | 1112.0184 | 93.7480 | 37.7510 | 6.7490 | 3/4 |
| 362 | 233.0017 | 128.9983 | 4627.080 | 390.0833 | 157.0810 | 28.0825 | 1/12 |
| 449 | 288.9994 | 160.0006 | 5739.099 | 483.8313 | 194.8321 | 34.8315 | 5/6 |
| 1260 | 811.0006 | 448.9995 | 16105.277 | 1357.746 | 546.7451 | 97.7456 | 3/4 |

The next approximation after the 14:9:5 ratio, 87:56:31, is then very accurate, and the fourth trio of numbers, 449:289:160, achieves an exactitude that can scarcely be bettered by the next series. When these values are compared with the ratios of the Jupiter/Saturn relationship, we see that the pure numbers also speak for a mysterious link between Neptune's orbital period and the pulse of our solar system:

$$\begin{array}{ccc} 149\ \text{Ju} = & 89\ \text{J/S} = & 60\ \text{Sa} \\ (300 + 149)\ \text{J/N} = & (200 + 89)\ \text{J/S} = & (100 + 60)\ \text{S/N} \end{array}$$

The final column in Table 12.2 shows the second aspect of the fractional resonance that leads to the 12-pointed star-formations. In all the consonances of their synodic periods, the three planets complete twelfths, fourths or sixths (after the decimal point) of their orbits. The degree of accuracy increases here, too, as in the left-hand part of the Table. We have listed only those combinations which in each instance give more exact equalizations than the previous ones. In between there are further trios of numbers which also give approximations depending on the exactitude on which they are based. In these, too, one sees that the planets cover fairly accurate multiples of one twelfth of their orbits in the relevant spans of time. In the interplay this leads to a spatial ordering of the conjunction and opposition positions which may be very significant for the long-term maintenance of the planetary system. In order to bring out this point with greater clarity we shall introduce a further means of depicting graphically the movements of the planets, namely a combination of the planet-centred depiction and that of the linklines. With the help of this new instrument let us look first at the planetary world from the viewpoint of Jupiter whenever it has a conjunction with Saturn. In the heliocentric view, Saturn and Neptune formed the 12-pointed star-figure made up of four triangles shown in the lower diagram in Fig. 12.3.

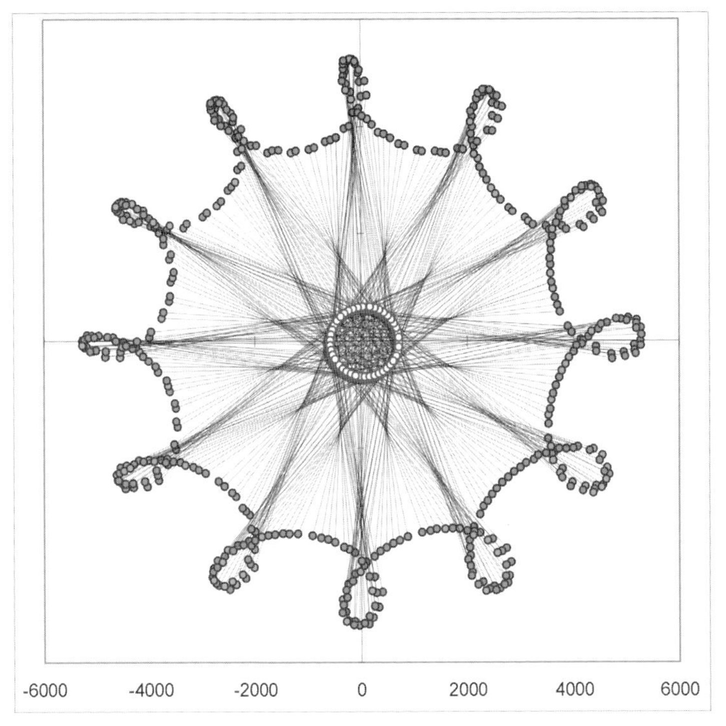


**Fig. 12.7** *Saturn-Neptune linklines as seen from Jupiter at Jupiter/Saturn conjunctions, 500 times, beginning 1 March 1494* BC, *period 9929.26 years*

At their conjunctions the distance between Jupiter and Saturn is always (almost) the same. Over the course of time, as seen from Jupiter, Saturn forms the circle shown in the planetary positions in the inner part of the diagram. Meanwhile Neptune forms a 12-fold figure with outward loops. The linklines between Saturn and Neptune trace, on a smaller scale, the same 12-pointed star-figure which has already appeared in the helio-centric view. This method of depiction also shows a further aspect of the interplay. The relatively tight loops show the opposition positions (or near-oppositions) of Jupiter and Neptune since the distance from the central point, where the former is located, is then at its greatest. In other words, with the Jupiter/Saturn conjunctions on which the picture is based, the oppositions of Neptune and Jupiter are only possible in the directions determined by the division of the circle into 12 sections, and always in very narrow regions.* The same applies in principle for Saturn/Neptune oppositions (when Saturn is in the centre), except that then the loops are considerably wider, as will be seen in one of the following pictures. But first let us show an opposite effect in a different constellation:

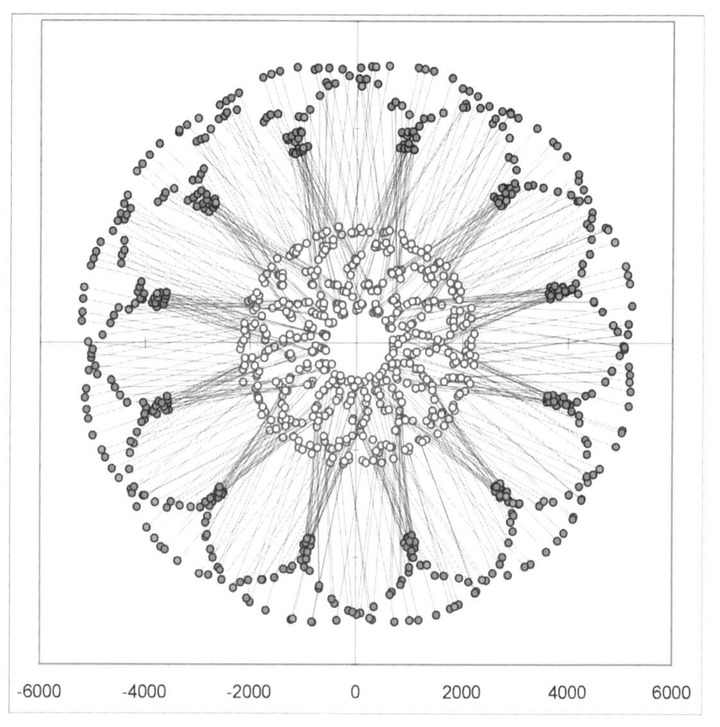


**Fig. 12.8** *Saturn-Neptune linklines as seen from Jupiter at Saturn/Neptune conjunctions, 500 times, beginning 9 October 6943* BC, *period 17,934.63 years*

---

*According to this, we have here a repetition of an effect of which we saw a different version in the Venus-Earth-Mars relationship (where at a conjunction with Mars the third planet cannot be in opposition; see explanation to Fig. 7.4).

We have already met with the outer loops of the figure in the lower diagram of Fig. 12.4. Even though not quite so clearly, a 12-fold order of loops also appears in the inner region of this picture. Here the linklines are all (approximately) of equal length because they mark the distances between Saturn and Neptune whenever these two are in conjunction. While, as seen from Jupiter, the directions of the Saturn positions and their distances from the centre are almost uniformly distributed, the shortest distances of Neptune are concentrated in its loops. These thus correspond to its conjunction positions (or those very close to conjunction) in relation to Jupiter. In the case of Saturn/Neptune conjunctions, therefore, Jupiter/Neptune encounters taking place at the same time are only possible in certain selected sections of Jupiter's horizon. When Saturn/Neptune and Jupiter/Neptune conjunctions occur at the same time, maximum gravitational force is exercised by them because all three planets are in line with the Sun.

The uniqueness of harmony in the interplay of those three distant divinities—a harmony arising in some measure from the truly remarkable details—is further illustrated from other angles in the following two diagrams. These bring us back to the main conjunctions already discussed.

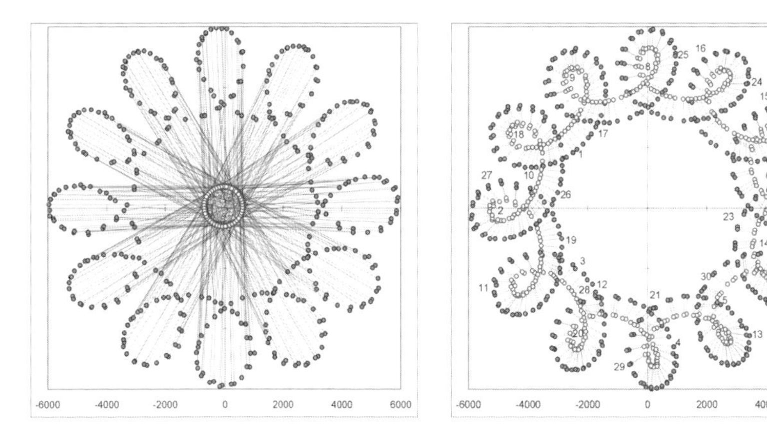


**Fig. 12.9** *Linklines from the planet-centred view at Jupiter/Saturn conjunctions, 500 times, beginning 11 May 1455* BC, *period 9929.26 years. Left: Jupiter-Neptune as seen from Saturn. Right: Jupiter-Saturn as seen from Neptune*

The left-hand picture is like Fig. 12.7 with Saturn in the centre and Jupiter therefore tracing the small circle of its positions. The right-hand picture shows the relationships as viewed from Neptune. The first 30 positions of the relatively tiny linklines (measuring *c.* 650 million km) of its inner companions are marked here to give an example of the chronological sequence of these lines and points being arranged in such a remarkable manner. Of course in principle this is no different from those of other planetary configurations shown, where in some cases a specific figure is traced quite rapidly while in others a long time has to pass before anything can be discerned. In this example the 12 loops show up when about 120 positions have been plotted. In summary we might add that the configuration of the three most massive planets must surely represent a factor for the stability of the planetary system as a whole which should not be underestimated. The fact that their interplay proceeds in the manner described while the structure of the solar system as such has lasted for many, many, many years without bursting asunder (despite calculations of susceptibility to chaos) demonstrates, we may suppose, that the invisible guiding hand of the Number of Perfection is at work here.

Let us now see whether the fourth of the great planets, Uranus, fits in with the order of the outer region by means of geometrical figures other than the hexagram it traces with Jupiter.

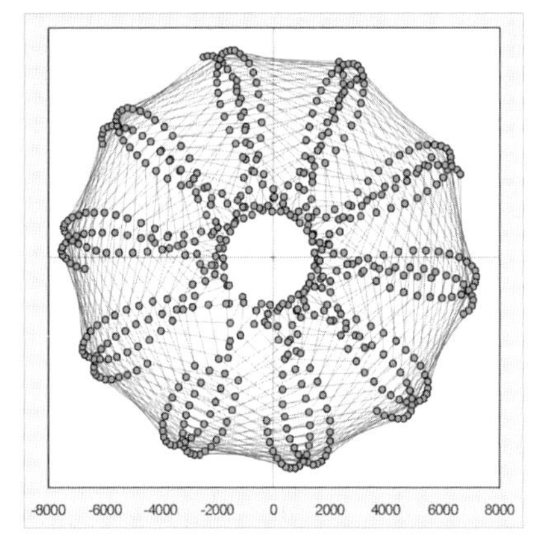


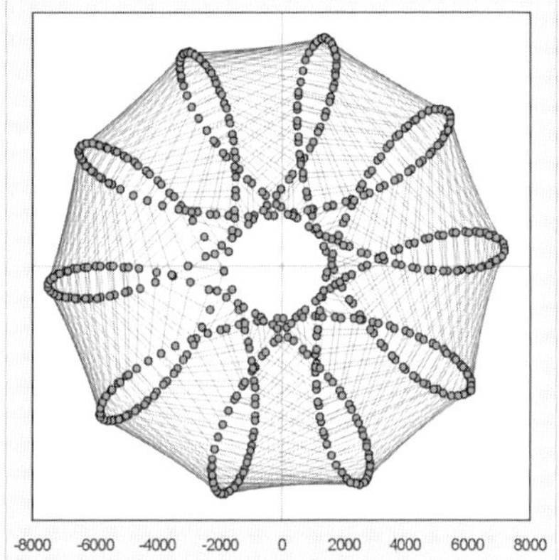


**Fig. 12.10** *Neptune as seen from Uranus, 600 times. Left: at Jupiter/Neptune conjunctions, beginning 4 March 2477* BC, *period 7669.18 years. Right: at every eighth Venus/Earth conjunction, beginning 22 May 3115* BC, *period 7673.58 years*

We can be brief now. There are no other truly resonant figures of small whole numbers in 3-planet constellations of the outer solar system. The figure showing most regularity is the left-hand one above. However, what initially appear to be ten loops turn out after a while to belong to a 51-looped figure. The reader may remember from Chapter 6 that the synodic period of the Jupiter/Neptune encounter corresponds to 8 Venus/Earth conjunctions except for a difference of 2.68 days in 12.8 years. If the same constellation is plotted at every eighth of these encounters, the result is a movement form that is as good as absolutely resonant. But one might naturally be tempted to question the purpose of relating the interplay of Uranus and Neptune with every eighth conjunction of very far distant inner planets. So the next thing we shall do is depict the corresponding movements in the rhythm of 4 Venus/Earth encounters which is so very characteristic for the inner solar system:

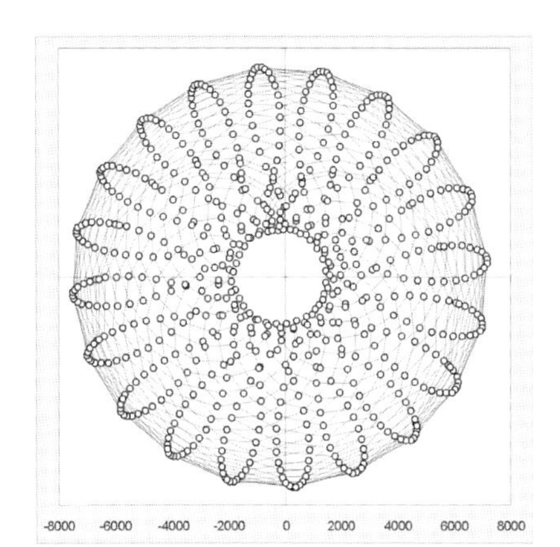


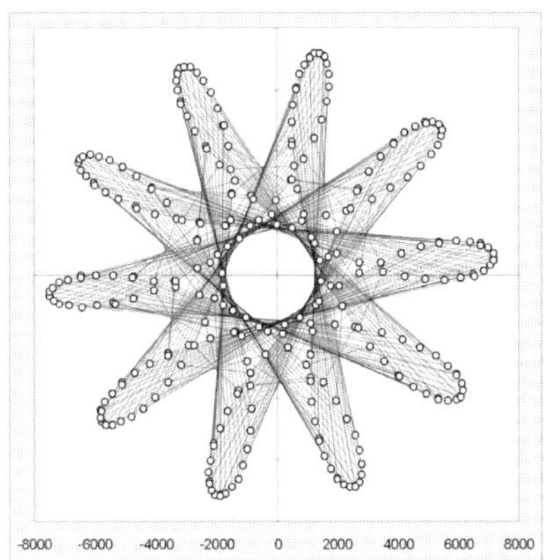


**Fig. 12.11** *Neptune as seen from Uranus. Left: at every 4th Venus/Earth conjunction, 750 times, beginning 20 April 557* BC*, period 4795.99 years. Right: at every 32nd Venus/Earth conjunction, 300 times, beginning 29 April 4995* BC*, period 15,347.15 years*

Instead of 10 loops we find 20; at every second Venus/Earth conjunction it would be 40, and finally at every encounter there would be 80, always with the same precision. This might lead us to expect five loops for every sixteenth conjunction in a similar depiction. But this does not happen. From the interval of 8 onwards the situation remains constant: from the planet-centred view Uranus and Neptune remain on the same 10 loops. (Or, when depicting the situation by means of the linklines, the corresponding star-shapes appear.) The connecting lines alone now draw the variations on the theme. At every sixteenth conjunction this would be a 10-pointed star-figure made up of two pentagons; and the star-figure we may admire in the right-hand diagram arises from the interval of 32. One could also say that the power and characteristics of the octave appear in their purest geometrical form in the celestial music arising from the movements of the planets involved here.

The fractional resonance leading to the number ten also answers the question as to why, beyond the interval of 8 Venus/Earth encounters, a further halving of the number of loops in the figure does not occur, leading to the appearance of a pentagram (or some other corresponding fivefold figure). Thirty-seven Uranus/Neptune synodic periods are approximately 496*8 Venus/Earth synodic periods. Here the outer planets complete 75.5007 or, respectively, 38.4997 orbits. The next approximation, 79:(1059*8), is extremely accurate, as are also the relevant orbits: 161.19998 for Uranus and 82.19996 for Neptune. Since there are very exact resonances for halves and fifths, this leads to the formation of figures ordered according to the number ten. This twofold resonance remains the same also when the interval of time is doubled, multiplied by four and so on. But a resonance based on half-orbits cannot cause a fivefold figure to come about. This is the arithmetical reason why the pentagram does not appear among the orbits of the outer planets. Instead, the not yet quite exact pentagram of Venus and Earth or its multiples—i.e. so to speak the higher octaves—generate a series of perfect 10-pointed star-figures in the movements of their outer opposites: Uranus and Neptune. The planets, then, are ordered as they are—and this order can be calculated—for the purpose of bringing something to expression. Arithmetic is the sibling of geometry and of music, and not the sovereign ruler it is made out to be by modern mankind in astronomy and in many other realms. It is hardly likely to enlighten us as to a possible inner link between the various pictures traced in the heavens.

So in order to move a little closer to what it is that the symphony of the celestial bodies is trying to tell us, let us now look at another configuration of the outer planetary system.

The meaning of a statement or of a work may sometimes be found in that which is not expressly formulated. The figure resembling a 5-pointed star seen here, highlighted by the lines joining the points, is the one which ought to appear. But planetary reality brings it about that the points showing the actual conjunction positions are to some extent wildly scattered around in the celestial landscape. To join them all up with connecting lines would render the image totally confusing. This is one of the very few examples where an expected geometrical formation, based on the synodic period, i.e. the mean value of the conjunction intervals, does not arise. Actually deviations from the mean value in Uranus/Neptune conjunctions are quite small, both in percentages and in days, at all events in the latter case as regards the scale of the outer planetary system (see Table in Appendix 6.3). In addition, the distance conjunctions shown opposite are those with still smaller deviations. But this, too, is no help. Just as in the previous example, the

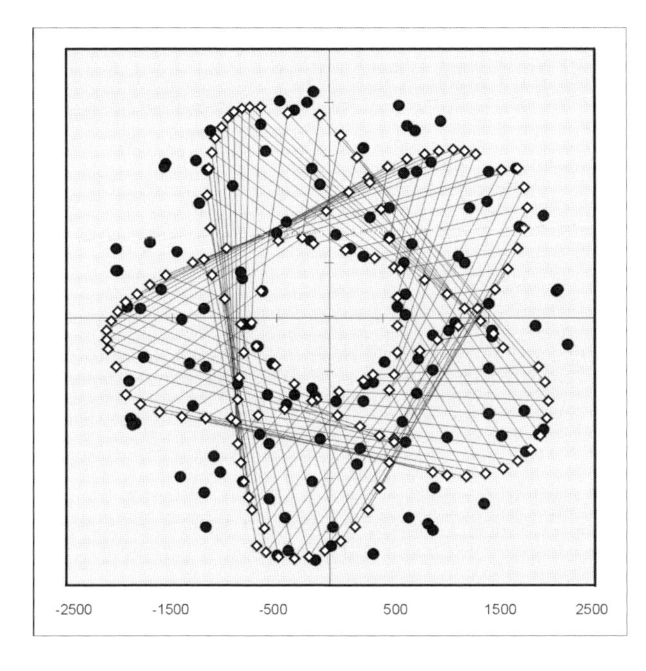


**Fig. 12.12** *Saturn as seen from Jupiter at Uranus/Neptune conjunctions, 130 times, period c. 22,287 years.** *Square marks: positions calculated from the mean values of the conjunctions. Round marks: actual positions at minimal distances between the planets*

ratios are such that the central figure of the inner planetary system cannot appear here. Evidently a pentagram is not required in that region, so far distant from human beings, where perfection bears the stamp of the number twelve.

Let us now look at the situation which is the opposite of the constellation just considered: what Uranus and Neptune do when Jupiter and Saturn are in line with the Sun. We saw that the magician who knows how to transform not only hexagrams into 5-pointed star-figures, but also 5-pointed star-figures into chaos, does not appear inclined to collaborate regularly in his relationships with, in each case, two other outer planets. But perhaps he will show himself from another aspect in the interplay of all four large planets (see Fig. 12.13).

In the left-hand picture the form in the centre is ordered according to the number nine, but all in all the figure does not at first appear to be particularly attractive aesthetically, nor does it show an especially high degree of order. The 15-pointed star-figure formed with Saturn in the

* This time-span actually exceeds the range for which it is possible to make exact calculations with the help of the software mentioned in Appendix 3.6. However, the positions that lie within the range of exact calculations show in principle the same picture, merely reduced by a few points and lines. So the overall message of the illustration is not in any doubt even though some positions may have shifted by a few degrees.

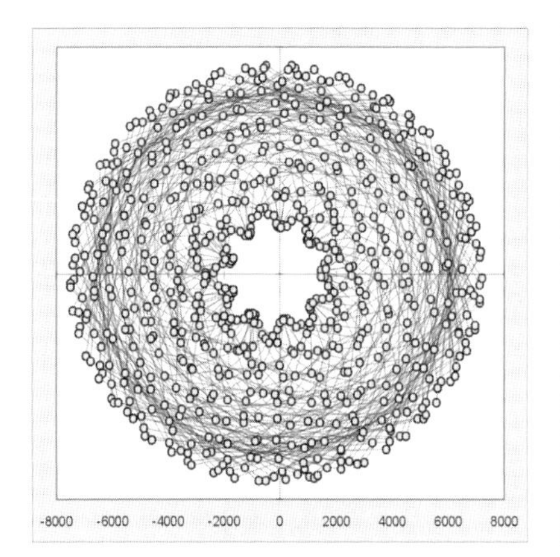


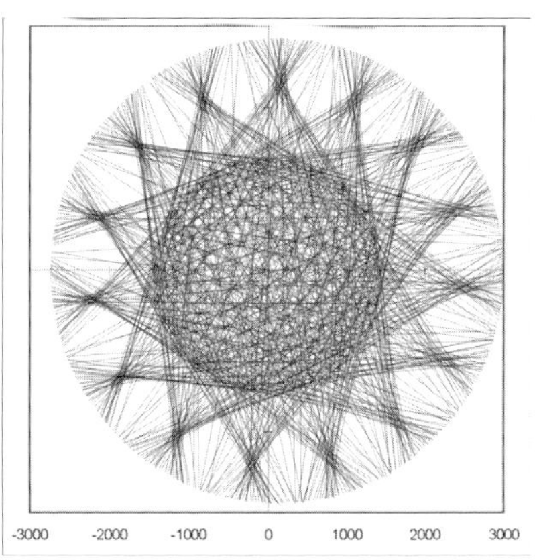


**Fig. 12.13** *Left: Neptune as seen from Uranus at Jupiter/Saturn conjunctions, 700 times, beginning 23 November 3958* BC, *period 13,900.97 years. Right: Saturn-Uranus linklines at Jupiter/Neptune conjunctions, 750 times, beginning 16 April 2605* BC, *period 9,586.47 years*

decisive Jupiter/Neptune encounter has a much clearer shape. And perhaps we are, after all, being shown another level when we see that the magician, too, can resonate in harmony in his interplay with the other three inhabitants of the realm of the number twelve where the number fifteen and even, fairly clearly, the number nine also appear. Since the Jupiter/Neptune conjunction in collaboration with Saturn evidently brings about the better-ordered figure, the obvious thing to do now is to follow the path of Uranus when the other three have one of their rare meetings (see Fig. 12.14).

Marvellous! Even though the web of spiralling threads is not yet quite complete, one can clearly see the flower with 12 petals coming into view. Thus are the four large planets united beneath the banner of the number twelve at the resonance level which embraces them all. A good approximation of their conjunction periods is evident in: 362 Ju/Ne = 335.011 Ju/Ur = 233.002 Ju/Sa = 128.998 Sa/Ne = 102.009 Sa/Ur = 26.989 Ur/Ne (= 4627.071 years).* Herewith they all circle the Sun relatively accurately x plus 1/12 times. (Regarding the period of time depicted, see the footnote to Fig. 12.12.)

Over and above this, the movements of the four large planets conceal another link so deeply hidden and so amazing that I once more receive the impression of the planets in our cosmic home dancing to a

* The next very accurate consonance is 912:844:587:325:257:68 in 11,657.15 years. Here, the 4 planets cover x plus $\frac{3}{4}$ orbits

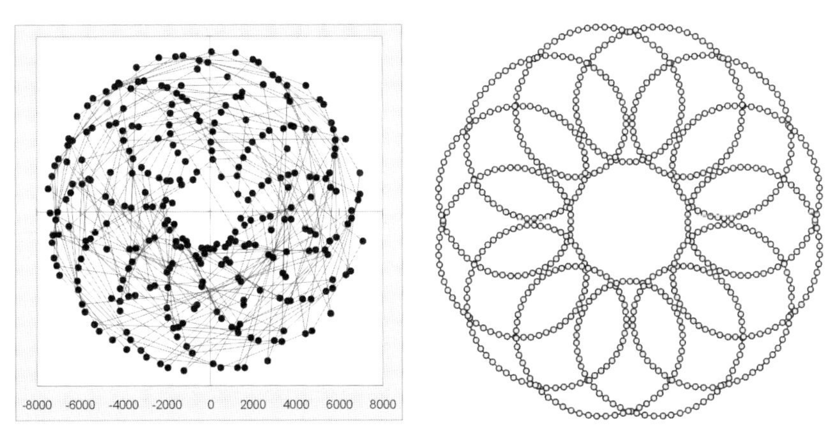


**Fig. 12.14** *Uranus as seen from Neptune at threefold conjunctions (<15°) of Jupiter, Saturn and Neptune, 300 times, period c. 45,478 years. Right: the archetype which accords with an extended epicycloid*[*]

choreography for a well-structured musical composition. The point of departure is the seemingly not well-ordered movement figure of Uranus and Neptune at Jupiter/Saturn conjunctions (Fig. 12.13 left). Here, too, the planet-centred depiction and the plotting of the middle points of the linklines yield an identical figure, except for the scale (see Chapter 6). Since Uranus and Neptune furthermore show a very similar mass ratio to that of Earth and Venus (0.843 as against 0.815), the difference in the tracks of middle points and centres of gravity initially appears to be insignificant (but a very meaningful deviation is about to come to light in the detail). In the sequence of pictures shown in Fig. 12.15, the centres of gravity of the Uranus-Neptune linklines are shown first, but this time not at conjunctions but at oppositions of the two largest members of the planetary system. The picture that arises when the conjunctions are shown is very similar, but here it is based on the oppositions for reasons which will shortly be revealed. Next, every third opposition is selected, and thereafter only every ninth. Finally only the positions at Jupiter/Saturn oppositions are marked when there is a simultaneous (approximate[†]) Jupiter/Neptune opposition. From the two opposition constellations it follows that there is also a simultaneous Saturn/Neptune conjunction.

---

[*] Mathematically: with the three parameters R = 12, r = 1, a = 7

[†] On p. 128 in Chapter 7 we explained that threefold conjunctions (or 2 simultaneous oppositions among 3 planets) never occur with 100 per cent accuracy. Since these constellations are also extremely rare, the positions plotted here were taken when the angle between Jupiter and Neptune is greater than 120°

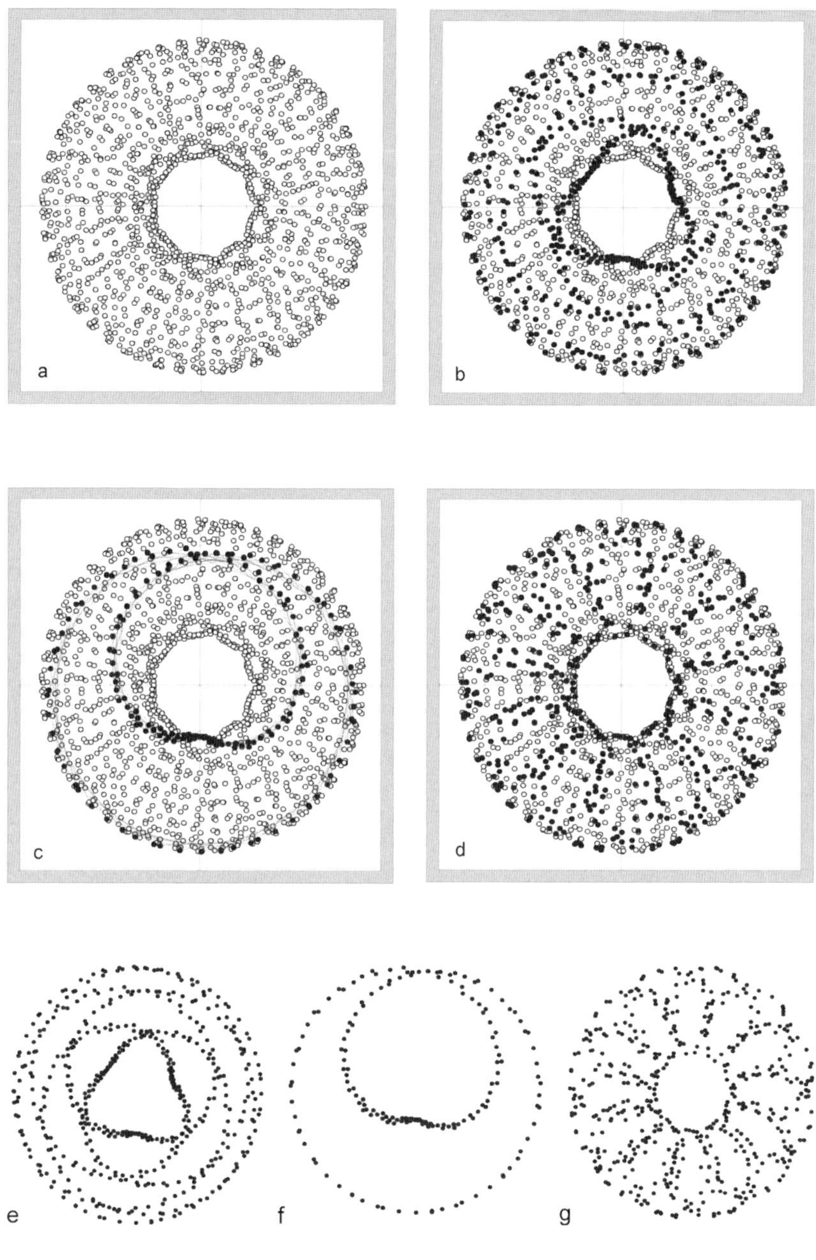


**Fig. 12.15 (a)–(g)** *The centres of gravity of the Uranus-Neptune linklines at Jupiter/Saturn oppositions, 1500 times, period* c. *29,788 years: (a) at every opposition; (b) emphasized at every third opposition; (c) emphasized at every ninth opposition; (d) emphasized at simultaneous Jupiter/Neptune near-oppositions (angle Ju–Ne >120°); (e) to (g) as in (b) to (d), for clarity without the background*

In the diagrams of Fig. 12.15 every black dot from the various sets covers one of the light ones in the initial totality. The threefold structure which arises during the course of several tens of millennia,* and which, moreover, reminds us strongly of the tracks of the centres of gravity of Jupiter and Saturn (see Fig. 9.1, bottom right), is thus present three times in the combined sets—in each case rotated by 40°. Then nine separate loops are crystallized out, since of course the one shown appears nine times—again at intervals of 40°. So what lies hidden in the initial tangle of dots, which despite the roughly ninefold form in the centre does not make a particularly orderly impression, is even at this early stage truly astonishing. But it is perhaps even more astonishing to find that within this near-chaos there is also an order involving the number twelve. This is revealed when one includes the other prominent constellations of the planets involved. In addition to their oppositions, a very similar order arises from the (once again approximate) Jupiter/Neptune conjunction positions—always together with the exact Jupiter/Saturn oppositions which provide the basis. In the case of Jupiter/Uranus conjunctions and oppositions, too, similar formations arise, though not quite so clearly. The conclusion to be drawn in summary is that at Jupiter/Saturn oppositions (and up to this point also at their conjunctions) all the other important constellations in the outer planetary system are ordered according to the number twelve in a hidden and very subtle way. To do justice to this amazing degree of subtlety, let us once again join Johannes Kepler in saying of the motion of the planets that: '*nothing is more admirable, nothing more beautiful, and nothing a better witness to the Creator's wisdom.*'[3]

Now why did we choose the oppositions of Jupiter and Saturn in the above depiction whereas before that in nearly every case we discussed the conjunction positions? Of course the former are more significant for the balance of the solar system as a whole since it is at least more nearly balanced when the two giant planets are situated on opposite sides of the centre. So to plot the positions of our Sun in relation to the centre of gravity at these Jupiter/Saturn oppositions over a longer period of time is another method that suggests itself (see Fig. 12.16).

Here in each case the lines depict the connections between two chronologically sequential positions. The distances show that the oppositions of Jupiter and Saturn do indeed maintain something like a balance in the solar system; in these constellations the Sun's mid-point is at most 700,000 km distant from the centre of gravity, whereas at other times it is possible for the distance to be as much as 1.5 million km. Although the

---

* Regarding the calculation time-scale, see above.

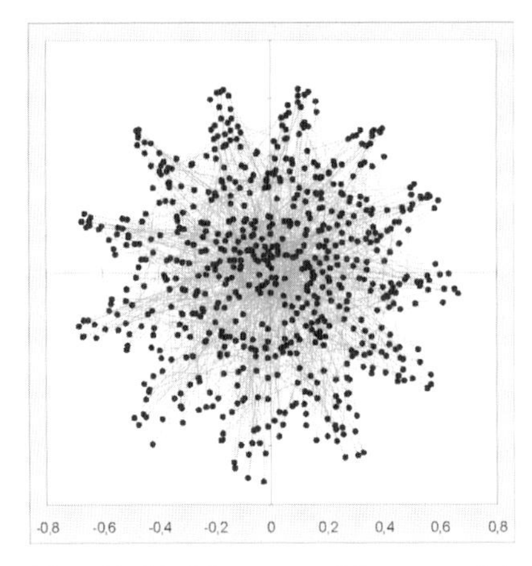


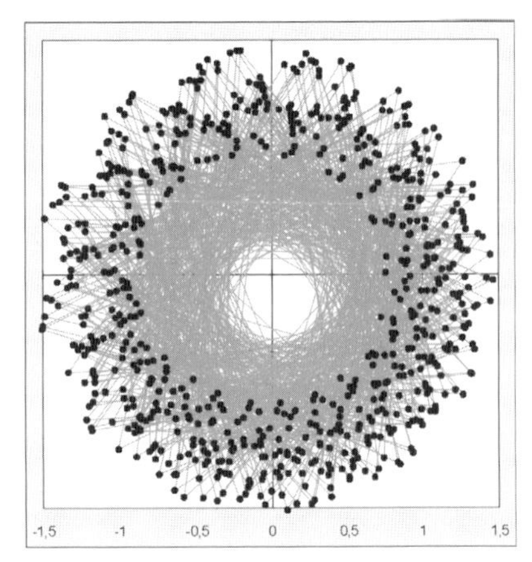


**Fig. 12.16** *Movement of the centre of the Sun around the barycentre at Jupiter/Saturn oppositions (left) and conjunctions, 700 times, period c. 13,881 years, beginning 2 December 4980 and, respectively, 7 April 4950* BC, *scale in millions of kilometres*

order revealed is not perfect, which could not be expected given the multitude of influences at work, nevertheless there is a clear ordering in accordance with the number twelve. It is also evident that there is almost no obvious structure in the conjunction pictures.

Given the varying influences of the four large planets on the movement of the Sun around the barycentre—as shown in Fig. 9.1—it is indeed astonishing that an ordered structure can arise and remain in place for longer periods of time. Basically, Fig 12.16 (left) shows a distorted variation of the movement structure 'Neptune as seen from Saturn at Jupiter/Neptune conjunctions' (see Fig. 12.4, top, and Plate 10). Hence we may perhaps regard this wonderful figure as the archetype on which the movement dynamic of the whole solar system is founded. The illustrations in this chapter show it in relation to the Jupiter/Saturn/Neptune relationship in a number of variations, but always purely geometrically, i.e. without taking the planetary masses into account. The structuring in accordance with the number twelve in the overall dynamic becomes much clearer when we include both Uranus and also the masses of the planets. This is shown in the following illustration.

Here we see that in the long-term interplay of Uranus and Neptune at Jupiter/Saturn oppositions shown in Fig. 12.15 the number twelve only appears distinctly when the centres of gravity of the linklines are plotted. Although, as already mentioned, there is relatively little difference in the masses, plotting the (mass-less) middle points leads to a much less clear figure. Not until the actual masses of the two planets

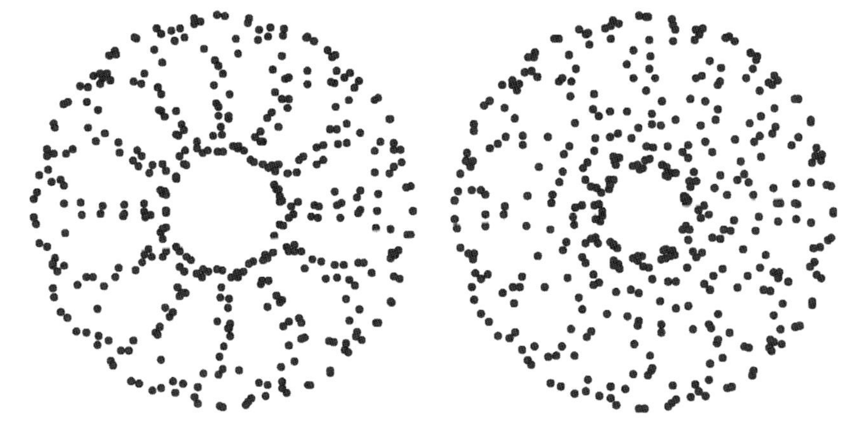

**Fig. 12.17** *The centres of gravity and middle points (right) of the Uranus-Neptune linklines at Jupiter/Saturn oppositions with simultaneous approximate Jupiter/Neptune conjunctions (angle Ju–Ne <60°), period c. 23,850 years*

are plotted does this lead to the 12 interlinking circles. Only because all the planets have the right mass at the right place are the movements of the whole system governed in a unique way by the Number of Perfection. I find it utterly impossible to imagine that none of this has any bearing on the long-term stability of our cosmic home. It speaks rather of the truth of something written down long ago in an ancient Book of Wisdom: '*But thou hast ordered all things by measure and number and weight.*'[4]

The question now remains as to whether the perfection attained by the distant gods also sends its influence into the inner realm where the Earth travels on its way between the polar opposite influences of its two neighbours, and where its human inhabitants so often have to witness and suffer circumstances that are so very far from perfect. The goddess of love gave them the number five while warlike Mars attaches them to the number four; but perhaps it is the task of the messenger of the gods to send a reflection into their world of something that appears to be unattainable. What could be a better starting point for this than the Jupiter/Neptune conjunction which forms the true backbone of the geometrical order in the realm of the number twelve? But since the threefold constellations introduced in Chapter 10 between inner and outer planets showed no sign of the required figures, Mercury will have to use the services of an ally to fulfil this current task. And since on the basis of its unusual rotation Mercury can be called the planet of the number three, it is probable that the planet of the number four will be the one to lend assistance in this mission:

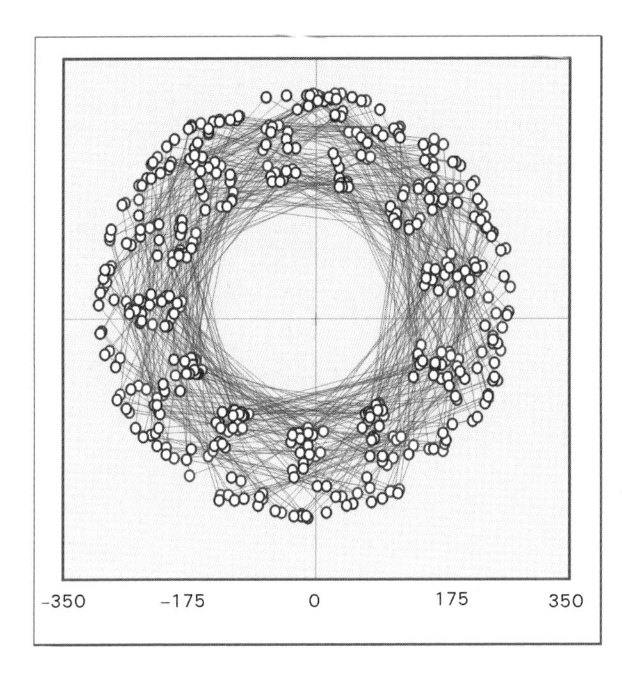


**Fig. 12.18** *Mars as seen from Mercury at Jupiter/Neptune conjunctions (minimal distances), 500 times, beginning 10 March 1199* BC*, period 6390.98 years*

Sure enough, Mercury and Mars succeed in bringing the number twelve into play also in the inner planetary system. While earlier on an expected figure failed to appear, here we have one of the figures that actually ought not to exist. The fluctuation range of intervals between Jupiter/Neptune encounters is *c.* 64 days; with the distance conjunctions applied here it is still about 35 days. This is *c.* 40 per cent of one Mercury orbit. Owing to this in connection with the high eccentricities of the inner planets (whose conjunction intervals diverge from one another by as much as 15 days) the looped figure which arises is quite surprising, and of course one cannot expect any very smooth contours. When conjunctions in ecliptic longitude are plotted the figure is even more blurred, although still quite recognizable. So here, once again, there must be some very fine tuning which even includes the various rhythms determined by the eccentricities. The fractional resonance of the ratios of the pure orbital periods is 29 Jupiter/Neptune synodic periods to 1342 Mercury/Mars synodic periods. This period of time coincides with x plus 1/12 orbits of Mercury and Mars. In principle the resulting movement picture closely resembles the lower diagram in Fig. 12.4 which is also formed by the participation of Neptune and Jupiter.

Looking back over the last few chapters we find that most of the figures shown were those which arise when the planetary conjunctions are related to one another by means of geometry. In seeking to establish whether specific structures can be discerned in the order of the planets, orientation according to the conjunctions yielded the greatest accord with the physical side of reality. The greatest mutual gravitational effects of the planets arise when the distance between them is shortest. Resonant ratios of orbital periods, and also of other orbital parameters, can lead to an accumulation of effects over long intervals of time. In an extreme case, irreversible orbital shifts could be caused by this. But on the other hand we still do not know which factors bring about the conditions that have evidently remained stable for hundreds of millions of years despite constant changes of the eccentricities, inclinations, and perihelion and nodal positions.

Aspects which would be missed by a purely arithmetical consideration are shown up here by the use of geometrical depictions of how the conjunctions of several planets are interrelated. For example we have discovered resonances that lie hidden in the rhythms of temporal fluctuations and we have shown whether and how a calculated approximate harmony based mainly on averages becomes visible in the reality of space in which the forces of gravity are at work. Over and above this the various modes of depiction have made it possible to detect, in the relationships of 3 or 4 planets, ordered structures that accord with an overall picture of their gravitational interaction. Although the fractional resonances which are thus brought to bear can also be arrived at by purely mathematical means, it is unlikely that they would have come into view had the instruments of geometry not been previously applied.

We have investigated the relationships within the inner and the outer planetary system, the interplay between the two regions and the rotations of the four slowly turning celestial bodies. The essence of the geometrically moving forms is, it seems, expressed in the figures of small whole numbers which occur repeatedly and/or manifest the strongest degree of resonance. In the direct conjunction ratios only two fairly resonant figures arise up to the number twelve (except for the presumed or actual 3:2 commensurability of Neptune and Pluto). In the inner realm this is the pentagram of Venus and Earth and in the outer realm the hexagon or, respectively, the hexagram of Jupiter and Uranus. There is a very strong resonance between the orbital period of Mars and the Venus/Earth synodic period. The result of this is a 20-pointed star-figure or also the square of Mars related to the cycle of 5 conjunctions.

The most noticeable relationship in the interaction of outer and inner realms is surely that of Jupiter and Uranus with Mars, Earth and Venus. This leads to very clear figures ordered according to the numbers five, ten and twenty. In the opposite direction we have seen an exactly formed sixfold looped figure in the relationship of Venus and Earth to Pluto. Furthermore, the rhythm of 4 or, respectively, 8 Venus/Earth encounters has a relationship with several pairs of planets in the outer system that leads to clear movement figures. Here above all the figure structured according to the number ten or its multiples formed by Uranus and Neptune deserves special mention. Among all the figures we have encountered, this is probably the one which manifests the highest degree of resonance.

Geometrical examination of the rotations involves both the relationships of the rotating bodies with one another and their relationships with the orbiting planets. Here the Moon occupies something of a middle position in that it can be regarded both as a rotating body and one which orbits Earth with the same period. Depiction of the geometrical relationships between rotations and orbits was made possible by investigating the orientation of a defined horizontal axis towards the moving planets. The ratios of the rotations can be accounted for by means of rotation conjunctions, i.e. the parallelism of the horizontal axes. The rotation of the Venus axis is related to all the planets up to Jupiter in accordance with the number five and its small multiples, whereby the 4*5 relationship to Earth is the most noticeable on account of its variability and because of the metamorphoses of the pentagram into the 10-pointed star-figure which are caused by the movements over longer periods of time. In addition, the rotating Sun forms a very resonant pentagram when seen at 'Venus-Sun-Views'. Here the rotation of the Sun is taken as the mean rotation period of sunspots, namely 25.38 days.

When the Moon's positions are plotted at moments when the solar axis is oriented towards the orbiting Earth/Moon system, the lunar orbit forms an utterly unexpected hexagon almost as though drawn with a ruler. The axis parallelism of Sun, Mercury and Venus provides with great accuracy all the lines possible in a decagon—depending on the degree of accuracy chosen, except for those forming the double pentagram. When Sun and Mercury are related via their axis parallelism and the Moon via its orbit, the positions show a 10-fold looped figure, whereby the connecting lines draw a continuous 10-pointed star-figure and the missing double pentagram.

Order in the outer planetary system is governed by the conjunction relationship of Jupiter, Saturn and Neptune. By means of their various configurations, and in very resonant ways, these planets form diverse 12-fold figures some of which are exceedingly beautiful. Uranus is included here, too, because at threefold conjunctions of the others its positions (in connection with Neptune) yield a geometrical sequence which is also regulated by the number twelve. Moreover, in the dynamic of the overall system, embodied by the movement of the Sun around the barycentre, this 12-fold order also emerges when the solar middle point is plotted at Jupiter/Saturn oppositions which are so decisive with regard to the balance of the system.

Even a mere enumeration of all these phenomena shows what exceedingly unusual conditions must have prevailed in our solar system when, firstly, the celestial bodies were taking their places and thereby establishing their orbital periods in accordance with the law discovered by Johannes Kepler and, furthermore, were beginning to tune their rotations in the measure we find in existence today. The future will show whether physical causes will ever be discovered for at least some of the evident preferences accorded to a few specific numbers in the various inter-relationships. We can, however, answer in some measure the question as to whether there is any degree of inner connection between the values that occur. To do this, let us now summarize the symphony of flowers and stars begun in Chapter 10 and brought to its conclusion here.

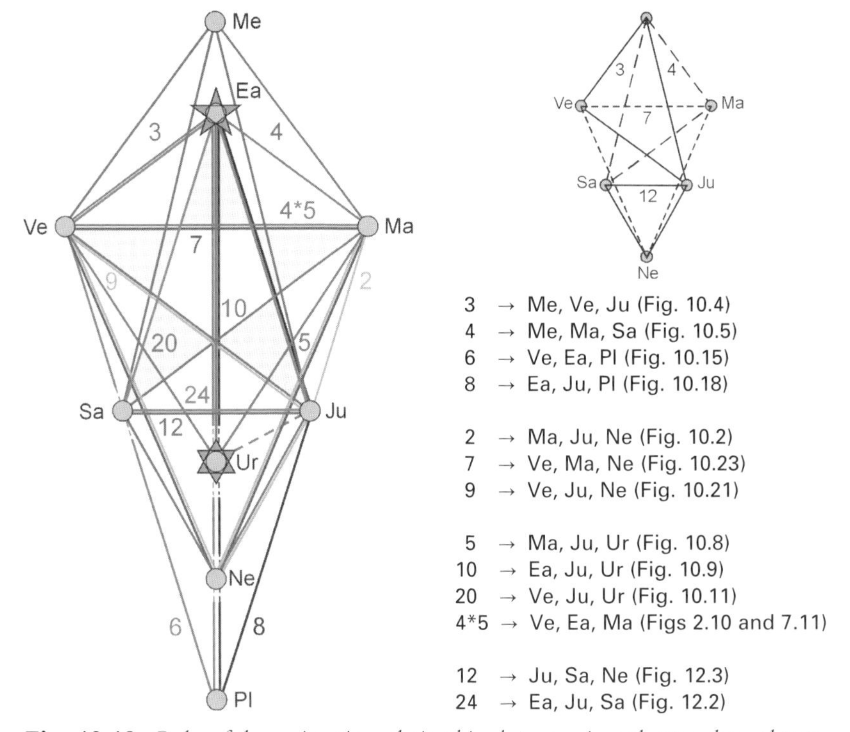


**Fig. 12.19** *Order of the conjunction relationships between, in each case, three planets, depicted by means of triangles and numbers of the same colour. (For reasons of simplicity, the formations arising from the Jupiter/Uranus hexagram are shown by one connecting line only. And the pentagram and hexagram is assigned to only one of the planets involved in each case.)*

→ Plate 12

All the numbers that have arisen are here linked with one another as coherently as possible in an overall picture—my best effort at discovering the logic that underlies the many and varied movement figures in the planetary system. If this is unsatisfactory, one may once more refer to Johannes Kepler's suggestion to his readers that they might like to 'construct a system that better depicts the celestial movements'. Not only does our diagram reveal the truly architectural beauty of planetary conjunction ratios; above all it gives a voice to the fundamental themes, the Cross and the Pentagram, which have appeared again and again during the course of our investigations.

As the small inset sketch shows, the planet closest to the Sun, Mercury, is linked to one inner and one outer planet through the number three and the number four. The integration of Neptune leads, in combination with the two inner planets, to the sum of those numbers, while in combination with the two outer ones it leads to their product. All the figures linked to the number five then arise through the inclusion of the third planet from the centre, Earth, and the third from the periphery, Uranus. And finally Pluto, the most distant planet from the Sun, plays its part by means of a doubling of the numbers brought to bear by its opposite, Mercury. The numbers two, nine and twenty-four are also evidently situated at their correct locations. One would surely have to investigate large numbers of fictional or yet to be discovered planetary systems before coming upon an equivalent degree of constant order. The oscillations which once upon a time fashioned and filled our solar system must indeed have been of a very special kind.

In another of Kepler's discoveries, which we met earlier in an overall view of the ratios of the limits of the velocities, distances and orbital periods, there is a further wonderfully simple image involving the most important of the numbers in our summary.

The 'Hedgehog' consists of 12 pentagrams, and it also possesses 12 points. It can be depicted from two angles (see opposite). The one on the right here shows its profile in the form of a 10-pointed star-figure while that on the left gives a different view: a hexagram. The right-hand figure is a variation discovered by the French mathematician L. Poinsot.* This figure consists of 20 equilateral triangles which interpenetrate one another.

In the dodecahedron-star the numbers five, six, ten, twelve and twenty are thus linked with one another in a unique manner—in what one might

* There is a further variation, discovered in 1810 also by Poinsot, which consists of 12 interpenetrating pentagons. Only four regular stellar solids exist: the dodecahedron-star, the icosahedron-star, and the two Poinsot-stars.

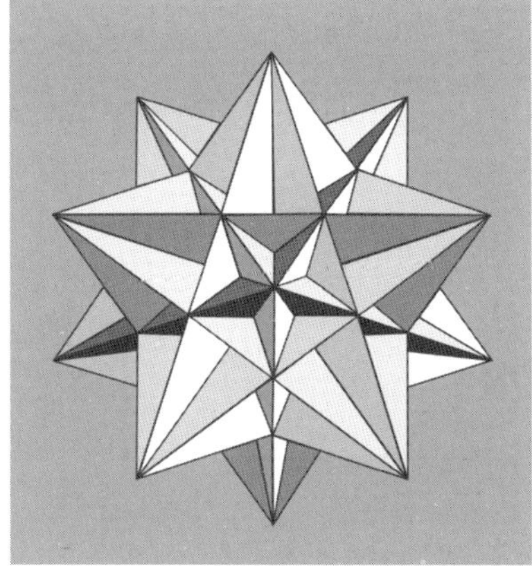

**Fig. 12.20** *Kepler's stellar solid, the 'Hedgehog'; right: a variation according to L. Poinsot*

call the world of pure spirit. The corresponding geometrical figures are those that appear most frequently in the solar system and, above all, in specific parts of it. This enables us to follow up the question as to whether the order we find there has any meaning. It is a question involving two presuppositions. The first of these is that one can only ask it in relation to the human being. Without the existence of humanity it would be meaningless to seek for meaning, quite apart from the fact that without humanity there would be no one who could ask the question. The other presupposition is that one can only think deeply about these things if one acknowledges the symbolism that has come to be attached to those basic geometrical figures during the course of human history. Since this involves many and varied associations, I shall here restrict myself to the core aspects—in the hope that I have recognized these correctly. I have thus applied some measure of general comprehensibility, without claiming any binding validity. The essence of the movement and rotation figures and the way in which they are arranged can be expressed in the following picture. The number seven, which shows in the interplay of Venus with various planets in different constellations (Figs 6.19 and 10.23), appears to be missing. We will allow it to remain hidden to symbolize the mystery which will never be grasped in its entirety.

The basic figure in the inner realm of the planetary system is the pentagram, the sign of the human being. Like his symbol, he has through his internal orientation the possibility or the freedom to raise his head up to the stars or else turn and fall downwards. In opposition to this, the hexagram of the outer planetary system represents the order of the external or material world which is polar in its structure.

Fig. 12.21

The pentagram is placed inside a quadrangle or, expressed in another way, one could say that the human being is nailed to the cross of this world. (In Fig. 10.14 the planets give us a very clear picture of this.) The rotations, which can be viewed as inner movements of the planets in contrast to their outer orbitings, unite with the Sun, the light, to form a 10-rayed star-figure. The nature of this figure is communicated to the Earth via the Moon. Inside the corresponding movement figure the double pentagram arises; it has overcome the polarity of the 5-pointed star-figure. Surely this can be seen as a symbol of the perfect human being—an ideal that would appear to be common to all religions.

When the hexagram is linked with the inner planets (Jupiter/Uranus conjunctions to Mars, Earth and Venus) the 5-pointed star-figure, or its multiples, comes into being. The natural human being grows out of the polarity of the material world or, expressed differently, during the course of evolution the human being has grown out of the animal kingdom which in its turn also has a long pre-history. That which has come into being out of nature returns once more to nature. Thus, in its relation to Pluto—significantly the planet of the underworld—the pentagram of the Venus/Earth constellation conversely gives the number of polarity back to the outer realm.

But in his inner life the perfect human being can be brought to birth by the light shining into him. The pentagram of Venus and Earth is only to be found in the form of a 10-pointed star-figure at the location in the outer planetary system which corresponds to their positions (Uranus-Neptune at 8, 16, 32.. Venus/Earth conjunctions). Only the symbol of the ideal human being enters into the perfection of the cosmos which is characterized by the number twelve. But since all the planets are most intimately interconnected through their interplay one might equally correctly say that the perfection of the cosmos can only be attained through the ultimate perfecting of the human being.

So let us close this chapter with a final glance at the already attained perfection in the outer region of our cosmic home.

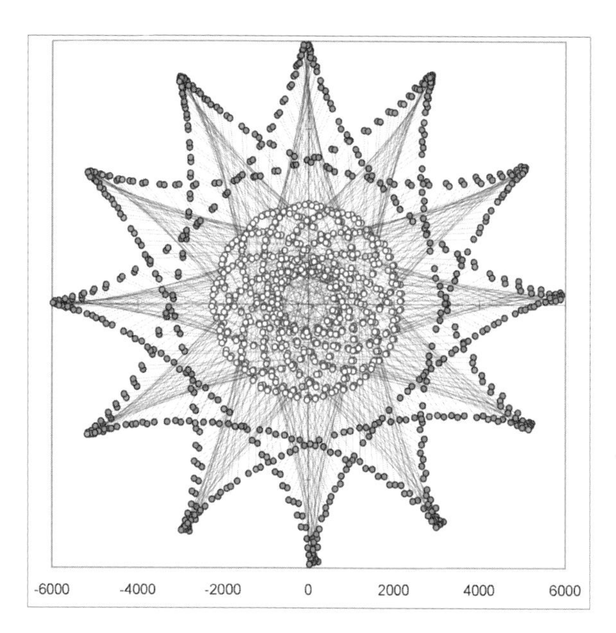


**Fig. 12.22** *Jupiter-Neptune linklines as seen from Saturn at Jupiter/Neptune conjunctions, 750 times, beginning 16 April 2605* BC, *period 9586.47 years*

→ Plate 13

# 13. The Signature

Planetary orbital periods and mean distances from the Sun are inextricably linked with one another through Johannes Kepler's Third Planetary Law: $T^2 = a^3$. But it is nevertheless justifiable to ask which of these two quantities was there first. To try and find an answer, let us return to what is now regarded as the age when the planetary system was born. In the protoplanetary disc, myriads of small objects circle around their centre at specific distances with which their velocities and orbital periods are necessarily linked. With constant collisions taking place, larger bodies gradually form. The impetus of every collision between blocks that are more or less of equal size should cause the two—whether still separately or now combined—to leave their previous orbit and circle around the Sun at a different, even if only slightly altered, distance from it. Therefore the resulting change in the orbital period is the secondary event. It is to this long series of spatial displacements that a fully fledged planet owes the position it finally comes to occupy in the system. One cannot imagine a reverse procedure leading to a deceleration or acceleration that would bring about a change in distance from the Sun based on some inevitable law of nature. And if we add the theory of shock-waves influencing those primordial events, the picture still remains the same. The waves as it were flush the planetesimals to a different position where the corresponding orbital velocity then takes over. Even the oscillations of some conceivable 'creation music'—such as we might imagine by looking at Fig. 5.4—would no doubt primarily serve to accumulate initially unformed matter into a specific spatial region.

The harmonies of planetary velocities elaborated in Chapter 4 with reference to characteristic points in their orbits are caused by the eccentricities and the distances from the Sun. The geometrical movement forms shown are the result solely of the mean distances; here the eccentricities at most distort (or in extreme cases prevent) the appearance of a regular figure (with the exception of the examples shown in Chapter 6 where it is the eccentricities that do actually cause the regularity). In this connection we should remind ourselves yet again that all the figures depicted come about simultaneously. In the figures in Chapter 10 each planet appears several times. Jupiter and also Venus are incorporated into the web of relationships through all* their

* The 11-petalled flower of Mercury, Venus and Neptune (Fig. 10.20) also appears in a similar way at Mercury/Venus conjunctions (not shown in the illustrations) when the Venus-Neptune linklines are applied.

conjunction periods. The calculation of these measures of time includes the orbital periods, the consequence being that the mean distances from the Sun must be very carefully selected if the varied relationships to the other planets are to come about.

Right at the outset in our search for the harmonic and geometrical ratios in our solar system we met with the semi-minor axes. Although these differ from the mean distances by a maximum of only 3.3% (in the case of Pluto), it was these relatively small deviations which showed up the basic structures of the spatial ratios. Here is an overview of the structure of the order, which is the same from inside outwards and from outside inwards:

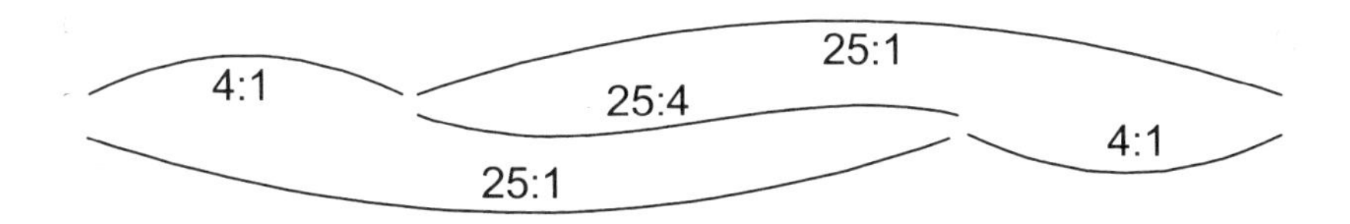


Taking our departure from this structure we shall now approach what in August 1998 I decided to call 'the signature of the celestial spheres' when the basic aspects became clear to me. As I proceeded with the work I realized that this signature involved a good many more levels than I had anticipated. In other words, what we are uncovering here is a theme with variations. Or, expressed more precisely: several geometrical depictions, which are to a large extent mathematically identical, are variations on the principal theme given by the actual astronomical conditions. As with music, what matters in the last resort is that which is present between the notes, as Gustav Mahler very aptly put it. Therefore it may turn out that the true 'signature of the celestial spheres' lies concealed somewhere among the images which we shall now investigate more closely.

In addition to the ratios shown in the above diagram we have also seen that, in the region most distant from the Sun, the Pluto/Uranus ratio is almost 2:1. Neptune, Venus and Jupiter were not integrated in that order, and Earth was only related to Mars with a relatively large deviation via a 3:2 ratio (see Fig. 1.5).

We shall now examine in detail the relationships of the semi-minor axes especially with regard to the planets not yet mentioned, beginning with the outer planetary system. The 4:1 ratio of Pluto and Saturn with its deviation of 0.34% is subdivided into two octaves by Uranus, whereby the differences are 0.60 and 0.27% (the underlying figures are shown in the Table in Appendix 6.5). If we remember the depiction of the double octave by means of the area ratios of the inner and outer circles of a triangle, or of two squares (see Fig. 2.2), the parallel is immediately obvious. Having seen towards the end of Chapter 6 that since the ratios of the overarching distances, as far as perihelion and aphelion intervals are concerned, correspond to geometrical area ratios (in that case those of circles applied to a pentagon), one might assume that a similar phenomenon could hold good here and include Neptune as well. In the picture of the double octave, the latter would have its orbit somewhere between the outer and the middle (2:1) circle. In the drawing this is where the square is situated through which the two circles are related. The interval of the semi-minor axes of Pluto and Neptune arises as 1.2719. Perhaps we shall be able to track down the origin of this figure by consulting the ratio that is offered to us by, as it were, the nature of geometry itself. The ratio of the outer circle to the square within it is $\pi/2 = 1.5707..$ (for the calculation see Appendix 1.3). This in turn corresponds closely to the value of the Neptune/Uranus relationship of 1.5683. The completion, the relation of the square to the inner circle, then has to be the same as the Pluto/Neptune ratio. This works out to be $4/\pi = 1.2732..$, i.e. apart from deviations of 0.11 or, respectively, 0.16%, the planetary ratios are thus approximated. This can be depicted geometrically as follows:

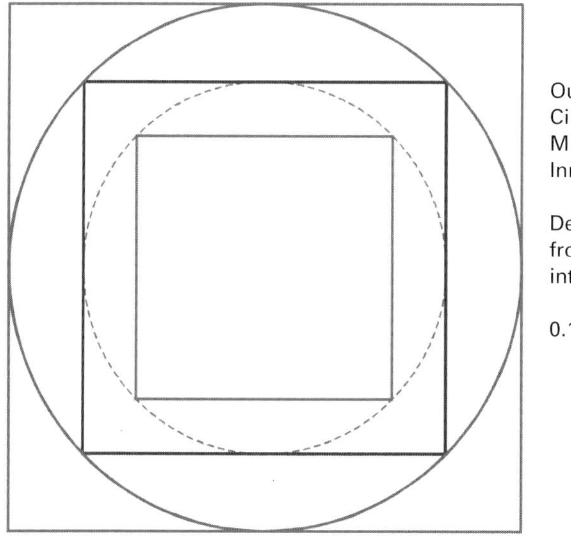

Outer square: Pluto
Circle: Neptune
Middle square: Uranus
Inner square: Saturn

Deviations of the planetary from the geometric intervals:

0.11–0.60%, average 0.29%

**Fig. 13.1** *Geometrical approximation of the ratios of the semi-minor axes in the outer planetary system*

The semi-minor axes of the four outer members of the planetary system are thus ordered in the same way as the areas in one of the simplest geometrical constructions. On the scale of this drawing (converted to lengths, i.e. to the length of one side of the middle square), the maximum difference from the actual ratios (0.60 for Uranus/Saturn) is *c.* 0.15 mm, i.e. about half the thickness of the line, so that one can indeed speak of quite a close approximation. It may be surprising to find three of the planetary orbits depicted as squares. But we are here concerned with showing the governing principle of distances, i.e. the semi-minor axes. Their ratios are related to one another by showing the areas of squares or, in the one case, of a circle. It does seem, then, as though the planetary intervals obtain their structure from a higher dimension. So let us now investigate whether these mathematical and at the same time mystical laws also apply to the other planets.

The four inner planets, too, are embraced by a specific ratio (like that expressed in the outer planetary system by the paler inner and outer squares in Fig. 13.1), namely the Mars/Mercury ratio of 4:1. As already mentioned, Mars forms a rather approximate ratio of 3:2 (instead of 2:1 between Pluto and Uranus) with Earth, the planet which correspondingly mirrors Uranus in the outer order. The ratio of Venus (the planet in the inner region which mirrors Neptune) and Mercury amounts to 1.9094. In the geometrical depiction this value would be some distance away from the 2:1 circle or, respectively, square. So how might the queen of heaven be included in the geometrical order which has, up to now, been perfectly simple? As we have shown, together with all the other planets Venus traces clear movement patterns in the firmament in specific configurations, so its position ought to fit especially well into the overall construction. If harmonious, musical considerations have a part to play in an architecture that underlies the ratios of the semi-minor axes, then the fifth ought to be the next interval to come to the fore with, moreover, a much higher degree of accuracy than that of the Mars/Earth interval. Perhaps this interval of the fifth is to be found in relation to the structure shown by the outer planets. The Pluto/Neptune ratio corresponds to $4/\pi$, so in this connection what is the situation with regard to Venus and Mercury?

$$\frac{b_{VE}}{b_{ME}} = \frac{108.2064(km * 10^6)}{56.6717(km * 10^6)} = 1.90936 \qquad \frac{4}{\pi} * \frac{3}{2} = \frac{6}{\pi} = 1.90985..$$

$$\frac{1.90985..}{1.90936} = 1.00027$$

And here we have it! With a deviation of only 0.027%, i.e. an accuracy enhanced to about a power of ten, the interval of the fifth in relation to the corresponding ratio in the outer square is realized by the two inmost planets. If we make Mercury's semi-minor axis the basic circle, then the semi-minor axis of Venus corresponds to a square which is $1\frac{1}{2}$ times the one circumscribing the basic circle. It is easy to construct this square. Finding that the proportions which arise are those of very simple geometrical forms, we shall here—and in what follows—assume that in constructing these ratios only those tools are needed which people have always used for geometry: compasses and ruler. In this way a square with an area of 3/2 can easily be obtained via the hexagon. A hexagon (pale grey, dotted, in Fig. 13.2) is inscribed within the 2:1 circle (thin line). The ratio of the incircle of the hexagon to the basic circle is 3:2 (see Fig. 2.3). The square around this incircle (thicker grey line) thus has the area $3/2 \star 4/\pi = 6/\pi$. So now let us add the ratios of Earth to this depiction. To do this we shall initially use the relatively inexact 3:2 ratio to Mars (deviation 1.15%) which corresponds to a similarly approximate 8:3 ratio of Earth and Mercury. We shall see later which construction is actually the basis for the orbit of Earth, or for its semi-minor axis, and why the value arising here is 1.5172 and not 1.5.

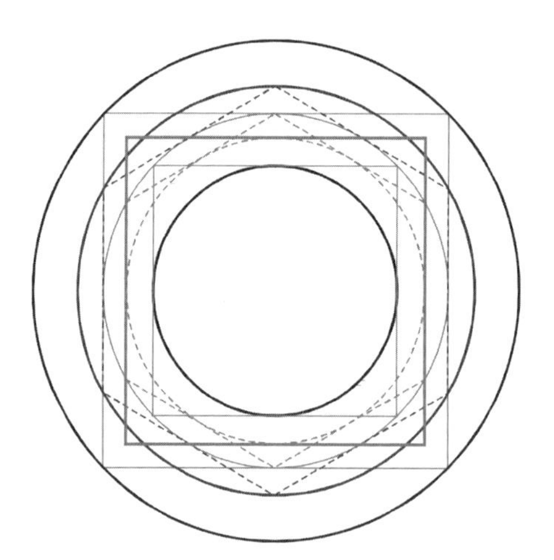

| | |
|---|---|
| Outer circle: | Mars |
| Middle circle: | Earth |
| Square: | Venus |
| Inner circle: | Mercury |

$$\frac{b_{EA}}{b_{ME}} \approx 2 * \frac{4}{3} = \frac{8}{3}$$

(basic circle-square-circle-hexagon-circle)

Deviation at 4:1 ratio: 0.114%

Mars/Venus ratio =
Ma/Me: (Ve/Me) $\cong 4/(6/\pi) = 2/3\ \pi$
(Deviation 0.140%)

**Fig. 13.2** *Initial geometrical approximation of the ratios of the semi-minor axes in the inner planetary system*

We now turn to the connection between the inner and outer region. The interval between Saturn and Mars was about 6.25 or 25:4. The exact figure is 6.281, i.e. there is a deviation of 0.49% which, in comparison with the differences just mentioned, must be seen as being too high. So in the way which has proved useful let us construct the ratio of the semi-minor axes of the two planets in order to find out whether this provides us with a better result:

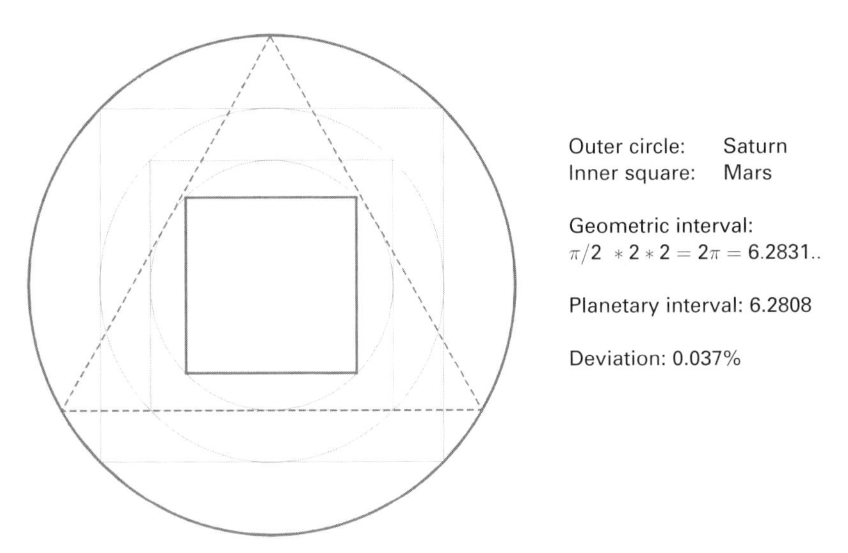


**Fig. 13.3** *Geometrical approximation of the ratio of the semi-minor axes of Saturn and Mars*

The range between closeness to and distance from the Sun is here also spanned with sufficient accuracy by the interlocking of squares (or the corresponding, dotted, triangle) and circles. From this we glean that the ratio which spans the middle and the inner or, respectively, the outer region is not 25:1 but $8\pi = 25.1327...$ The real intervals of Saturn/Mercury and Pluto/Mars are 25.1519 and 25.2079 (deviations of 0.076 and 0.299%). According to this, at least seven of the nine planets (since Earth and Jupiter are as yet unaccounted for) are included in the structure of the simplest possible geometrical constructions. This is indeed truly remarkable. As we shall see, the probability of such an order coming about purely by chance is much lower even than in the case of the harmonic probabilities of the velocity intervals as shown in Chapter 4. The figures certainly show this. Were we to find them traced in the sand we would realize that the natural forces known to us could not at all easily have depicted them thus. 'Friends,' said the Greek philosopher Aristippus when he espied those geometrical figures in the sand, 'we may permit ourselves to entertain some hope.'

**Fig. 13.4** *William Blake, 'The Ancient of Days', illustration in* Europe, a Prophecy, *1794*

*'To-day there is a wide measure of agreement, which on the physical side of science approaches almost to unanimity, that the stream of knowledge is heading towards a non-mechanical reality; the universe begins to look more like a great thought than like a great machine. Mind no longer appears as an accidental intruder into the realm of matter; we are beginning to suspect that we ought rather to hail it as the creator and governor of the realm of matter . . . We discover that the universe shews evidence of a designing or controlling power that has something in common with our own individual minds—not, so far as we have discovered, emotion, morality, or aesthetic appreciation, but the tendency to think in the way which, for want of a better word, we describe as mathematical.'*

Sir James Jeans (1877–1946)[1]

However, there is still something lacking in the three geometrical depictions (and I do not here mean the two planets, Earth and Jupiter, which remain to be discussed). Looking closely, we find that in Fig. 13.2 Mars is represented by a circle and in Fig. 13.3 by a square. With Saturn the reverse applies. Arithmetically this is not a problem. For example one can calculate the ratio of Neptune/Mars to $2\pi^2$ (Ne/Sa⋆Sa/Ma = $\pi$⋆$2\pi$) or that of Saturn/Venus to $4/3\ \pi^2$ (Sa/Ma⋆Ma/Ve = $2\pi$⋆$2/3\ \pi$). However, in order to bring all the various realms together in a single construction it would be useful to be able to do something in the geometry of the heavens which many mathematicians had been attempting for over two thousand years until in 1882 Ferdinand von Lindemann proved it to be impossible, namely to square the circle by using only compasses and ruler. One might initially be forgiven for asking why people had the ambition in the first place to make something completely round into a right-angled shape of the same area. And why did they restrict themselves in the matter of which tools they might use for the purpose? As an archetypal geometrical form the circle may be seen as an image of the prime cause. Perhaps—either consciously or unconsciously—they wanted to prove mathematically that all variety had arisen out of a single first principle by using only the possibilities arising from that. Mathematicians call the number π a transcendental number, i.e. one that transcends whatever is rationally comprehensible. In a certain sense they perhaps wanted to inject the one force which had generated all the other forms into the world that has come to be symbolized by the square.

Whatever the case may be, the planets ought to be able to have command over something which is beyond the abilities of the human mind. Or they ought to be at least as clever as the human mind, for in searching for the square of the circle one discovered procedures which made good or indeed very close approximations possible. As early on as 1700 BC the Egyptians had discovered a construction which enabled the arithmetical value

$$\pi \cong \frac{256}{81} = 3.160..$$

to be worked out. The architecture of the pyramid of Cheops, built even earlier, is said to be founded on a quadrature which is based on the very much more accurate correspondence of 22/7 = 3.1428.. with π (deviation 0.04%).[2] In the heavens which surround Earth, the planet of human beings, it ought to be possible to approach an even greater degree of precision in order to enable not only the Earth but also all its companions to circle around the Sun for a sufficiently long stretch of time. In addition one might surmise that once the square and the hexagon had come into play the number of the human being, the number five, would also be required. And it did indeed appear, above all in connection with its neighbouring planet, the goddess of love. So let us recall the ratio of the semi-minor axes of Earth and Venus discussed towards the end of Chapter 1. Its value is close to that of the silver section: 1.3819... Since it is not possible to square the circle, let us try to make use of this figure that also plays a part in the construction of the pentagon which can be attained by using compasses (see Appendix 1.2).

$$1.38196..^2 = \left(\frac{5-\sqrt{5}}{2}\right)^2 = \frac{5}{2}\left(3-\sqrt{5}\right) = 1.90983..$$

$$\frac{b_{VE}}{b_{ME}} = 1.90935 \qquad \frac{1.90983}{1.90935} = 1.00025$$

This slays two birds with one stone. On the one hand it shows that the ratio of the semi-minor axes of Venus and Mercury depicts the squaring of that of Earth and Venus, and on the other that the former can be expressed not only by the value 6/π, but also with almost the same deviation by the square of the silver section. The ratio of Earth to Mercury, therefore, is:

$$\frac{b_{EA}}{b_{ME}} = 2.63936 \cong 1.38196..^3 = 2.63932.. \cong \left(\frac{6}{\pi}\right)^{\frac{3}{2}} = 2.63938..$$

The deviations here are a mere 0.0015 or 0.0008%. Although the right-hand part of the equation is even closer to being exact, it does not help the construction since this would necessitate taking the square root of 6/π by using compasses which, according to Lindemann, cannot be done.* So let us stay with the silver section. We came across it in Chapter 2, in the discussion of the number five's 'rights of citizenship' in the world of music. The semi-divine proportion—if one may put it like this—made it possible for the major third to shine out as an intermediate step in the depiction of the harmonic intervals as area ratios of simple geometrical figures (see Fig. 2.7 and Appendix 1.6). Before the interval 5:4 was able to emerge from the sphere of the irrational through the combination of a pentagon and two decagon circles, the first of the two latter gave rise to an area with a content of 1.38196... And this construction fits the Earth/Venus relationship with 0.027% accuracy. When executed twice there is a deviation of 0.025% in the Venus/Mercury interval; and when executed three times, that of Earth and Mercury has a degree of accuracy that exceeds the above by a good power of ten. Here is the result in a geometrical depiction:

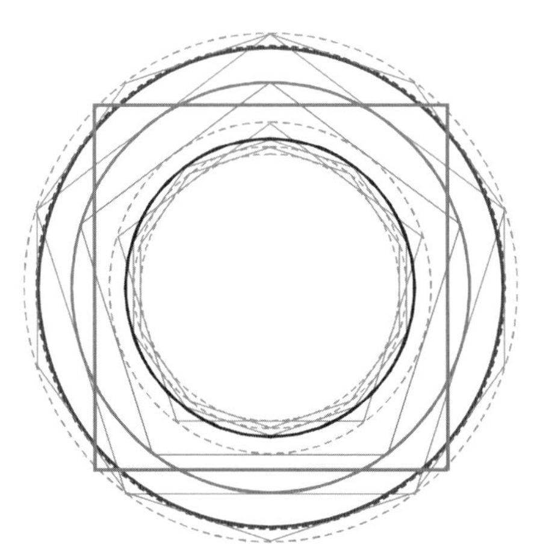

Outer black circle: Earth
(for comparison:
immediately adjacent dotted circle
using hexagon construction)

Grey: Venus
(a) Square: using hexagon construction
(b) Circle: via pentagon construction

Difference between procedures (a) and (b):
$6/\pi/1.38196..^2 = 1.0000153$
or 0.00153% deviation

Inner black circle: Mercury

**Fig. 13.5** *Geometrical approximation of the semi-minor axes of Mercury, Earth and Venus as a squaring of the circle*

---

* The square root of π as the side of a square would lead to it having an area of π. Starting from a circle with r = 1, this would realize the squaring of the circle.

Two decagons are inscribed inside the basic Mercury circle. (One could alter the sequence, but this way round is perhaps the clearest arrangement.) Beginning at the inmost decagon circle, two pentagons then lead to the grey circle with its area of 1.90983.., i.e. the squared value of the silver section. This shows a difference of 0.00153% from the area of the square gained in Fig. 13.2 via the hexagon. Related to a length, for example the distance of Earth from the Sun of 149.6 million km, this amounts to a distance of 2289 km, i.e. just under one fifth of Earth's diameter; related to the length of a football pitch it would be 1.5 millimetres. So we can talk of a nicely accurate approximation to the solution of the old conundrum of squaring the circle.*

The squaring of the circle obtained with the help of the silver section and the value $6/\pi$ will later enable us to bring together in an overall picture the constructions, shown above, made up of circles and squares. However, the geometrical construction of our planetary system is so astonishing and varied that it will initially be worth our while to continue our investigation with the help of the ratios arising from the pentagon and decagon. We have seen that the ratio of the Venus/Mercury semi-minor axes corresponds to the square of that of Earth/Venus. (In what now follows, mention of a planet or its orbit should be taken as referring to the semi-minor axis unless otherwise stated.) By now looking at the next planet, Mars, we shall arrive at an initial explanation as to why the semi-minor axes are so relevant to the spatial architecture of the solar system. We shall begin by relating its perihelion distance to the semi-minor axis of its inner neighbour:

$$\frac{Ma_{pe}}{Ea_b} = \frac{206.65692\ (km * 10^6)}{149.57718\ (km * 10^6)} = 1.38161 \qquad \frac{1.381966..}{1.38161} = 1.00026$$

Once again the silver section emerges very accurately. From this we can deduce that the Mars-perihelion/Venus interval is equal to the Venus/Mercury interval. Thus the orbit of Venus represents the mean proportional (which is the same as the geometric mean) between the closest point of Mars and of the orbit of Mercury to the Sun. The deviation in this case is only 0.013%. Furthermore Earth's orbit forms the mean proportional of the Mars perihelion and Venus.

---

* The possibility of this geometrical approximation is known at least in principle. It can also be related to the following relationship between $\pi$ and the value of the golden section ($g = 1.618..$): $\pi \cong 6/5{\star}g^2 (g^2 = 2.618.. = 4 - 1.3819..)$. In Schwentek, p. 62, you will find an approximate solution for constructing $\pi/2$.

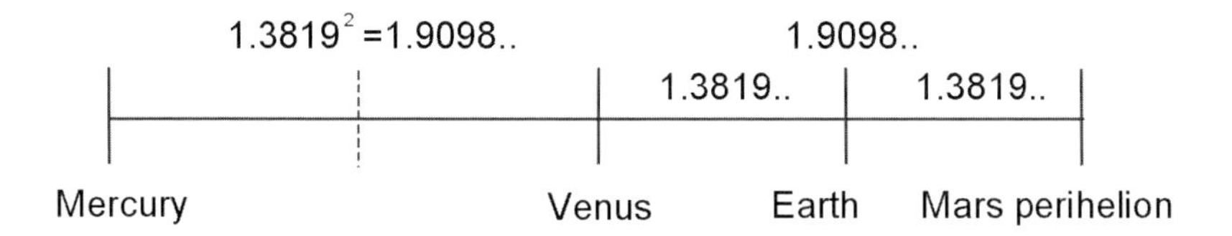


Let us remind ourselves (as explained in connection with Fig. 6.20) that the semi-minor axis b is the geometric mean between the aphelion and the perihelion distance. This—and thus b—can be simply constructed in a triangle with Euclid's Height Theorem (see Appendix 1.4). So with the help of a composite depiction of the intervals of the inner planetary system we can now see how the semi-minor axis of Mars fits in with what has thus far been a well-proportioned ordering:

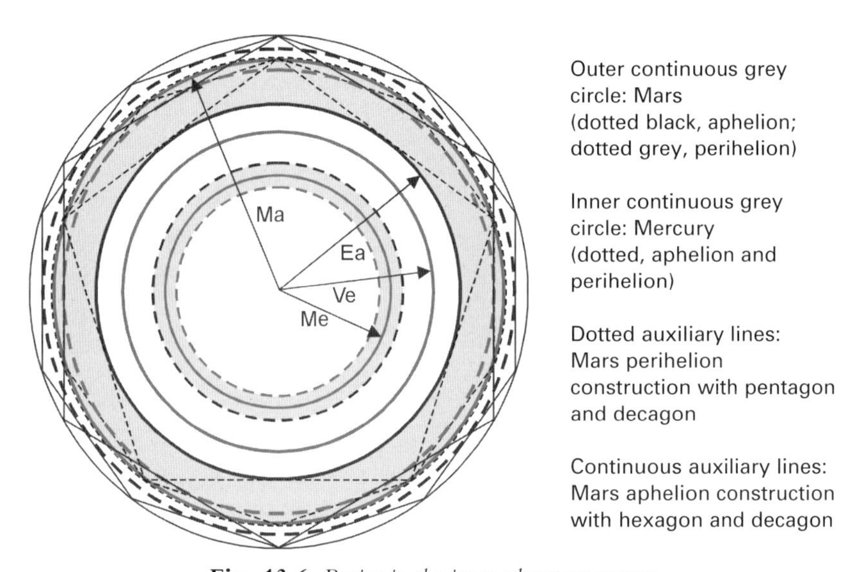


**Fig. 13.6** *Ratios in the inner planetary system*

As mentioned, the Mars-perihelion circle arises out of the pentagon and decagon by means of the silver section construction. If we now draw a hexagon around this circle, add the circumcircle (outer auxiliary lines) and then inscribe a decagon, we arrive at the black dotted aphelion circle. By means of the construction of the mean proportional, shown in Appendix 1.4, Fig. 14.5, the grey circle of the semi-minor axis can then arise (it almost coincides with the dotted auxiliary circle). Correspondence with the actual ratios is getting better and better:

$$\frac{\text{Hexagon ratio}}{\text{Decagon ratio}} = \frac{1.33333..}{1.105572..} = 1.20601..; \quad \frac{Ma_{ap}}{Ma_{pe}} = \frac{249.2242}{206.6569} =$$

$$1.20598; \quad \frac{1.20601}{1.20598} = 1.000026$$

In another sequence one reaches the Mars aphelion circle from the Earth circle via a pentagon and a hexagon, reduced by two decagons. The area ratio which arises is 1.38196.. ⋆ 1.20601.. = 1.66666... This gives us the construction which leads to the musical interval of the major sixth, i.e. from note C to the 'natural' chamber note A (see Fig. 14.8). The planetary ratio is 1.66619; i.e. here once again the difference is minimal (0.029%). In addition the Mars/Earth interval (semi-minor axes) can also be calculated from the above construction since the semi-minor axis b as a geometric mean is the square root of the aphelion distance multiplied by the perihelion distance.

$$\frac{Ma_b}{Ea_b} \cong \frac{1.66666..}{\sqrt{1.20601..}} = 1.38196.. * \sqrt{1.20601..} = 1.3196.. * 1.09818..$$

$$= 1.51765..$$

Once again the actual value (1.51724) deviates minimally from this (by 0.03%). This figure is reached even more accurately via the Mercury-aphelion/perihelion ratio of 1.51770. (The two almost equal proportions are shaded grey in Fig. 13.6.) So according to this all the spatial ratios of the inner planetary system can be built up very accurately from the silver section (the pentagon reduced by the decagon) and the construction of the geometric mean (the hexagon reduced by the decagon); or, respectively, they can be calculated from the two number values 1.38196.. and 1.09818... (The only exceptions are the aphelion and perihelion distances of Earth and Venus which because of the small eccentricities differ only very slightly from the measure of the semi-minor axes.) If the first value is A and the second B, the ratios can be depicted as follows:

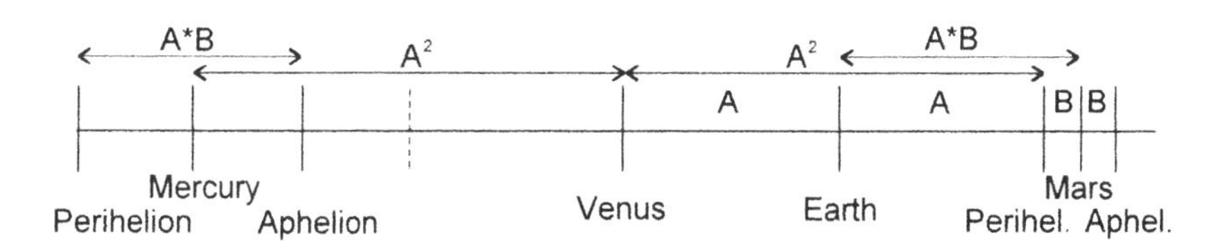

The reader will presumably agree that this result is startling. The inner planetary system represents a geometrical totality bound together by the pentagon, the hexagon and the decagon in which the two larger eccentricities of Mercury and Mars (or the extreme distances resulting from these) are organically incorporated. The semi-minor axis forms the mean proportional of the extreme values in the orbit of a planet, and this special ratio evidently also plays an important part in the intervals between the various planets. Thus the spatial structure of the whole can be logically reconstructed by using the ratios of the semi-minor axes.

So we can now ask whether it is also possible to find a geometrical foundation for the combination of intervals in this way, or at least for those of the semi-minor axes. In this connection let us look at the Platonic solid which, according to former interpretations, was seen to symbolize the celestial substance or ether.

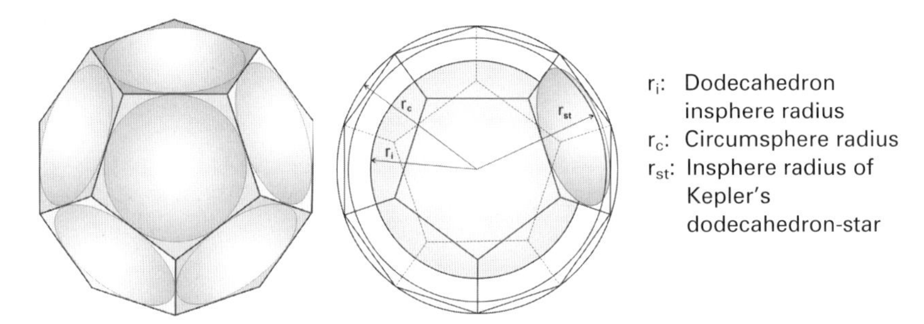


**Fig. 13.7** *Dodecahedron with circum- and in-spheres*

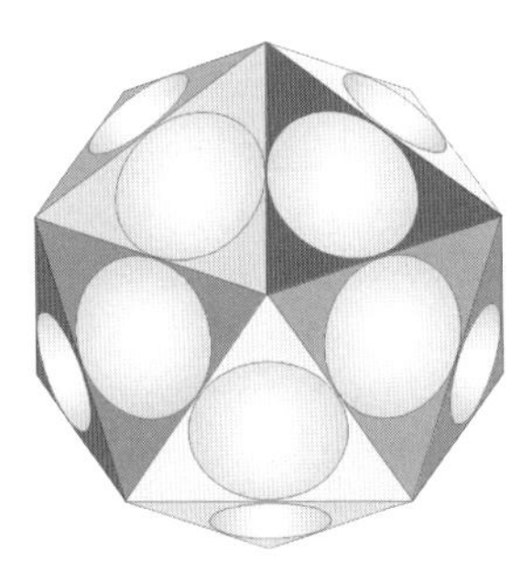

**Fig. 13.8** *Icosahedron with the insphere of the star-figure*

As we have hitherto found that the ratios of the distances of the planets can best be approached via ratios of areas we shall here, too, pay special attention to these. The sphere which in Fig. 13.7 shows between the edges of the dodecahedron represents the insphere of the dodecahedron-star to which Johannes Kepler gave the name 'Hedgehog'. The area of this sphere arising in the side-view has the ratio of our value A of the silver section in relation to the insphere of the basic dodecahedron (for the calculation, see Appendix 2.4). The interval B arises directly if one relates the surface area of the dodecahedron to that of an icosahedron which circumscribes the same insphere. Furthermore, the areas of the insphere of the 'Hedgehog' and those of Kepler's second stellar solid, the icosahedron-star, have the ratio $B^2 = 1.20601\ldots$ In this figure the circumsphere of the icosahedron has the ratio of the silver section to the insphere of the star. Out of this arises the value of 1.20601.. for the areas of the inspheres of Kepler's two stellar solids (for the calculation see Appendix 2.4 again). But we shall soon see that B can also be found indirectly in the dodecahedron-star.

In Fig. 12.20 we met the two possible side-views of the 'Hedgehog', one as a combination of the pentagram and a continuous 10-pointed star-figure, and the other as a hexagram. If we imagine a combination of these two possibilities, we discover a circle—corresponding to the area of the sphere surrounding the 'Hedgehog'—within which are inscribed a 5-pointed, a 6-pointed and a 10-pointed star-figure. Thus, united here in a unique geometrical figure—the dodecahedron-star—we have those stars which are formed from the polygons that permit the above geometrical constructions of the ratios in the inner planetary system. In this depiction we can plot the circles given by the star-figures and calculate the ratios of the corresponding areas.

**Fig. 13.9** *Combination of the side-views of the dodecahedron-star. The little 'Hedgehog' shows the derivation of the 5-pointed and 10-pointed star-figures, and a rotation of the star around 18° takes the side-view over into a hexagram. The two (inner) continuous blue circles arise from the side-views of the inspheres of the dodecahedron, on which the drawing is based, and of the dodecahedron-star. The red (middle) circle is determined by the hexagram (the circle which surrounds the hexagon inside it), and the black (outer) circles by the points of intersection of the lines of the 10-pointed star-figure.* → Plate 14

The inner (continuous) blue circle—the side-view of the incircle of the dodecahedron—is the basic circle. Based on this, the following area ratios arise:

**Table: 13.1** *Area ratios in the dodecahedron-star in comparison with the intervals in the inner planetary system (colours refer to Plate 14; A = 1.38196.. , B = 1.09818..)*

| | Geometric ratio | As a combination of | Planetary ratio of the semi-minor axes | | Deviation (%) |
|---|---|---|---|---|---|
| Basic circle | 1 | | — | — | — |
| Outer blue circle | 1.38196.. | A | Earth/Venus | 1.38233 | 0.0265 |
| | | | Mars perihelion/Earth | 1.38161 | 0.0260 |
| Red circle | 1.66666.. | $A \star B^2$ | Mars aphelion/ Earth | 1.66619 | 0.0285 |
| Inner black circle | 1.90983.. | $A^2$ | Venus/Mercury | 1.90935 | 0.0250 |
| Outer black circle | 2.63932.. | $A^3$ | Earth/Mercury | 2.63936 | 0.0015 |

The Mars/Earth ratio or the value B = 1.09818.. could, as before, be obtained by means of the construction of the mean proportional of the middle (red) circle and the circle inside it (the outer blue one) so that all the above ratios of the inner planetary system are established in Johannes Kepler's favourite star-figure in a conclusive, i.e. hardly otherwise possible manner. Earth represents the basic circle since all its ratios with the other inner planets are shown to be taking their departure from it. So one can say that Earth is surrounded by the celestial ether of the dodecahedron or by its stellar extension and that its semi-minor axis represents the starting point for all the ratios. Kepler's notion that the Platonic solids, or his 'Hedgehog' star, are closely bound up with the ratios of the solar system—the measure of all things herein being the Earth's orbit—is thus borne out in a most beautiful albeit metamorphosed manner.

Before our final approach to the mystery of the solar system, let us now enjoy a brief pause during which another aspect of the harmony of the spheres may be heard. A regrettably little-known poet here expresses in verse something ever and again surmised by human beings, namely that the human soul, too, is most profoundly related to the cosmos.

*soundlessly o music of the spheres*
*open up the cosmic inner space*
*let the distances bring forth themselves*
*for our longing for our dream*

*emanations from arcturus*
*lightly skim all depths all heights*
*greater yet than light and rise:*
*inaudible world symphony*

*soundlessly o music of the spheres*
*open up the cosmic inner space*
*let bring forth the stars themselves*
*for our vision for our dream*

David Colombara
'Solitude'[3]

Despite all the aspects we have thus far discovered concerning the marvellous spatial ordering in our solar system there is still more that needs to be completed if we are to unravel it fully. In Chapter 12 we saw that the 12-pointed star-figure finally brings to perfection the symphony arising from the orbital periods and the rotations, giving it a deeper meaning in the overall view. So we must still ask whether the Number of Perfection also lies hidden in the intervals of the semi-minor axes. The kingly planet Jupiter will be our point of departure in this quest. In the—so to speak geocentric—depiction of the inner planetary system created with the help of Kepler's dodecahedron-star (Fig. 13.9) we saw that the most appropriate procedure was to relate the ratios of the semi-minor axes to Earth (as a basic circle). So let us now look at the corresponding ratio of Jupiter and Earth and, by way of comparison, that of the semi-major axes:

$$\frac{Ju_b}{Ea_b} = \frac{777.416\ (km * 10^6)}{149.577\ (km * 10^6)} = 5.19742$$

$$\frac{Ju_a}{Ea_a} = \frac{778.322\ (km * 10^6)}{149.598\ (km * 10^6)} = 5.20276$$

The difference between the two values is a mere 0.1%. And yet it has evidently been this minimal difference which has up to now concealed the correspondence between the Jupiter/Earth relationship and a very simple geometric construction. This is because the 2 after the decimal point in the interval of the semi-major axes hides the fact that 5.19615.. is the operative number. This is the decimal expression for three times the square root of three or, expressed differently, three to the power of one and a half. As a ratio this value occurs, for instance, between the volume of two cubes of which the second is inscribed in the insphere of the first. The same applies to the octahedron. However, since the distances in the planetary system are not regulated two dimensions higher but only one, we shall stay with the ratios of areas. The value $3^{\frac{3}{2}}$ can then be depicted as shown on next page (Fig. 13.10).

$$\frac{F_{tr}}{F_{sq}} = \frac{\pi}{2} * 2 * \frac{3^{\frac{3}{2}}}{\pi} = 3^{\frac{3}{2}} = 5.19615..$$

(To calculate the triangle see Appendix 1.3.)

The difference between this and the interval of the semi-minor axes is only 0.024% which means we have found an excellent correspondence. Since this is perhaps even easier to see if one imagines the planetary orbits as circles, we should mention in addition that with the help of a somewhat

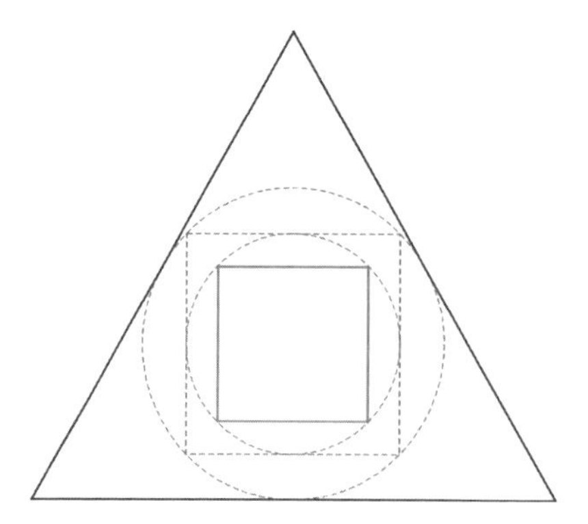

**Fig. 13.10** *Geometrical approximation of the ratios of the semi-minor axes of Jupiter (triangle) and Earth (square)*

*'Philosophy is written in that great book—I mean the universe—that forever stands open before our eyes, but you cannot read it until you have first learned to understand the language and recognize the symbols in which it is written. It is written in the language of mathematics and its symbols are triangles, circles, and other geometrical figures without which one does not understand a word, without which one wanders through a dark labyrinth in vain.'*

Galileo Galilei[4]

more complicated construction the interval $3^{\frac{3}{2}}$ can also be depicted as a ratio between two circles.* Furthermore, studying other planetary intervals one finds that the Neptune/Venus ratio can also be depicted by means of a sequence of figures that is in principle the same:

$$\frac{Ne_b}{Ve_b} = \frac{4498.000\ (km * 10^6)}{108.206\ (km * 10^6)} = 41.56872 \quad 8 * 3^{\frac{3}{2}} = 41.56921..$$

$$\frac{41.56921}{41.56872} = 1.000012$$

The Neptune/Venus ratio is thus almost eight times that of Jupiter and Earth. It can be reached via Fig. 13.10 by means of the additional circumscription of one of the dotted circles by a 4:1 circle (via a triangle) and a 2:1 circle (via a square). We shall see later that with the help of the Neptune/Venus ratio in the planetary system the direct transformation of a circle into an equilateral triangle (in addition to the squaring of the circle described above) can be achieved. But let us first follow up the indication pointing to the number twelve hidden in the ratios, for this is what one might surmise from the two appearances of a triangle in connection with a square. To do this we have drawn the

* Earth would be the basic circle which is first circumscribed by a square and then by the 2:1 circle. Then comes an equilateral triangle which is transformed into a square of the same area. This can be done with the aid of the geometric mean because the side s of this square is: s = square root(h*a/2) with h = height, a = side of the triangle. The square arrived at by this means is circumscribed by a circle which has the area ratio of 5.19615.. to the basic circle, thus corresponding to Jupiter's semi-minor axis.

constructions shown at the beginning of this chapter in combination with that given in Fig. 13.10:

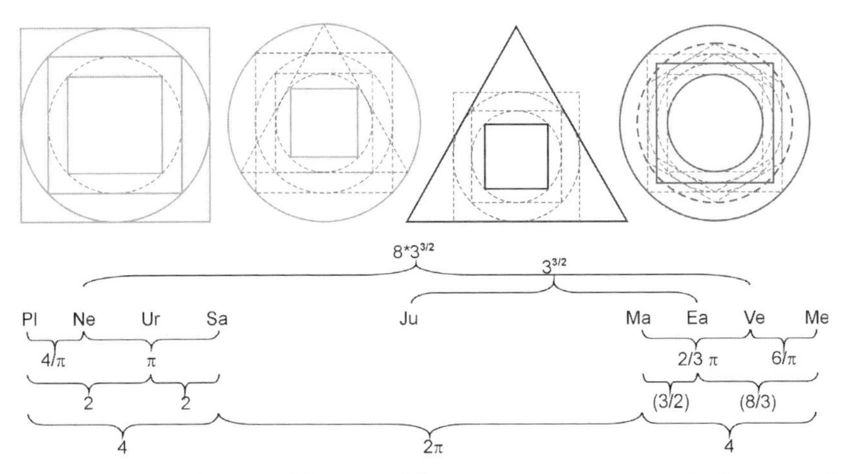


**Fig. 13.11** *Combination of the ratios of the semi-minor axes, construction by means of circle, square and triangle. From left to right: outer region, Saturn/Mars, Jupiter/Earth, inner region*

The values of all the relevant intervals are shown below the drawings; 3/2 and 8/3 are bracketed because the ratios Mars/Earth and Earth/Mercury are only rather approximate. Surprisingly, all the other ratios shown also appear in the area ratios of the 12-pointed star-figure, as we can see from Fig. 13.12 (calculation in Appendix 2.3). Here the inner region of the continuous 12-pointed star-figure is shown which by means of its rays brings about the derived star-figures made up of 3 squares, 4 triangles and 2 hexagons.

The ratios in the inner planetary system are shown by the continuous grey lines on the left side and those of the outer system by the ones on the right side (light grey). Venus, for example, is shown by the dodecagon within the 2:1 circle; this has the same area as that of the square obtained via the hexagon construction in the right-hand drawing in Fig. 13.11: $2/(\pi/3) = 6/\pi$. The Saturn/Mars value $\cong 2\pi$ turns up here as π—as it were in the lower octave, but it could just as easily be derived via a circumscribed square and circle. Correspondingly, the existing ratio of triangle D and square Q ($= 3^{\frac{3}{2}}/4$) can be transformed into that of Jupiter and Earth—by inscribing two further squares into square Q (see Fig. 13.11, second figure from right).

So—if we regard the architecture built on the ratios of pentagon,

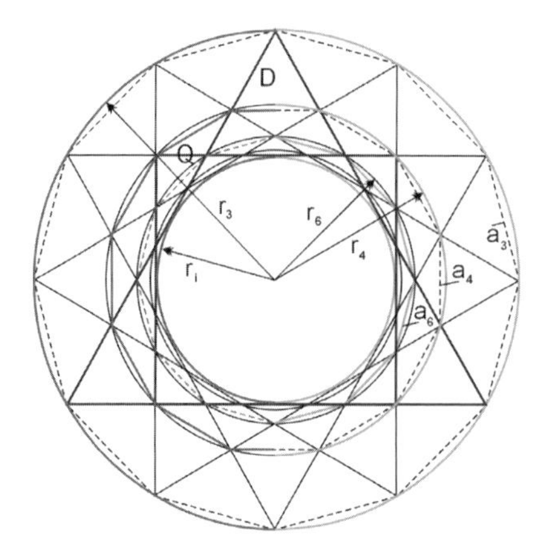


The indices stand for the polygons formed by the rays of the 12-pointed star-figure, i.e. 3, triangle; 4, square; 6, hexagon. r denotes the radii of the circumcircles (except for ri), and a the sides of the dodecagon. F denotes the relevant areas of the circles and A those of the inscribed dodecagons. From this the following arises:

$$\frac{A_6}{F_i} = \frac{4}{\pi}\ (Pl/Ne) \quad \frac{F_4}{A_6} = \frac{\pi}{2}\ (Ne/Ur)$$

$$\frac{F_4}{F_i} = 2\ (Ur/Sa) \quad \frac{F_3}{F_i} = 4\ (Pl/Sa,\ Ma/Me)$$

$$\frac{A_4}{F_i} = \frac{6}{\pi}\ (Ve/Me) \quad \frac{F_3}{A_4} = \frac{2}{3}\pi\ (\mathrm{Ma/Ve})$$

$$\frac{A_4}{A_6} = \frac{3}{2}\ (Ma/Ea) \quad \frac{F_3}{A_6} * 2 = 2\pi\ (Sa/Ma)$$

**Fig. 13.12** *The area ratios in the 12-pointed star-figure which correspond to the intervals of the semi-minor axes*

hexagon and decagon as the side view (of the inner region)—we now have before us the front elevation, so to speak, of the building almost in its full glory. The geometrical origin of its ratios arising out of a unity, the star-figure of the Number of Perfection, is revealed. It is hard to imagine a more excellent correspondence with the story recounted to us by the movements of the celestial bodies. The interplay of the planets also, however, included the Sun in a very significant role. So the final question to be asked is whether the central star of our solar system might not also be involved as the source of the physical emergence of the planets in the order which is blazing forth ever more wondrously. In order to answer this question we shall first have to follow a train of thought involving physics. This can also provide yet another indication (not a logical explanation, which I am unable to provide) as to why the dimensions of the semi-minor axes are so crucial in the structure of the planetary system.

The latter could be linked to the angular momentum of the elliptical orbits. Angular momentum is defined as the product of mass, velocity and distance of an orbiting body (or, respectively, the mass, velocity and radius of a rotating body). This momentum plays a considerable role, for example, in the theories concerning the coming into being of a solar system. As mentioned in Chapter 5, young suns rotate very fast. The explanation for this is that according to a fundamental physical law the angular momentum always remains constant (at least within a closed system). A sun emerges from a rotating cloud of gas which condenses more and more. So in accordance with the conservation of angular momentum the rotational velocity must be increasing. The best example of this, shown in all books on this subject, is that of a pirouetting ice skater who automatically spins ever more rapidly the closer the arms are held to the body. Our Sun, however, no longer rotates as rapidly as does a young star. So at some later point during the phase of its coming into being a large part of the angular momentum must have been transferred to the planets or to their original material constituents; or it may have been removed from the solar system altogether, since now only 2 per cent* of the momentum of the total system can be found in the Sun. According to current theories, the transfer came about through a transport of matter outwards from the Sun which takes place particularly efficiently in connection with magnetic fields.[5] The conclusion one can draw from this is that the loss of matter and the accompanying reduction of angular momentum would have continued until the Sun reached more or less its current size. So let us now relate this value to the structure of the semi-minor axes (for example to that of Mercury, the planet closest to the Sun) even though we might be tempted to ask what the one has to do with the other.

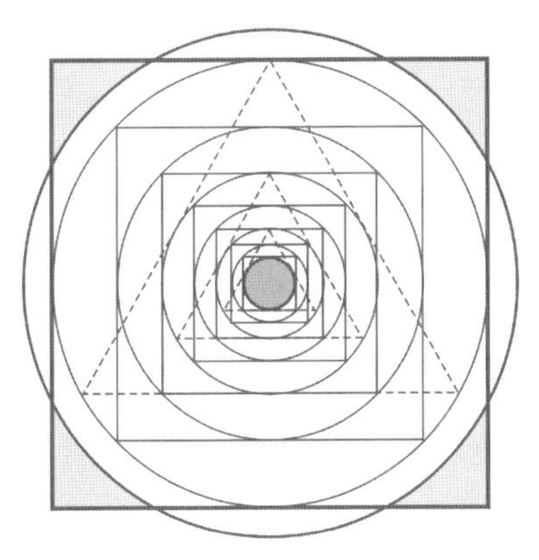

**Fig. 13.13** *Geometrical approximation to the ratio of the semi-minor axis of Mercury and the solar radius*

$$\frac{b_{ME}}{\text{Solar radius}} = \frac{56.6717 * 10^6\ km}{0.69626 * 10^6\ km} = 81.3945$$

$$2^6 * \frac{4}{\pi} = \frac{2^8}{\pi} = 81.4873..$$

$$\frac{81.4873..}{81.3945} = 1.00114$$

corresponding to 0.114% deviation

$$\frac{b_{PLUTO}}{\text{Solar radius}} = \frac{5720.8}{0.69626} = 8216.48 \cong 2^{13}$$

(Deviation: 0.299%)

Taking our departure from the solar circle, having circumscribed squares and circles, we arrive with the seventh square (outer dark line) exactly (except for almost one thousandth) at the semi-minor axis of

---

* Data in the literature vary somewhat, with some sources mentioning *c.* 0.5%

Mercury. The slightly thinner line represents the circle with the same area that was the point of departure for the ratios in which Mercury was involved (Figs 13.2 and 13.5). The grey shading in the four corner sectors symbolizes the necessary squaring of the circle. Of course the first six squares in the construction could be replaced by three triangles. But we shall stay with the squares. If we were now to circumscribe the Mercury square by more circles and squares, the new seventh circle would correspond to the semi-minor axis of Pluto (the sixth to that of Uranus, the fifth to that of Saturn). In this way the total measure of the planetary system from Sun to Pluto amounts to 13 interlocking circles and squares or, mathematically, $2^{13}$.

All this makes it obvious that our central star is indeed included among the planetary ratios. And it is therefore perfectly possible to imagine that the foundation stone for the structure we have been discussing was already laid when the angular momentum of the young Sun was transferred to the planets or to their primordial substance, because the semi-minor axis b, among other parameters, is included in the calculation of the angular momentum of an elliptical orbit.* So in accordance with this a specific arrangement of the b-axes corresponds to the distribution of angular momentum in the planetary system. In this sense the part played by the semi-minor axes must be seen as fine tuning, since the momentum of an orbit is primarily determined by the mass and the mean distance from the Sun of the planet in question. Furthermore, by far the greatest part of the angular momentum results from the orbit, whereas in the case of the Sun it arises solely from the rotation. In this sense it is therefore not arbitrary to relate the radius of the Sun to the semi-minor axes of the planets, since in the Sun's case it is its size which determines the angular momentum. Quite the contrary: in this sense the diameter of the Sun has proved to be the geometrically and possibly even the physically fundamental basic measure of the harmonious order of the ratios—referred to elsewhere as the progression of distances. Having once arrived at this conclusion it need not take us long to exclaim: Of course! What other measure, if not that of the Sun, can we expect to provide the foundation for the ratios of the planetary system?

So let us now combine the above separate aspects in a composite depiction:

---

* Orbital angular momentum: $L = b * m * \sqrt{G * M/a}$ with m, planetary mass; G, constant of gravitation; M, solar mass; a, semi-major axis.

Fig. 13.14 *The signature of the celestial spheres* → Plate 15

We have allocated the common structural principle laid down in the 12-pointed star-figure to the Sun (inmost circle) as being the source of all things in our planetary system. For reasons of space it has been enlarged by the amount of two triangle-circles. Venus is shown as a dodecagon, which is one of its possibilities as demonstrated in Fig. 13.12. For Earth we chose the less accurate but on this scale more easily visible construction involving a hexagon (as in Fig. 13.2). In order to combine the various regions depicted in Fig. 13.11 in the construction, Mercury, Earth, Mars, Saturn and Pluto have been included both as circles and as squares of the same area. The quadratures needed are shown in each case by the white gaps at the edges and by dotted lines. Neptune appears as a circle and a triangle (green, see Plate 15); this accords with the direct transformation put forward in Fig. 13.10. The circle—related to Venus—can be calculated via multiplication of the intervening ratios as being $4/3\pi^3$ ($Ma/Ve \star Sa/Ma \star Ne/Sa = 2/3\ \pi \star 2\pi \star \pi$), whereby the deviation from the actual interval is 0.549%. The triangle stems from the already mentioned ratio $Ne/Ve = 8 \star 3^{\frac{3}{2}}$ (with a very much smaller difference of 0.001%); for the sake of clarity the auxiliary lines leading from the Venus square to the

Neptune triangle have not been included. The outer dark blue ring marks the region—bordering on the stars—of Pluto's orbital sphere from the semi-minor axis to the aphelion. That is to say that the whole of Pluto's sphere (from aphelion to perihelion) can also be approximated by the interval 1.666.. constructed via decagon, hexagon and pentagon. In other respects I hope this depiction speaks for itself on the basis of all that has gone before.

Circle, square and triangle, then, are the invisible symbols written in the sky which so impressively embody the mysterious order that exists in our solar system. Making our way through the world with open eyes we ever and again discover combinations of these three figures in eloquent symbolism. In scientific modes of thought people link them with the idea of a generally valid and rationally comprehensible cosmic order, while in the religious sense they connect them with the certainty of their faith in a creative force which has ordered the world in ways that will, in the end, make everything turn out for the best. Our celestial home bears these three archetypal images given to us by geometry—and united in the Number of Perfection—as a mark, a signature almost, but perhaps also as a challenge and a promise.

It remains for us to examine the evidence as to the probability of such structuring having come about by chance. Once again I shall put forward only basic ideas (details in Appendix 4.5). Only the simplest pattern can be taken into account in this appraisal, namely the one which involves a geometrical circumscription by squares and circles. In the opposite extreme one could formulate a specific condition for each interval. This would in each case be achievable in a good approximation, from which a small degree of probability would result, leading to a microscopically tiny but completely unrealistic overall probability. Related to the Sun as the basic circle (or its radius), the pattern of circle and square is covered with a very good probability by 5 of the 9 ratios under consideration. Using a standard procedure of probability calculation, one arrives at a chance of *c.* 1:40 million (i.e. this is the approximate probability that the depicted order of the semi-minor axes could have arisen purely by accident). Even if one generously reduces the already carefully chosen presuppositions in the calculation, a probability of about 1:140,000 remains. Whichever value one prefers, the probability always remains so small that it is all the more likely that some physical cause for the order of the semi-minor axes must exist or must once have existed.

So, finally, we should investigate the permanence of the ratios we have discovered, since of course even the measurements and ratios of the semi-minor axes also change over the course of time. However, the structure we have described and ascertained via the planetary data valid at present will not change noticeably, at least during the coming few thousand years. Over 5000 years into the future or the past the measure of the b-axes varies in the case of Saturn by a maximum of 0.3% and that of the other planets in some instances by a great deal less, so that here, too, there will only be minimal variations in the relevant intervals. Admittedly in relation to the lifespan of the solar system 10,000 years do not amount to much. Nevertheless, one would only know for certain how the order of the planets was changing over 100,000 or a million years if one were to carry out the same calculations with the orbital data valid when that time came, something which, as mentioned in Chapter 5, certain scientific research institutes would not find too difficult. Even with simple means one can reach at least a partial overview regarding the longer-term tendency of how the semi-minor axes are ordered. The calculations to ascertain them include the mean distances from the Sun and the eccentricities of the planets. The latter, in particular, are changeable, while the semi-major axes remain as good as stable. In keeping with this, the measure of fluctuation of the eccentricities is what determines the changes in the minor axes and thus also in the corresponding ratios. By a considerable margin Mercury has the widest variability, and this also has the greatest effect in the variations of the ratios between the planets. (Pluto is not included in this consideration.) We saw in Chapter 6 that the numerical eccentricity of Mercury varies by *c.* 0.1 and 0.3% over a period of 400 million years, and that at present it is 0.206%, i.e. around the average. As an example, one can calculate the intervals which arise on the basis of the highest limits of Mercury. Compared with today's values, this shows deviations of just under 1.7% (at an eccentricity of 0.1) or, respectively, 2.6%. This is sufficient to cause a considerable blurring of the conditions, but the means obtained nevertheless deviate from the current figures by only 0.47%.* (These averages are not identical with those that arise in the case of a Mercury eccentricity of exactly 0.2; in this case the values of the ratios differ from those of today by only 0.12%.) So there are a number of reasons for believing that the structure of the semi-minor axes at the present time is identical to, or comes very close to, showing an average of the conditions that fluctuate somewhat in the longer term.

All this will nevertheless have to remain as a preliminary estimation, so

---

* In the case of the Earth/Mercury ratio, for example, Mercury eccentricities of 0.1 and 0.3 yield values of 2.596 and 2.708 respectively, assuming all other conditions are equal. The mean here is 2.652. An eccentricity of exactly 0.2 leads to a ratio of 2.636. The current value is 2.639.

we shall abstain from going into the details of the calculations. But what we can claim for certain is that the ratios have been as they now are ever since human beings have consciously looked up to the stars, or at least ever since the day when records about this began. And thus they will remain for at least as long again. This goes not only for the signature of the celestial spheres as such but equally for the planetary harmonies in the velocities and for the movement forms depicting the conjunction relationships. If it were to turn out that in a few hundred millennia or millions of years from now nothing or little of the order at present prevailing in every aspect could be expected to remain, it would be all the more astonishing to find that during the very period when human beings are present (however long this may be) the planets move in the heavens in such an apparently very special way.

More than 200 years ago it was already possible to encapsulate the approximate spatial sequence of the orbits they follow in the formula of the Titius-Bode law which concerns the mean distances from the Sun. On the other hand one can only partially arrive at a mathematical formulation concerning the order of the semi-minor axes. Although this is sufficient for an estimation of the probabilities, it would be difficult to attribute all the existing ratios to a uniform formula or law. In imitation of Galileo's almost prophetic words one could—with regard to the structures discussed here—say that although the signature of the celestial spheres is written in the language of mathematics and although its symbols are triangles, circles and other geometrical figures, nonetheless this is not all there is to it. Those who view it solely with the eyes of arithmetic, or should we say of a law, may have thrown some light into the dark labyrinth while nevertheless continuing to wander about aimlessly in its entanglements. The order that is revealed in its spatial architecture is more like that of a musical composition or melody. There are relatively few geometrical possibilities (i.e. building blocks that obey specific laws) which can be likened to musical themes or the notes of a musical scale. In the case of music these are combined freely and creatively with the purpose of realizing the intentions of the composer. On the whole, however, scientifically inclined individuals do not believe that the phenomena of the universe—i.e. above all the cosmos, life and human beings—have been brought into existence intentionally by some creative power, be it God, a divine force or whatever else one might call that which is too great to be named. The scientific view is often thought to rest on the overarching perception of the various branches of research that one thing emerges from another, suns from the cosmos or its clouds, life from the inanimate, etc. Yet a more intensive wrestling with the various aspects of evolution tends rather to lead to the conclusion that in all the essential transitions which have taken place a complete riddle remains as to how, concretely, something that is more highly developed can have emerged from what was less complicated or, indeed, how something new can have arisen out of what did not exist before.

In the final analysis it is a matter of belief to infer that a more or less incomprehensible existence and what looks like a highly improbable ordering of things points to the work of a creative force. So of course the order in our planetary system, too, does not prove that God or anything similar exists. All that it reveals in its various aspects is that the arrangement of the celestial bodies is built up in a way that looks like the work of a conscious, creative force. To imagine this as something impersonal (whereby to think the opposite does not involve imagining something that resembles a human person) reduces it to the level of a natural law. In this case the most one could ascribe to the creator would be a role as the originator of the laws of nature, the one who has arranged them in such a way that they as it were inevitably bring forth the productions of nature. But surely, a force that is capable of doing this would also have the ability to handle these laws in a creative way.

Always at loggerheads with the strict adherence to the law of his religiously dogmatic contemporaries, Christ said he had come not to abolish the law but to fulfil it (Matthew 5:17). One might also say that although founded on the law, creative freedom begins where the law is at its end. If the laws of nature alone existed, even creations fashioned by human beings would be unlikely to come into existence. As the example of Kepler's stellar solids shows, such creations are anyway often more akin to discoveries of existing realities already present in some mysterious realm of being. In this sense I am grateful to that God who is too great to be named for permitting me to discover all that has been presented in this book.

# Appendix

## *1. Basic geometric and harmonic data*

### 1.1 The ellipse

The path of an ellipse passes between the circles drawn around its centre with the radii of the semi-major and the semi-minor axis. In addition, Fig. 14.1 shows a circle with the radius of the eccentricity e because, surprisingly, it has the same area as that of the ring formed by the two larger circles.

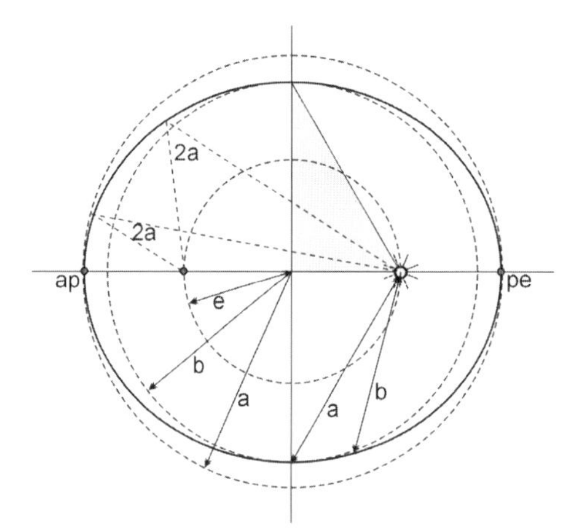


The ellipse in Fig. 14.1 has a numerical eccentricity $\epsilon$ of *c.* 0.5. The extreme values of $\epsilon$ = 0 arise in the case of a circle, those of $\epsilon$ = 1 in that of a parabola and those of $\epsilon$ > 1 in that of a hyperbola. All four figures represent conic sections.

**Fig. 14.1** *Ellipse with circles with the radii of the semi-major axis a, the semi-minor axis b and the linear eccentricity e (ap, aphelion; pe, perihelion)*

The form of an ellipse is defined by the ratio of the linear eccentricity e and the semi-major axis a. This value is designated as the numerical eccentricity $\epsilon$ (epsilon):

$$\epsilon = e/a \quad \textit{resp.} \quad e = \epsilon * a$$

The straight lines a, b and e give rise to the shaded right-angled triangle, so that the relationship between these three values (or $\epsilon$ after conversion) is given by the Theorem of Pythagoras:

$$a^2 = b^2 + e^2 \Rightarrow a = \sqrt{b^2 + e^2} \Rightarrow b = \sqrt{a^2 - e^2} \text{ or } b = a * \sqrt{1 - \epsilon^2} \text{ etc.}$$

In addition, perihelion and aphelion are defined by:

$$pe = a - e, \quad ap = a + e \Rightarrow pe + ap = 2\ a$$

So any two of the parameters indicated suffice for the exact definition of an ellipse. In astronomical tables the values $\epsilon$ and a are given to mark the elliptical planetary orbits (with additional parameters being necessary for an exact definition of their positions in space).

The semi-major axis a represents the average distance of a planet from the Sun. That this is indeed the case can be clearly shown by the exceedingly crafty construction of the ellipse that gardeners also use (indicated in Fig. 14.1 by the two sets of dotted lines). Hammer a post into each focus of the ellipse and attach a length of string between them which is longer than their separation. A stick is used to hold the string taught, and is then dragged round, scoring the ground, to create the ellipse (i.e. the gardener's flower bed). At every point along the resulting curve, the lines joining the stick to the two posts (focus points) have a constant sum. Astronomically, since this amounts to 2a both at the perihelion and the aphelion it must necessarily have the same value everywhere else. The mean distance of the stick from one of the posts or, respectively, of a planet from the Sun, is thus half of 2a, i.e. the semi-major axis.

Let us now see what a planetary orbit looks like over long periods of time. Depending on the planet, the line of apsides, i.e. the positions of perihelion and aphelion, turns full circle over a period of tens of thousands or hundreds of thousands of years. This is also termed the precession of the perihelion.

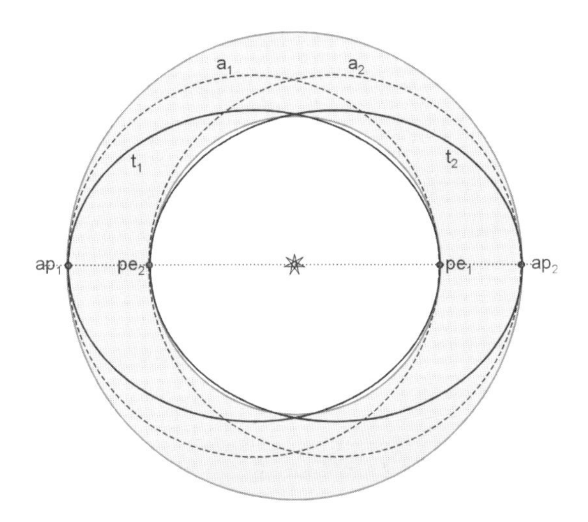


**Fig. 14.2** *The orbital sphere of a planet showing the ellipses at points in time $t_1$ and $t_2$ corresponding to a precession of the perihelion of 180°; $a_1$ and $a_2$ are the circles around the central points of the ellipses with the radius of the semi-major axis at those times. (For the sake of clarity, the ratio of the eccentricity to the shape of the ellipse is not entirely to scale.)*

During the course of aeons, aphelion and perihelion appear at first glance to describe perfect circles around the Sun. In doing so they mark out the space (shaded)—one might also call it the sphere—within which a planet can be located. In this respect the ancient Platonic idea[1]—which probably harks back to the Pythagoreans—of perfectly circular planetary orbits shows its validity at least partially. However, the actual situation is much more complex. Thus over similarly long periods of time the eccentricity also oscillates between values that vary, depending on the planet in question (for more see Fig. 6.15). In consequence the orbital sphere is not the equivalent of a fixed circular ring (shaded) or spherical shell such as one might envisage by elaborating on the ideas of antiquity, but rather it pulsates—for want of a better comparison—like the heart or the breath of a living being.

From the point of view of physics, the long-term changes are determined by the interplay of the gravitational influences of the planets (technical term: perturbations or orbital perturbations). Chapter 5 deals with this in more detail. But we shall close this section with a different view, quoted from a book by Thomas Ring in which he describes the solar system as a living organism:

> *Seen from a distant location in space skew to the ecliptic, and in the same timescale, the planetary system of bands would be a system of interlocking zones delineated by threads of light* (Ring means the light-trails which a timeless observer would see following the planets) *around a central forward-moving body. In shape these would not be the rigid circle-like ellipses we draw to illustrate the planetary orbits. With slight vibrations propagating from zone to zone the whole set-up would in our eyes resemble a pulsating organism. If we were able to slow the timescale down we would see not the threads of light but individual bodies circling round, and we would notice how the movement of one body very slightly interferes with that of the other, and conversely, how the second body also affects the first; and as our observation progressed we would have before our very eyes the effect of each body on all the other bodies in the system . . . So when seen overall, together with the very orderliness of the movement processes, what in the isolated case appears to be no more than a mechanical perturbation would come to appear as an expression of quivering life in an elemental being.*[2]

## 1.2 The golden section

The golden section can be constructed geometrically in various ways. I have chosen the method which is directly connected with the pentagon and thus also with the decagon:

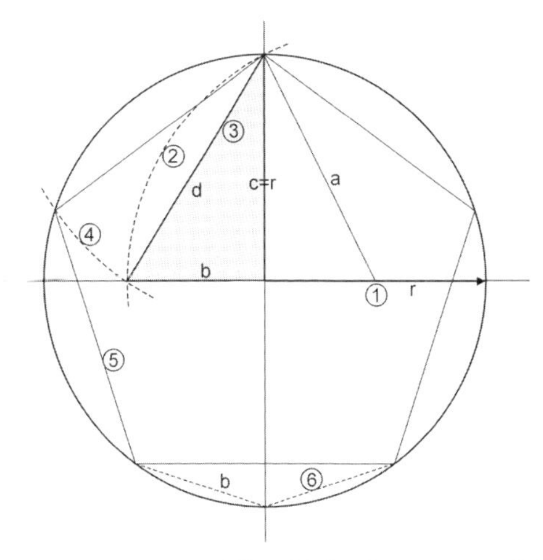


The shaded triangle which makes this construction possible is known as the triangle of Eudoxus (after one of Plato's pupils) or also the golden triangle.

**Fig. 14.3** *Construction of the golden section*[3]

Construction:

1. The radius r is halved to give point 1.
2. With the resulting line a as the new radius, a circle is drawn centred on that point.
3. The marked triangle with sides b, c and d arises via the intersection of this circle with its horizontal diameter.
4. A second circle with radius d is drawn, centred on the apex of this triangle.
5. The intersection with the outer circle gives the side of the pentagon. Perhaps rightly, Walther Bühler writes: '*This remarkable triangle contains within it a fact which in its turn may be described as one of the great wonders of geometry. Its hypotenuse* [side d, my comment] *can be marked off exactly five times around the circle.*'
6. The ratio between side b and the radius is exactly that of the golden section. At the same time b is a side of the decagon (mathematical derivation see Appendix 2.2).

Calculation (with r = 1):

$$a = \sqrt{1^2 + \left(\frac{1}{2}\right)^2} = \sqrt{\frac{5}{4}} = \frac{\sqrt{5}}{2}$$

$$b = a - \frac{r}{2} = \frac{\sqrt{5}}{2} - \frac{1}{2} = \frac{\sqrt{5}-1}{2} = \underline{\underline{0.618034..}} \Rightarrow \frac{r}{b} = \frac{1}{0.618..} = 1.618..$$

$$d^2 = b^2 + c^2 = \left(\frac{\sqrt{5}-1}{2}\right)^2 + 1^2 = \frac{6-2\sqrt{5}}{4} + \frac{4}{4} = \frac{5-\sqrt{5}}{2}$$

$$= 1.381966..$$

$$\Rightarrow d = \sqrt{\frac{5-\sqrt{5}}{2}} = 1.175570..$$

The golden section is equally intimately connected with the pentagram:

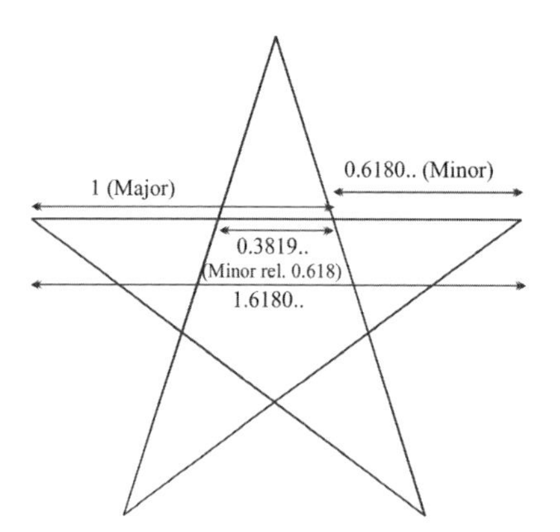


$$\frac{1.618..}{1} = \frac{1}{0.618..} = \frac{0.618..}{0.3819..}$$

*'... the pentagon star [or pentagram], a figure called Witch's Foot in German, and by Paracelsus the sign of health.'*[4]

**Fig. 14.4** *The golden section in the pentagram. The larger part arising in the division is termed 'Major', the smaller one 'Minor'*

## 1.3 Square, triangle and hexagon

### *Square*

We shall begin with the square since this is mathematically the simplest. The radius of the incircle, also in the following diagrams, is taken as 1.

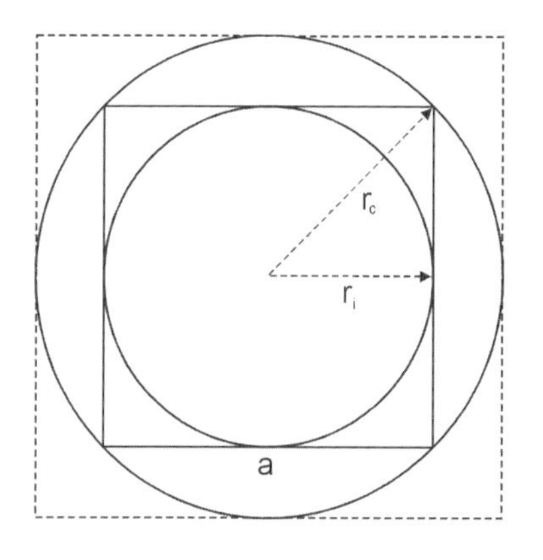


With $r_i = 1$, $a = 2$, $\pi = 3.14159..$
Areas of circles: $F_i$, $F_c$.
Area of square: A
Area of the incircle and the square:
$F_i = \pi * r_i{}^2 = \pi$ *and* $A = a * a = 4$
The radius of the circumscribed circle, using Pythagoras' Theorem, is:

$$r_c = \sqrt{\left(\frac{a}{2}\right)^2 + \left(\frac{a}{2}\right)^2} = \sqrt{1^2 + 1^2} = \sqrt{2}$$

*so* $F_c = \pi * r_c{}^2 = 2\pi$
(The subscript c refers to the circumcircle, i to the incircle.)

Thus the following area ratios arise:

$$\frac{F_c}{F_i} = \frac{2\pi}{\pi} = 2; \quad \frac{F_c}{A} = \frac{2\pi}{4} = \frac{\pi}{2}; \quad \frac{A}{F_i} = \frac{4}{\pi}$$

The length of the sides of the next square, traced around the outer circle, would be:

$$a_2 = 2 * r_c = 2 * \sqrt{2} \Rightarrow A_2 = 8 = 2 * A$$

So the ratio of the circle areas divided by a square and that of the square's area bounded by the circle is the same. This proposition can be applied to all regular polygons.

***Triangle***

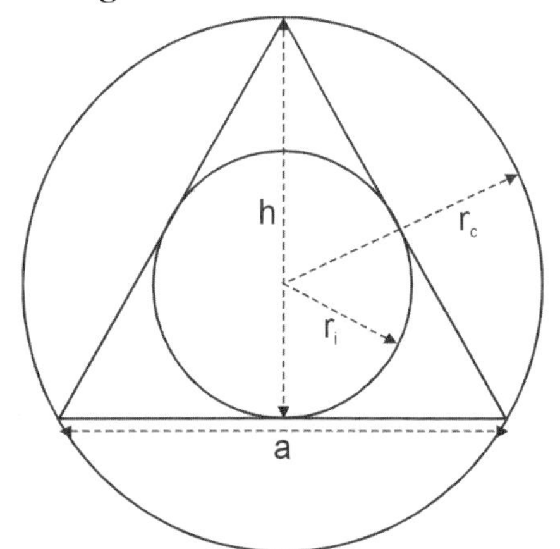


First determine the height and area of the equilateral triangle:

$$h = \sqrt{a^2 - \left(\frac{a}{2}\right)^2} = \sqrt{\frac{3}{4}a^2} = \frac{\sqrt{3}}{2} * a$$

$$A = \frac{g * h}{2} = \frac{a}{2} * \frac{\sqrt{3}}{2} * a = \frac{\sqrt{3}}{4} * a^2$$

(g: base of triangle, which here is the same as a)

Related to the incircle radius $r_i = 1$, a can now be found according to the formula:*

$$r_i = \frac{A}{s} \ \textit{with} \ s = \frac{a + b + c}{2} = \frac{3}{2}a$$

where s is half the sum of the sides; b and c are the other sides of the triangle, here equal to a:

$$r_i = \frac{A}{s} = \frac{\sqrt{3}}{4}a^2 * \frac{2}{3a} = \frac{\sqrt{3}}{6}a = \frac{a}{2 * \sqrt{3}} \Rightarrow a = 2 * \sqrt{3} \Rightarrow$$

$$h = \frac{\sqrt{3}}{2} * 2\sqrt{3} = 3 \ (* \ r_i)$$

* Refer to mathematical textbooks for the derivation of this and other basic formulae.

The area of the triangle A is thus:

$$A = \frac{\sqrt{3}}{4} * (2 * \sqrt{3})^2 = \frac{\sqrt{3}}{4} * 4 * 3 = 3^{\frac{3}{2}}$$

The circumcircle radius of the triangle can be calculated using the formula:

$$r_c = \frac{b * c}{2h_a} = \frac{(2\sqrt{3})^2}{2 * 3} = \frac{4 * 3}{6} = 2 \Rightarrow F_c = 4\pi$$

The ratios of the areas are therefore:

$$\frac{F_c}{F_i} = \frac{4\pi}{\pi} = 4; \quad \frac{F_c}{A} = \frac{4\pi}{3^{\frac{3}{2}}} = 2.418399..; \quad \frac{A}{F_i} = \frac{3^{\frac{3}{2}}}{\pi} = 1.653986..$$

### *Hexagon*

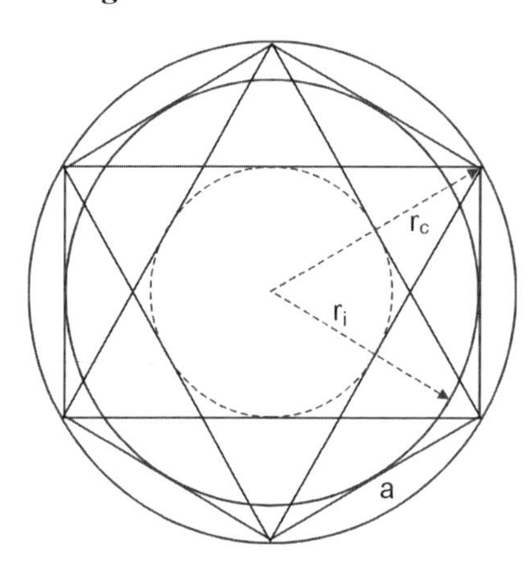


Let $r_c$ = the radius of the circumcircle and $r_i$ that of the incircle. Now a = $r_c$. The general formulae for the relationship between a, $r_i$ and $r_c$ and for the area of a polygon A are:

$$r_i = \frac{1}{2}\sqrt{4{r_c}^2 - a^2}$$

$$A = \frac{n * a * r_i}{2}$$

Thus (for $r_i$ = 1) we obtain:

$$r_i = \frac{1}{2}\sqrt{3{r_c}^2} = \frac{\sqrt{3}}{2} r_c \Rightarrow r_c = \frac{2}{\sqrt{3}} \Rightarrow F_c = \frac{4}{3}\pi$$

$$\text{and} \quad A = \frac{6 * 2}{\sqrt{3} * 2} = 2\sqrt{3}$$

The ratios of the areas are:

$$\frac{F_c}{F_i} = \frac{4\pi}{3 * \pi} = \frac{4}{3}; \quad \frac{F_c}{A} = \frac{4\pi}{3 * 2\sqrt{3}} = \frac{2\pi}{3^{\frac{3}{2}}} = 1.209199..; \quad \frac{A}{F_i} = \frac{2\sqrt{3}}{\pi} = 1.102657..$$

Briefly, regarding the hexagram: Its inner hexagon can be divided into 6 equal triangles. Thus the star consists of 12 of these smaller triangles. The ratio of its area to that of the hexagon is 2:1. The large triangle is made up of 9 smaller ones. So the ratio of hexagram to triangle is 4:3.

The incircle radius of the small hexagon which is formed by the hexagram is, as the drawing shows, half of the circumcircle radius of the original figure. Thus one also notices that the ratio of the areas of the circumcircle of the large hexagon and the incircle of the small hexagon is the same as that of the circum- and incircles of the triangle, namely 4:1. Since in the hexagon $F_c/F_i = 4/3$, one can further calculate that the areas of the two hexagons have a ratio of 3:1, and those of the large hexagon and the hexagram 3:2.

## 1.4 Euclid's Height Theorem

This theorem states that the square on the height of a right-angled triangle (b in Fig. 14.5) has the same area as the rectangle ap⋆pe formed from the divided hypotenuse. The geometric mean $\sqrt{ap * pe}$ can then be constructed in combination with the Thales circle (dark semicircle). By applying this construction to the sphere of a planetary orbit bounded by aphelion and perihelion (see Fig. 14.2) one arrives at the semi-minor axis b.

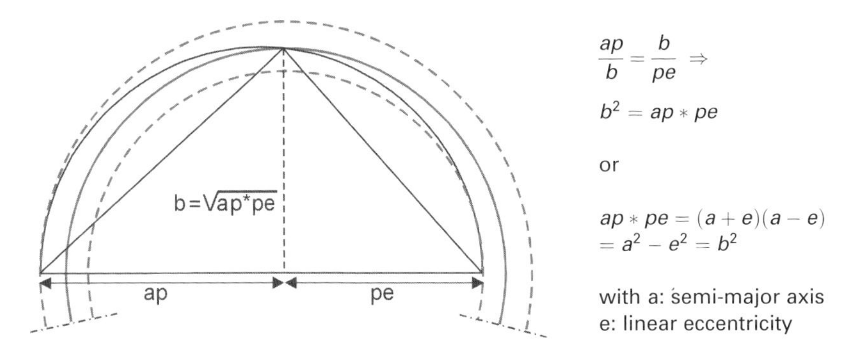


**Fig. 14.5** *The geometrical connection between aphelion distance, semi-minor axis and perihelion distance*

## 1.5 Octavation, intervals and scales

***Octavation***

By the method of octavation any interval can be converted into a number between 1 and 2 and thus compared in a simple way with harmonic proportions. The basis for this is that the space of all musical notes is structured by the octave. This corresponds to double the frequency or, respectively, half the length of a string which allows the keynote and, when halved, the octave to sound. In terms of physics the principle is based on the overtones.*

The mathematical method is based on continuous division or multiplication by 2, depending on whether the interval to be transposed is larger than 2 or smaller than 1. Here are some examples:

$$7.5 \rightarrow \frac{7.5}{2*2} = 1.875; \quad 88 \rightarrow \frac{88}{2*2*2*2*2*2} = \frac{88}{2^6} = 1.375;$$

$$0.19 \rightarrow 0.19 * 2^3 = 1.52$$

***Intervals and scales***

But what are the other building materials belonging to the art that has the most beautiful task of reflecting the human soul while leading it into regions far and away more lofty than our everyday earthly world? What is the basic structure of the musical order of notes capable of doing this, an order also present in a similar form in the movements of the planets?

Fig. 14.6 shows all 12 possible basic intervals† arising within the space of an octave. They are divided into three groups according to their origin or, respectively, their significance. At the top are the intervals that result from the structure of the Greek scales which are based on tetrachords. Then come the thirds and sixths which give harmonious European music its character; and finally the remaining semitones. We give examples for the seven-step scales together with the positions of the semitones that are characteristic for major and minor, and also the ratios and names of the intervals. An explanation follows as to why the order of notes conforms to this particular structure.

In the Greek scale consisting of two tetrachords, the octave is divided into the fourth 4:3 and fifth 3:2. Geometrically this can be depicted by intersecting circles that mirror one another (as in Fig. 3.2). The whole note is divided off as 9:8 in the middle part while the actual centre remains

---

* Every musical note is accompanied by whole-number multiples (1-fold, 2-fold, 3-fold, 4-fold, etc.) of the fundamental frequency.

† By mentioning basic intervals I imply that further intervals can be made possible in the various scales by augmenting or diminishing the notes by means of accidentals. The additional intervals thus gained are then very close to the basic intervals, e.g. F♯ (as the basic interval related to C) and G♭, of which the ratios are 45/32 = 1.40625 or, respectively, 64/45 = 1.4222... The intervals thus derived are, however, not relevant in our present context and will therefore not be further dealt with here.

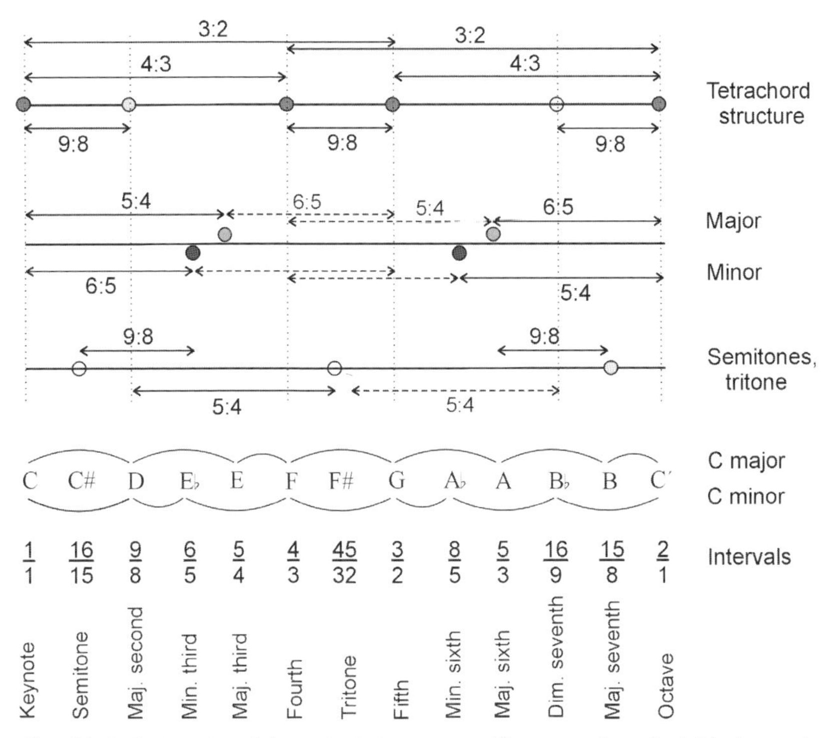


**Fig. 14.6** *Basic ratios of the major/minor system (dim. seventh = diminished seventh which is distinct from minor seventh 9/5, though these terms are not always uniformly applied)*

unoccupied. As described in Chapter 2, in Greek times and right up to the Middle Ages thirds were not regarded as consonant intervals. In keeping with this view, further division of the scale was achieved with the help of the ratios mentioned. When applied to musical instruments this procedure is described as Pythagorean tuning.

The tetrachord structure continues on into the further development of music and the tonal differentiation connected with it. In the system of major and minor, the thirds and sixths are added; these characterize the harmonic tuning. Of course these intervals also occurred in Greek musical theory, albeit in somewhat deviating proportions, for example with the third as 81:64. In major scales the interval 5:4 determines the next step, i.e., beginning with the keynote, the second whole note. The two circles of fifths get their structure (5/4 ⋆ 6/5 = 3/2) in association with the ratio 6:5 which results in their being very much more strongly emphasized in comparison with the Pythagorean order. Fig. 14.7 shows the integration of the ratios formed by the number five into the ratios of the basic order:

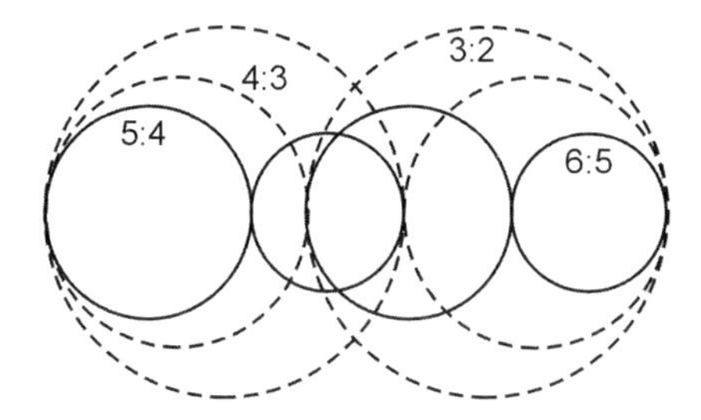


Fig. 14.7 *Basic structure of the major scale*

The overreaching importance of the major and also the minor triad in European music thus gains its geometric correspondence which is founded in the mathematical logic of form. As shown in Fig. 14.6, the minor scales arise simply from the fact that the sequence within the fifth is reversed: 6/5 ⋆ 5/4 = 3/2. Mathematically the result is identical; yet when the relevant chords are sounded one often experiences them as being as different as light and shade.

The seven intervals—octave, fifth, fourth, major and minor third, major and minor sixth—were arrived at through a gradation in two steps; since the time of Johannes Kepler at the latest these intervals have been experienced as being consonant. The notes at either end of the seven-step major or, respectively, minor scale can only be accommodated in the remaining larger spaces at the left-hand and right-hand edges of the depicted order. In relation to major, the left-hand side shows the range of two whole notes which are sensibly subdivided by the major second which is equal to the Pythagorean whole note interval 9:8 (thus the second whole note now has to be 10:9, the minor second, so that the sum can again amount to 5:4). The right-hand space is equal to one and a half notes. Beginning at its lower end (the note A in our example of C major) the major seventh is attained by inserting another whole note. The two missing intervals arise in a similar manner in the minor scale (see Fig. 14.6).

To complete the sequence of intervals for a total of 12 possible semitones one can subtract the proportion of a whole note from the major or minor third; but there are also other ways of doing this. In the middle, at the tritone, the arrows coming from below (i.e. from the left) and from above (i.e. from the right) do not meet and the one coming from the right gives us G♭. The centre remains unoccupied both in the Pythagorean and in the harmonic tuning. On each of the notes arising it is now possible to erect one major and one minor scale which follows the structural principle shown and yet possesses its own sphere of expression. So twice 12 scales each with seven steps are needed, and are also sufficient, for the creation of a whole musical world. (As chance will have it, this is also the number of hours in one week, and therefore the period of time during which—so an ancient account tells us—the actual world is supposed to have been created.)

In summary we notice that the structure of the scales with their steps of whole tones and semitones in certain positions, which is so difficult to understand and even sometimes thought to be arbitrary, thus receives its natural geometrical explanation. The inner logic of the form that has developed over long periods of time, which enables it to be described in some ways as a creative deed of the whole of humanity, thus possesses great clarity. The hierarchical structure, which in its differentiation follows the simple series of intervals 2:1, 3:2 . . . up to 6:5, implies that this is the only possibility. By building on this it has been possible for the mysterious creative genius of a number of great composers to give to us, during the course of only a few centuries, one of our most precious treasures: their music, of which the overall import and depth has not as yet been truly plumbed. In bringing this theme to a close, let us hear what one such composer has had to say on the subject:

> *When I feel in myself the urge* [to compose] *I direct my thoughts first of all to my Creator, asking him the three most important questions that relate to our life on this earth—whence, wherefore, whither?*
>
> *Immediately after doing this I sense vibrations which permeate my whole being. These are the Spirit which lights up the inner forces of my soul, and in this state of rapture I clearly see what in my ordinary frame of mind is obscured. Thereupon I find myself able, as was Beethoven, to allow inspiration to enter into me from on high. In such moments I become aware above all else of that most lofty revelation given to us by Jesus: 'I and the Father are one.' Those vibrations take on the shape of specific mental images once I have formulated the desire I have, which is to be inspired to compose something that can console and encourage humanity—something that is of lasting value.*
>
> Johannes Brahms[5]

## 1.6 The musical intervals as ratios of areas

In Chapter 2 we introduced the three basic procedures by means of which the musical intervals can be represented as the ratios of areas of simple geometrical figures. Multiplication here implies that each subsequent figure is circumscribed, while division indicates that it is inscribed:

1) Octave 2:1 — circle ⋆ square ⋆ circle (Fig. 2.1)
2) Fourth 4:3 — circle ⋆ hexagon ⋆ circle (Fig. 2.3, left)
3) Major third 5:4 — circle ⋆ pentagon ⋆ circle / decagon / circle / decagon / circle (Fig. 2.7)

The major third in Fig. 2.7 has the following mathematical background.

The ratio of the areas of the circum- and incircles of the pentagon (see Appendix 2.1 and 2.2 for the derivations of this) is:

$$6 - 2 * \sqrt{5} = 1.52786.., \text{ or, in the decagon, } 8/(5+\sqrt{5}) = 1.10557..$$

The ratio of the incircle in the first decagon to the basic circle arises from the division of these two ratios:

$$\frac{1.52786..}{1.10557..} = (5-\sqrt{5})/2 = 1.381966..$$

(The silver section, as the reader will no doubt recall.)

And then the second decagon circle causes the whole irrationality to vanish into thin air:

$$\frac{1.381966..}{1.105572..} = \frac{(5-\sqrt{5})(5+\sqrt{5})}{2*8} = \frac{25-5}{16} = \underline{\underline{1.25}}$$

whereupon the major third is revealed in all its glory.

The next two intervals arose out of a combination of the first and second constructions (Fig. 2.3, right).

| | |
|---|---|
| 4) Fifth 3:2 | circle ⋆ square ⋆ circle / hexagon / circle |
| 5) Whole note 9:8 | circle ⋆ square ⋆ circle / hexagon / circle / hexagon / circle |

Another example, that of the major sixth 5:3, shows how all the musical intervals can be obtained via the three simple basic procedures:

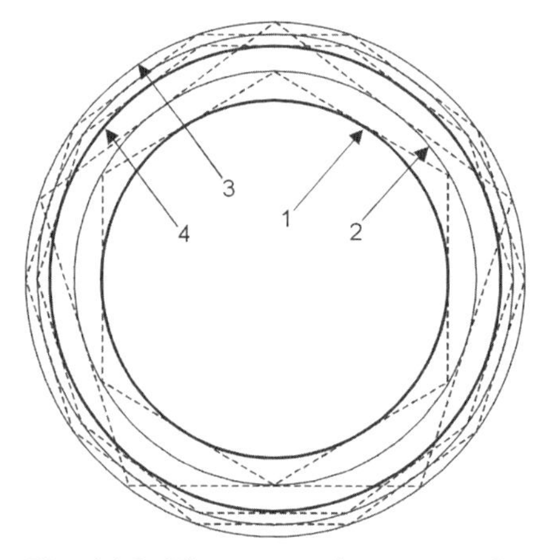


Construction:

1. Basic circle (= 1)
2. Circumscribed: hexagon and circle
3. Circumscribed: pentagon and circle
4. Inscribed: two decagon circles

**Fig. 14.8** *The major sixth as a ratio of areas*

The construction shows a combination of those required for the intervals of the fourth and the major third. The calculation arises from the ratios in Appendix 1.3 (hexagon), or 2.1 (pentagon) and 2.2 (decagon):

$$6) \quad \textit{Maj. sixth} = \frac{4}{3} * \frac{5}{4} = \frac{\textit{hexagon} * \textit{pentagon}}{2\ \textit{decagons}} = \frac{1.33333.. * 1.52786..}{1.10557.. * 1.10557..} = 1.66666..$$

The remaining intervals can be similarly obtained via the following combinations (for the sake of simplicity, the relevant intermediate circles are not specifically mentioned here):

| | |
|---|---|
| 7) Minor sixth 8:5 | circle / pentagon ⋆ 2 decagons ⋆ square ⋆ circle |
| 8) Minor third 6:5 | circle / pentagon ⋆ 2 decagons ⋆ square / hexagon / circle |
| 9) Dim. seventh 16.9 | circle ⋆ hexagon ⋆ hexagon ⋆ circle |
| 10) Major seventh 15:8 | circle ⋆ square / hexagon ⋆ pentagon / 2 decagons / circle |
| 11) Semitone 16:15 | circle ⋆ hexagon / pentagon ⋆ 2 decagons ⋆ circle |
| 12) Tritone 45:32 | circle ⋆ square / hexagon / hexagon ⋆ pentagon / 2 decagons / circle |

## 2. *Star-figures*

We shall here depict only those star-figures that arise in connection with the planets in especially conspicuous ways, either as though appearing to be traced in the firmament by movement or because the closest possible correspondences can be found via the geometrical ratios. And since our investigations have yielded results mainly in connection with area ratios, this is where I shall place my emphasis.

Our point of departure will be the formulae, given in any mathematical textbook, for calculations concerning the regular polygons. All our calculations will be related to the radius of the circle inscribed in the polygon in question, which will be set at 1.

### 2.1 The pentagram

Before discussing the mathematics of the pentagram, let us ask once again whether this star-figure is indeed something like an archetypal image of the human soul, not by putting forward various more or less clever treatises on the subject but simply by placing before the reader two plain pentagrams (Fig. 14.9). Those who wish to do so might contemplate these for a while in order to find out whether they notice any particular impact on their own inner feelings and whether there is in this respect any difference between the left-hand and the right-hand drawing.

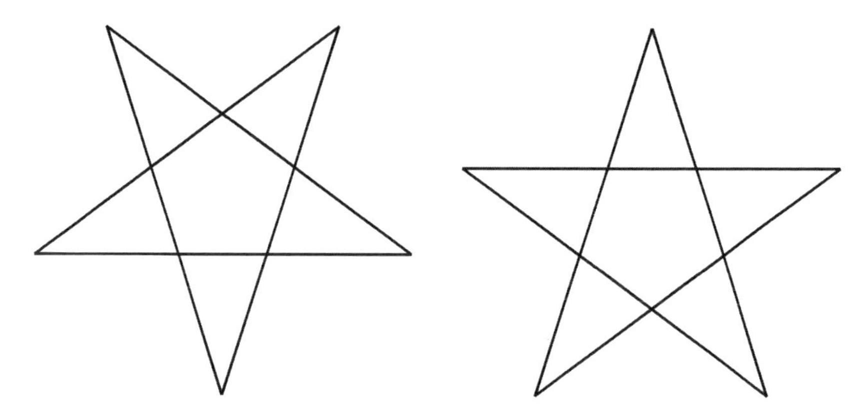

**Fig. 14.9** *Two versions of the pentagram*

At the end of this section I shall return to this pictorial or indeed perhaps archetypal aspect of the pentagram with a brief glance at the symbolism attributed to this star-figure. But let us first now turn to the numerical ratios which arise in connection with both pentagon and pentagram.

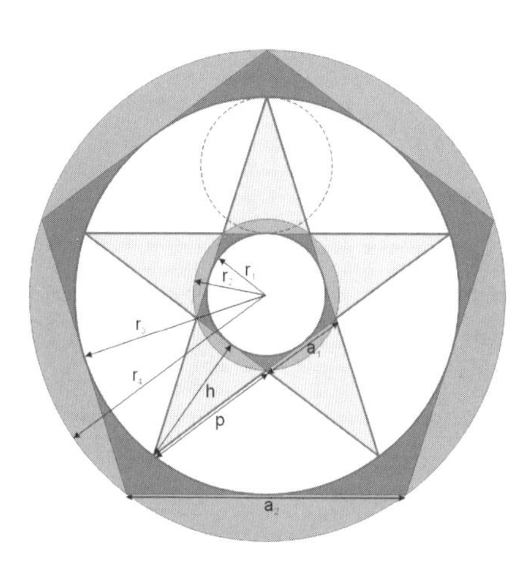


**Fig. 14.10** *Pentagon and pentagram*

Terms:
$r_1$ (= $r_i$): incircle radius
$r_2$ (= $r_c$): circumcircle radius
a: side of the pentagon
h: height of the pentagram triangle
p: side of the pentagram triangle
s: side of the pentagram = $a + 2 * p$
$F_1$: area of the incircle
$F_2$: area of the circumcircle
$F_3$: area of the pentagram circle
$F_4$: area of the outer circle
$A_1$: area of the inner pentagon
$A_2$: area of the outer pentagon
D: area of the pentagram triangle
P: area of the pentagram = $A_1 + 5 * D$
g: ratio of the golden section = 1.61803..
(see Appendix 1.2)
$\pi$: 3.14159..

Formulae for calculating the pentagon:

$$r_c = \frac{4}{\sqrt{5}+1} r_i \qquad a = \frac{10}{\sqrt{25+10\sqrt{5}}} r_i \qquad A = \frac{5a}{2} r_i$$

With $r_i = 1$ we arrive at:

$$r_1 = 1 \quad r_2 = 1.23606.. = \sqrt{5}-1 \quad a_1 = 1.45308.. = 2\sqrt{5-2\sqrt{5}}$$
$$A_1 = 3.63271..$$

and $F_1 = \pi {r_1}^2 = \pi;\ F_2 = \pi\ {r_2}^2 = \left(\sqrt{5}-1\right)^2 * F_1 = 1.527864.. * F_1$

Therefore the basic ratios of the areas of the pentagon, incircle and circumcircle are:

$$\frac{F_c}{F_i} = \frac{F_2}{F_1} = 1.52786..\,; \quad \frac{F_2}{A_1} = \frac{1.52786 * \pi}{3.63271} = 1.321306..\,;$$
$$\frac{A_1}{F_1} = \frac{3.63271}{\pi} = 1.156328..$$

It is easy to calculate the pentagram triangle via the side-length p which is in the ratio of the golden section to a:

$$p = g * a = 2.35114.. \quad h = \sqrt{p^2 - \left(\frac{a}{2}\right)^2} = \sqrt{5} = 2.23606..$$
$$D = \frac{a * h}{2} = 1.62459..$$

$$\Rightarrow \quad s = 2 * p + a_1 = 6.15536.. \quad P = 5 * D + A_1 = 11.75570..$$

Thus the square calculated from the height of the pentagram and the square circumscribing the inmost circle (of which one side is $2 \star r_1 = 2$) have areas in the ratio of 5:4. Since the ratios remain the same this also applies to the circles inscribed within these squares (shown in Fig. 14.10 by the dotted circle surrounding the upper pentagram triangle).

So now the radius $r_3$ arises from $r_1$ plus the height h:

$$r_3 = \sqrt{5}+1; \quad F_3 = \left(6+2\sqrt{5}\right) * F_1 = 10.47213.. * F_1$$

**Table 14.1** *Area ratios in the pentagon*

| Ratio | $F_1$ | $A_1$ | $F_2$ | P |
|---|---|---|---|---|
| $A_1$/.. | 1.15633 | 1 | | |
| $F_2$/.. | 1.52786 | 1.32131 | 1 | |
| P/.. | 3.74196 | 3.23607 | 2.44914 | 1 |
| $F_3$/.. | 10.47214 | 9.05637 | 6.85410 | 2.79857 |

The ratios of all the areas calculated thus far are shown in Table 14.1. The intervals between the circles and pentagons arise by taking the number π into account, e.g. $F_3/A_1 = 10.472..\star \pi / 3.632.. = 9.056..$

The values for the area which reaches beyond the pentagram can be derived from those already obtained since the ratios of the large pentagon to its in- and circumcircles are equivalent to those of the small one:

$$a_2 = 1.45308.. * \left(\sqrt{5}+1\right) = 4.70228.. \qquad A_2 = \frac{5a_2}{2} r_3 = 38.04226..$$

and $$r_4 = \left(\sqrt{5}+1\right)\left(\sqrt{5}-1\right) = \underline{\underline{4}} \quad \Rightarrow \quad F_4 = \underline{\underline{16}} * F_1$$

We thus find that it has been well worth doing all these calculations because now we once more encounter the phenomenon we already discovered in the case of the major third of the pentagon and decagon (see Fig. 2.7). After a specific point the irrationality of the golden section, or rather of the numbers related to it, resolves into simple whole-number ratios. From an initial pentagon we make a pentagram; then we join its points by means of a circle; this is circumscribed by a second pentagon; and the latter is then surrounded by another circle. The areas of the incircle of the first pentagon and of the outermost circle have the ratio 1:16. In this respect the construction leads to the same result as that of the sequence circle-triangle-circle-triangle-circle:

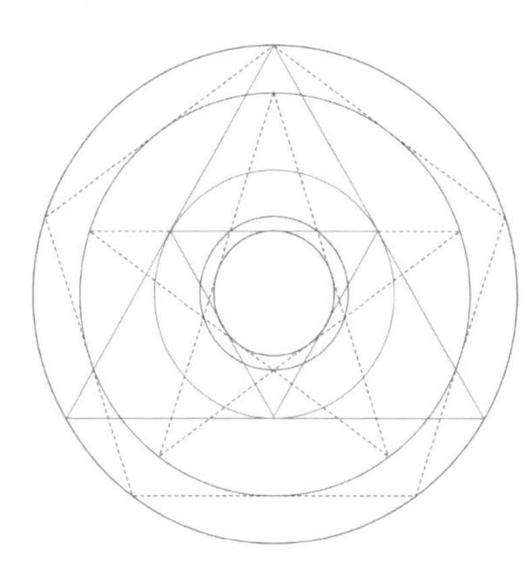

**Fig. 14.11** *Triangle and pentagram*

Furthermore, the radii of the pentagram circle $r_3$, of the middle triangle circle ($r_m = 2 * r_1$) and of the pentagon circumcircle $r_2$ are in the ratio of the golden section:

$$r_3/r_m = \frac{\sqrt{5}+1}{2} = 1.61803..$$

and $$r_m/r_2 = \frac{2}{\sqrt{5}-1} = 1.61803..$$

Accordingly, their areas have the ratios

$$F_3/F_m = F_m/F_2 = g^2 = 2.61803..$$

or, respectively,

$$F_3/F_2 = g^4 = 6.85410..$$

Although there are a number of pleasing connections, the various numerical ratios may thus far appear to be rather confusing. Or, in other words, one might say that the polarity expressed in Fig. 14.9, showing the pentagram standing either on two legs or else on its head, has not yet been resolved in any synthesis. Perfect synthesis will only be attained by the pentagram when we find it in the decagram.

Could it be that this not-yet-perfection is one of the reasons why in past ages the pentagram was often linked with the figure of the human being? Think of the example given by the natural philosopher and occultist Agrippa von Nettesheim (1486–1535) who depicted the human frame with outstretched arms as a right-angled cross, but also within a circle, in the shape of a pentagram, showing the arms inclined downwards and the legs apart. The reader's sensitivity and judgement will surely lend meaning to whatever he or she may feel about the symbolism of the pentagram or 5-pointed star.

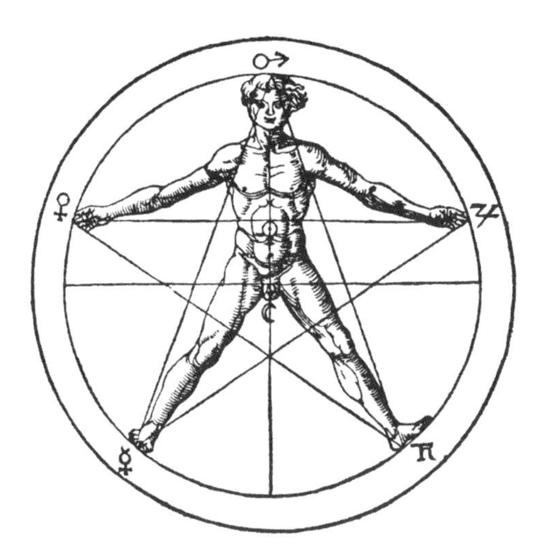

**Fig. 14.12** *Drawing by Agrippa von Nettesheim*

## 2.2 The 10-pointed star

The geometry of the decagon together with its star-figures is so marvellous that it is worth a moment's contemplation to find out what effect this artwork of nature (if it is indeed nature that makes such shapes possible) might have on heart and mind (see Fig. 14.13).

The outer circle surrounds both the large decagram and the light-coloured double pentagram. The circle within the inner decagon is also the inner circle of the two pentagrams. Therefore the ratio between this

**Fig. 14.13** *Decagon and star-figures*

inner circle and the outer circle bordering the whole figure is the same as in the case of the pentagram. In order to calculate further ratios we shall use the following terms, where the radius of the inner circle is once again set at 1:

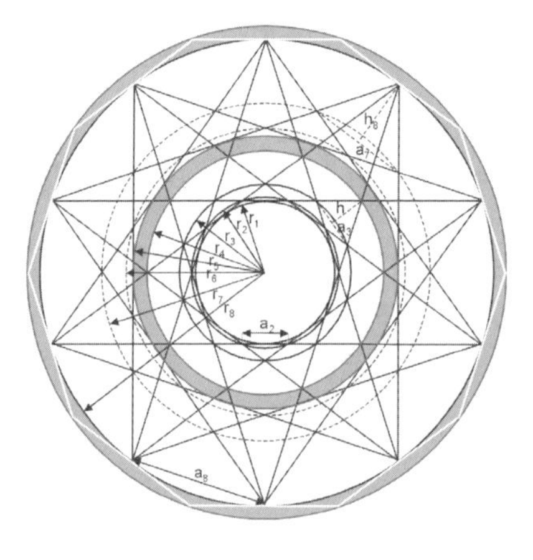

**Fig. 14.14** *Decagon and star-figures II*

Terms:
$r_1$ (= $r_i$): incircle radius
$r_2$ (= $r_c$): circumcircle radius (decagon circle)
$r_3$: inner pentagon circumcircle radius
$r_4$: inner decagram circumcircle radius
$r_5$: outer decagram incircle radius
$r_6$: $r_5$ * decagon circle radius
$r_7$: $r_5$ * pentagon circle radius
$r_8$: outer decagram circle radius
a: one side of the decagon (e.g. $a_2$ is inscribed inside the circle with the radius $r_2$)
$F_1$: incircle area
$F_2$: circumcircle area ... etc.
A: area of the decagon (e.g. $A_2$ is the decagon with sides $a_2$)
Z: area of the (outer) decagram
g: ratio of the golden section

Here are the basic formulae pertaining to the decagon to be found in standard mathematical textbooks:

$$r_c = \frac{4 * r_i}{\sqrt{10 + 2 * \sqrt{5}}} = 1.051462.. \Rightarrow F_c = r_c^{\,2} * \pi = 1.1055728.. * \pi$$

$$a = \frac{r_c}{2}\left(\sqrt{5} - 1\right) = 0.618034.. * r_c = 0.649839..$$

$$A = \frac{5a^2}{2}\sqrt{5 + 2 * \sqrt{5}} = 3.2491969..$$

Thus the side of the decagon relates to the radius of the circle surrounding it as in the ratio of the golden section. (Therefore the decagon inscribed into the inner circle has side length $a_1 = 0.618.. \star r_i$.) The ratios of the areas of the decagon and its outer and inner circles are:

$$\frac{F_c}{F_i} = \frac{F_2}{F_1} = 1.1055728..; \qquad \frac{F_2}{A_2} = \frac{1.105573 * \pi}{3.249197} = 1.068959..;$$

$$\frac{A_2}{F_1} = \frac{3.249197}{\pi} = 1.034251..$$

Radius $r_3$ is that of the outer circle of the double pentagon arising inside the double pentagram. Radius $r_4$ of the inner decagram that coincides with the lines of the pentagram can be calculated with the help of the triangle which comes about through the dotted decagon side $a_3$ and the height h constructed upon it. The ratio of side $a_3$ to $a_1$ is the same as the ratio $r_3$ to $r_1$, so that it is easy to calculate its length with the help of the circumcircle radius of the pentagon (see Appendix 2.1). The distance h can then be calculated via the tangent of the angle of the decagram point which (without being proven here) is 72°:

$$a_3 = r_3 * a_1 = \left(\sqrt{5} - 1\right)\frac{\left(\sqrt{5} - 1\right)}{2} = 0.763932..$$

$$h = \frac{a_3}{2 * \tan\,(36^\circ)} = 0.525731..$$

$r_4$ is now the radius of the (not named here) inner circle of the decagon with side $a_3$ plus the height h. This radius arises by dividing $r_3$ by the decagon-circle ratio $r_2$ (= $r_c$):

$$r_4 = \frac{r_3}{r_2} + h = \frac{\sqrt{5} - 1}{1.051642} + h = 1.7013016.. \quad \Rightarrow \quad F_4 = 2.894427.. * \pi$$

$r_5$ can now be found by dividing the outer-circle radius $r_8$ by $r_4$, since the same ratio arises as that between $r_4$ and $r_1$; $r_6$ and $r_7$ then result by multiplying $r_5$ by the decagon or, respectively, the pentagon ratio ($r_2$, $r_3$):

$$r_5 = \frac{\sqrt{5}+1}{1.7013016} = 1.902113.. \Rightarrow r_6 = 2 \quad and \quad r_7 = 2.351141..$$

The relationship to the rationality of the triangle—which appears in connection with the pentagon and its pentagram, where it first arises as though approaching from the outside by passing over the pentagram circle—is here found in the inner part of the decagram. The harmony of decagram and triangle is of course also valid for the double triangle, i.e. the hexagram:

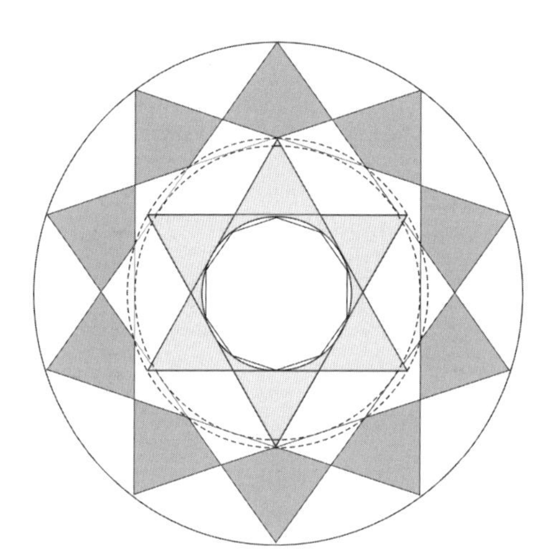

The ratios valid for the radii or, respectively, the circle areas are:

$$\frac{r_8}{r_6} = \frac{\sqrt{5}+1}{2} = 1.618..$$

$$\frac{F_8}{F_6} = \frac{(\sqrt{5}+1)^2}{4} = 2.618..$$

(see Fig. 14.11)

The inner decagon is the one with the radius $r_i = 1$ inscribed within the inner circle.

**Fig. 14.15** *Decagram and hexagram*

The principles of the number ten and the number three with its multiples as it were interpenetrate one another in the dotted decagon circular ring formed by $r_5$ and $r_6$. They are linked with the help of the golden section ratio in that the circles with the radii $r_8$ and $r_6$, which delineate the two figures, are related to one another. The decagon side $a_6$, which in the same way has this special ratio with radius $r_6$, measures $a_6 = 0.618.. * 2 = \left(\sqrt{5}-1\right)$, from which we can deduce that $r_8$ and $a_6$ have the same interval as the areas $F_8$ and $F_6$, namely double that of the golden section (2.618..).

**Table 14.2** *Ratios of decagon and decagram*

| | Radius | Circle area (... * π) | Decagon area A | $F_2/F_1$, $F_3/F_2$ ... |
|---|---|---|---|---|
| $r_1$, $F_1$ | 1 | 1 | 2.938926 | — |
| $r_2$, $F_2$ | 1.051462 | 1.105573 | 3.249197 | 1.105573 |
| $r_3$, $F_3$ | 1.236068 | 1.527864 | 4.490280 | 1.381966 |
| $r_4$, $F_4$ | 1.701302 | 2.894427 | 8.506508 | 1.894427 |
| $r_5$, $F_5$ | 1.902113 | 3.618034 | 10.633135 | 1.25 |
| $r_6$, $F_6$ | 2 | 4 | 11.755705 | 1.105573 |
| $r_7$, $F_7$ | 2.351141 | 5.527864 | 16.245985 | 1.381966 |
| $r_8$, $F_8$ | 3.236068 | 10.472136 | 30.776835 | 1.894427 |
| $r_9$, $F_9$ | 3.577709 | 12.8 | 37.618256 | 1.222291 |

A summary of the radii and areas which we have worked out shows that other rational ratios can also result. The circular ring marked in the middle of Fig. 14.14 (formed by the incircle of the outer decagram and the circumcircle of the inner decagram) is another way of geometrically depicting the major third 5:4 (see Fig. 2.7). The radius $r_9$ denotes the circle which arises by means of two further decagons drawn around the outer circle of the decagram (only one of these is shown in the diagram, and the occupied space is emphasized). The ratio to the inner circle of 12.8 corresponds to the musical interval of the minor sixth augmented three times. The ratio of the outer pentagon circle $F_7$ to the inner decagon circumcircle $F_2$ is 5:1. And there is an additional golden section ratio between the radii $r_4$ and $r_2$ (1.701../1.051.. = 1.618..). The most pleasing relationship, however, is probably that of the decagram and the pentagram. The area of the pentagram has already been mentioned on page 311 (equation $P = 5 \star D + A_1 = 11.75570..$). It is the same as that of the decagon $A_6$. By how much does the area of the decagram exceed that of the pentagram?

It arises from the area of the decagon $A_7$ and the ten points which stand upon it. These triangles (let us call them D) are calculated from the side $a_7$ and the height $h_8$. The segment $a_7$ is arrived at (as above) via the ratio $r_7/r_1$ and $h_8$, here shown with the help of the tangent:

$$a_7 = r_7 * a_1 = \frac{(\sqrt{5}-1)}{2} * 2.351141 = 1.453085..$$

$$h_8 = \frac{a_7}{2 * \tan(36°)} = 1$$

$$\Rightarrow \quad D = \frac{a_7}{2} * h_8 = 0.726542..$$

$$\Rightarrow \quad Z = 16.245985 + 10 * D = 23.511409..$$

Surprisingly, the height $h_8$ is the same as the radius of the innermost circle. The decagram area Z becomes the octave of the pentagram, i.e. it is exactly double its area. Twice five is of course ten, but it is not all that obvious that this calculation also applies to the areas in question. The area of the double pentagram (which will not be calculated here) is 17.01302.., and the ratio of the area of the decagram to it is the ratio of the silver section (1.381966..). Further harmonious area ratios are, for instance, that of the double octave 4:1 of the pentagram area to the inner decagon $A_1$, that of the octave interval of the double pentagram to the decagon $A_4$, and the 5:2 ratio of the decagon $A_7$ to the inner decagram (of which the area is 6.498394..). So it is correct to state that the polarity of the pentagram is resolved in the decagram both harmonically in the octave and also geometrically by the close kinship with the divine proportion.

In conclusion, the area ratios which arise through the decagram are here depicted in a way that reveals the interpenetration of the ratios from inside outwards and from outside inwards in a manner similar, for example, to the tetrachord structure shown in Fig. 3:2.

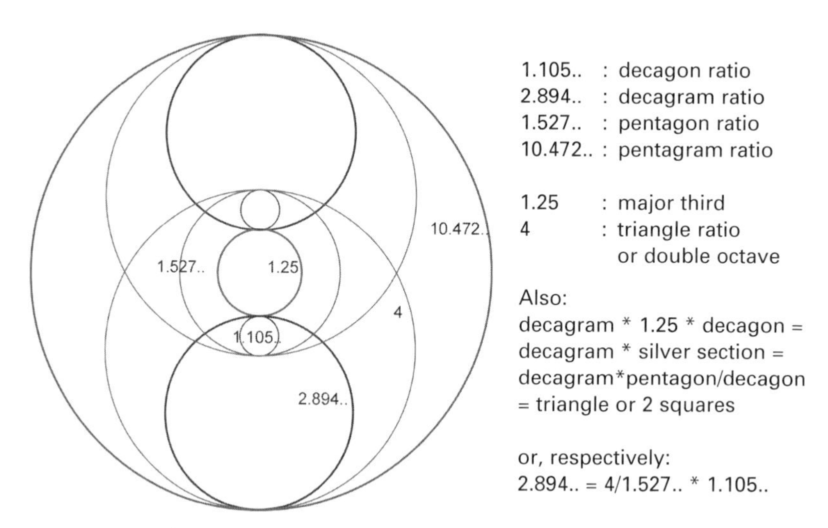


**Fig. 14.16** *Structure of the ratios of decagon and decagram*

Although this structure does not represent any specific peculiarity of the decagon or its descendants, since it occurs in similar ways with other star-figures, nevertheless perhaps the way in which the rational and irrational number ratios interact does reveal the basic structure most strikingly and clearly. There is an evident resemblance between the geometrical principle of form expressed here and the principles of form arising in music in the structure of the scales (see Fig. 14.7) and in astronomy in the ordering of the planets (see Fig. 1.5).

## 2.3 The 12-pointed star

The 12-pointed star encompasses the dodecagon, the hexagon, the square and the triangle. With it we have arrived, you could say, at the epitome of perfection in geometrical form. The figure as a whole has something about it that is free of clutter, radiant and majestic, something that does not appear in the same way in the rather more mysterious 10-pointed star.

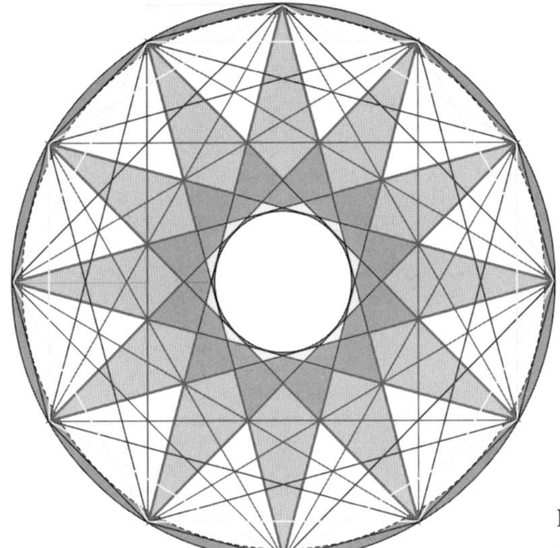

**Fig. 14.17** *Dodecagon and star-figures*

The grand and unique 12-pointed star contains composite star-figures, consisting of two hexagons, three squares or four triangles. These reappear when one skips the relevant points in the outer dodecagon. The basic ratio of circumcircle to incircle arises from a general formula for calculating the area of any polygon, while side a and area A of the enclosed dodecagon are derived from another calculation already mentioned in connection with the hexagon (Appendix 1.3):

$$r_c = \frac{1}{\cos(\alpha/2)} * r_i = \frac{1}{\cos(15°)} * 1 = 1.035276.. \Rightarrow$$

$$F_c = r_c^{\,2} * \pi = 1.0717967.. * \pi$$

$$a = 2\sqrt{r_c^{\,2} - r_i^{\,2}} = 0.535898.. * r_i$$

$$A = \frac{n * a * r_i}{2} = 6a = 3.215390.. * r_i$$

This yields the area ratios:

$$\frac{F_c}{F_i} = 1.0717967..; \quad \frac{F_c}{A} = \frac{1.071797 * \pi}{3.21539} = 1.0471975.. = \frac{\pi}{3};$$

$$\frac{A}{F_i} = \frac{3.21539}{\pi} = 1.0234905..$$

The list of radii and areas arising is shown here without any further calculation. The radii and circle areas (except for those of the 12-pointed star) result from the values for triangle, square and hexagon given in Appendix 1.3. The sides and areas of the dodecagon can be found via the relevant ratios relating to $r_c$ and $F_c$ (e.g. $a_6 = r_6 \star a_c / r_c$).

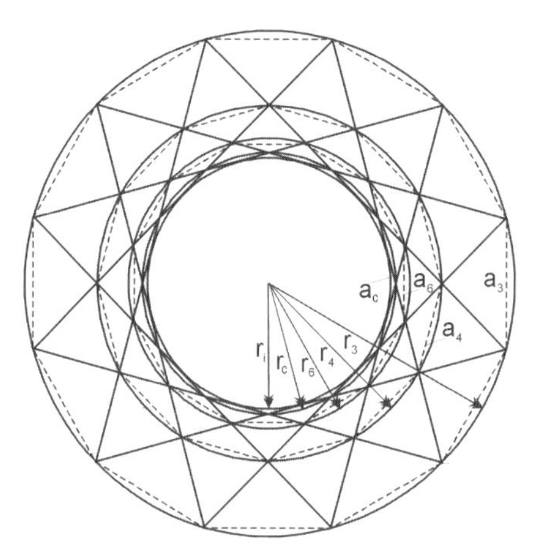


For the sake of clarity only the inner region of the 12-pointed star is shown here. The terms are given from the inside outwards:

- $r_6$: radius of the circumcircle of the hexagons
- $F_6$: area of the circle
- $a_6$: side of the dodecagon inscribed into this circle
- $A_6$: area of the dodecagon
- $r_4$: radius of the circumcircle of the squares etc.
- $r_3$: radius of the circumcircle of the triangles etc.
- $r_{12}$: radius of the circumcircle of the 12-pointed star etc.
- $r_{16}$: radius arising when a further dodecagon is added round the outside etc.

(Subscript c denotes the circumcircle, i the incircle.)

**Table 14.3** *Ratios in the 12-pointed star*

| | Radius r | Circle area F (.. ★ π) | Dodecagon side a | Dodecagon area A |
|---|---|---|---|---|
| $r_i$ | 1 | 1 | 0.517638 | 3 |
| $r_c$ | 1.035276 | 1.071797 | 0.535898 | 3.215390 |
| $r_6$ | 1.154701 | 1.333333 | 0.597717 | 4 |
| $r_4$ | 1.414214 | 2 | 0.732051 | 6 |
| $r_3$ | 2 | 4 | 1.035276 | 12 |
| $r_{12}$ | 3.863703 | 14.928203 | 2 | 44.78461 |
| $r_{16}$ | 4 | 16 | 2.070552 | 48 |

When a further dodecagon is added, the result—as with the pentagram—is: radius $r_{16}$ = 4 or circle area $16\pi$. From the areas of the dodecagons a number of intervals can be formed which are musical equivalents of octaves, fifths and fourths. Moreover the area of the continuous 12-pointed star also fits harmoniously into this sequence since its value is exactly 24. The 12-pointed star is composed of the area of the dodecagon $A_3$ (which is demarcated by the triangles) and its 12 points. Thus the area of one point of a 12-pointed star surprisingly measures exactly 1. It is also worth mentioning that the area of a dodecagon inscribed within a circle measures 3/2 of the area of a square within the same circle. The dodecagon $A_4$ inscribed inside the circle which surrounds the (inner) squares has, for example, the area 6. The square, i.e. one of the three squares, has the side-length 2 ⋆ $r_i$ and therefore the area 4.

We shall now investigate the ratios of the circle areas and those of the dodecagons since we noticed even during our discussion of the basic ratios that the dodecagon and π were related in whole numbers.

$$\frac{F_c}{A_c} = \frac{\pi}{3} \qquad \frac{F_4}{A_6} = \frac{\pi}{2} \qquad \frac{F_3}{A_4} = \frac{2}{3}\pi \qquad \frac{F_3}{A_6} = \pi \qquad \frac{A_6}{F_i} = \frac{4}{\pi} \qquad \frac{A_4}{F_i} = \frac{6}{\pi}$$

As we shall see, these remarkable ratios hidden within the perfection of the 12-pointed star also have a very important part to play in the architecture of our planetary system.

## 2.4 Kepler's stellar solids

As our basis for calculating the sizes and proportions in Kepler's two stellar solids we shall use the formulae for the two initial figures, the dodecahedron and the icosahedron, which are found in mathematical textbooks. Then all the calculations will be related to the radius of the inscribed spheres which is set at 1: $r_{i,D} = r_{i,I} = 1$. Thus:

**Dodecahedron:**

$$a = \frac{20}{\sqrt{10(25+11\sqrt{5})}} r_i = 0.898056..r_i$$

$$r_c = \frac{a\sqrt{3}\,(1+\sqrt{5})}{4} = 1.258409..r_i$$

$$A = 3a^2\sqrt{5\left(5+2\sqrt{5}\right)} = 16.650873..r_i{}^2$$

**Icosahedron:**

$$a = \frac{12}{\sqrt{3}(3+\sqrt{5})} r_i = 1.323169..r_i$$

$$r_c = \frac{a\sqrt{2\,(5+\sqrt{5})}}{4} = 1.258409..r_i$$

$$A = 5a^2\sqrt{3} = 15.162168..r_i{}^2$$

$$V = \frac{a^3\left(15 + 7\sqrt{5}\right)}{4} = 5.550291..{r_i}^3 \qquad V = \frac{5a^3\left(3 + \sqrt{5}\right)}{12} = 5.054056..{r_i}^3$$

with a: side-length of the pentagon in the dodecahedron and of the triangle in the icosahedron;
$r_c$: radius of the circumsphere; A: area; V: volume

For the next calculations one must first determine the radii of the circum- and inspheres. The necessary steps are explained here using the dodecahedron-star as the example.

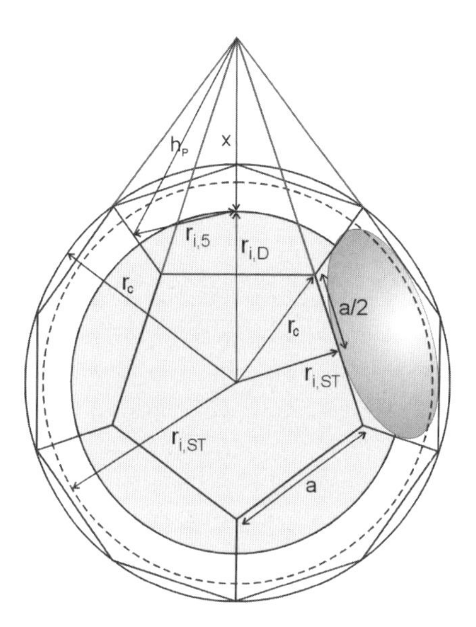


**Fig. 14.18** *Dodecahedron with star-point. For reasons of perspective, not all the sizes are to scale*

The insphere of the star with radius $r_{i,ST}$ is that which projects beyond the side areas of the dodecahedron. It touches the edges, i.e. the sides a of the pentagon, at their central points. The radius $r_c$ meets the corner points, so that $r_c$, $r_{i,ST}$ and a/2 form a right-angled triangle. In accordance with the Theorem of Pythagoras the result is:

$$r_{i,ST} = \sqrt{r_{c,D}^2 - (a/2)^2}$$

$$= \sqrt{1.258409^2 - (0.898056/2)^2}$$

$$= 1.175571..r_{i,D}$$

This is the square root from the silver section (= 1.38196..), which thus expresses the ratio of the relevant areas. The radius of the circumsphere of Kepler's dodecahedron-star is composed of $r_{i,D}$ and the length x. This is contained within a right-angled triangle of which the two other sides are formed by the height of the pentagram-triangle $h_p$ and the incircle radius of the pentagon $r_{i,5}$. We are already familiar in principle with these two numbers from the calculation of the pentagram (Appendix 2.1). Here they merely have to be related to the applied basic measure $r_{i,D}$.

$$r_{i,5} = \frac{a}{10}\sqrt{25 + 10\sqrt{5}} = 0.618034..r_{i,D} \quad h_P = \sqrt{5}\; r_{i,5} = 1.381966..r_{i,D}$$

$$\Rightarrow \ x = \sqrt{h_P^2 - r_{i,5}^2} = \sqrt{5} - 1 = 1.236068..r_{i,D}$$

$$\Rightarrow \ r_{c,ST} = 1 + \sqrt{5} - 1 = \sqrt{5} \ r_{i,D}$$

The radii of the in- and the circumsphere of the icosahedron-star arise accordingly. That of the insphere is:

$$r_{i,ST} = \sqrt{r_{c,I}^2 - (a/2)^2} = \sqrt{1.258409^2 - (1.323169/2)^2} = 1.070467..r_{i,I}$$

And the radius of the circumsphere can be found via the triangle (calculation, see Appendix 1.3) in the icosahedron. The triangle side a is also the pentagon side of the pentagram (see Fig. 3.9):

$$r_{i,3} = \frac{a}{2\sqrt{3}} = 0.381966..r_{i,I}$$

Since a = 1.453085.. $\star r_{i,5}$ in the pentagon, the ratio $h_p/a$ can be calculated in order to determine $h_p$:

$$h_P/a = \frac{\sqrt{5}}{1.453085..} = 1.538842.. \Rightarrow \ h_P = 1.538842..a = 2.036148..r_{i,I}$$

$$\Rightarrow \ x = \sqrt{h_P^2 - r_{i,3}^2} = \sqrt{2.036148^2 - 0.381966^2} = 2 \ r_{i,I}$$

$$\Rightarrow \ r_{c,ST} = 1 + 2 = 3 \ r_{i,I}$$

The ratios thus arising are shown in Table 14.4. The ratios of the areas are formed via the square of the radii, and those of the volumes via the cube, i.e. the third power.

The ratios specifically discussed at the end of Chapter 3 and in Chapter 13 are shaded grey. In Chapter 3 this is the sequence icosahedron-star—dodecahedron-star—icosahedron-star in which the overall ratios of velocities, distances and orbital periods can be discovered. Where there are metamorphoses of one of Kepler's stellar solids into the other the boundary sphere is the circumsphere of the dodecahedron or, respectively, of the icosahedron, both of which of course have the same radius. The former is also the circumsphere of the inner icosahedron-star and the latter correspondingly that of dodecahedron-star (see Fig. 3.10). Thus from the radii of the circumspheres of the two Platonic solids and their star-figures there arise not only the interval of the velocities Mercury/Pluto at 2.3839.. ⋆ 1.7769.. ⋆ 2.3839.. = 10.09863 (the difference from the planetary interval of 10.09960 amounts to 0.01%), but also the other ratios corresponding to the areas and volumes by raising the radii to the powers of 2 and 3.

**Table 14.4** *Ratios of radii, areas and volumes for dodecahedron, icosahedron and their star-figures. (Insph = insphere, I = icosahedron, D = dodecahedron, Circ.s = circumsphere, ST = star)*

| | Insphere Icosahedron, Dodecahedron | Insphere Icosahedron-star | Insphere Dodecahedron-star | Circumsphere Icosahedron, Dodecahedron | Circumsphere Dodecahedron-star | Circumsphere Icosahedron-star |
|---|---|---|---|---|---|---|
| | **Radii** | | | | | |
| Insph. I, D | 1 | | | | | |
| Insph. I.ST | 1.070466 | 1 | | | | |
| Insph. D.ST | 1.175571 | 1.098185 | 1 | | | |
| Circ.s. I, D | 1.258409 | 1.175571 | 1.070466 | 1 | | |
| Circ.s. D.ST | 2.236068 | 2.088873 | 1.902113 | 1.776901 | 1 | |
| Circ.s. I.ST | 3 | 2.802517 | 2.551952 | 2.383963 | 1.341641 | 1 |
| | **Areas** | | | | | |
| Insph. I, D | 1 | | | | | |
| Insph. I.ST | 1.145898 | 1 | | | | |
| Insph. D.ST | 1.381966 | 1.206011 | 1 | | | |
| Circ.s. I, D | 1.583592 | 1.381966 | 1.145898 | 1 | | |
| Circ.s. D.ST | 5 | 4.363390 | 3.618034 | 3.157379 | 1 | |
| Circ.s. I.ST | 9 | 7.854102 | 6.512461 | 5.683282 | 1.8 | 1 |
| | **Volumes** | | | | | |
| Insph. I, D | 1 | | | | | |
| Insph. I.ST | 1.226645 | 1 | | | | |
| Insph D.ST | 1.624598 | 1.324424 | 1 | | | |
| Circ.s. I, D | 1.992806 | 1.624598 | 1.226645 | 1 | | |
| Circ.s. D.ST | 11.180340 | 9.114567 | 6.881910 | 5.610351 | 1 | |
| Circ.s. I.ST | 27 | 22.011255 | 16.619491 | 13.548735 | 2.414953 | 1 |

The ratios specifically discussed in Chapter 13 arise between the areas of the inspheres of the 'Hedgehog' star and of the dodecahedron or, respectively, of the circumspheres of the icosahedron and the insphere of its star-figure (value A = 1.38196..). The insphere areas of the two star-figures thus show an interval of $B^2$ = 1.20601... We find the square root B = 1.09818.. not only as the geometric mean or, respectively, as the interval of the relevant radii but also in the ratio of the surface areas and the volumes of the two Platonic solids (16.650../15.162.. = 5.550../ 5.054.. = 1.09818..). The order in which the ratios of the areas of the four spheres mirror one another can be depicted as follows:

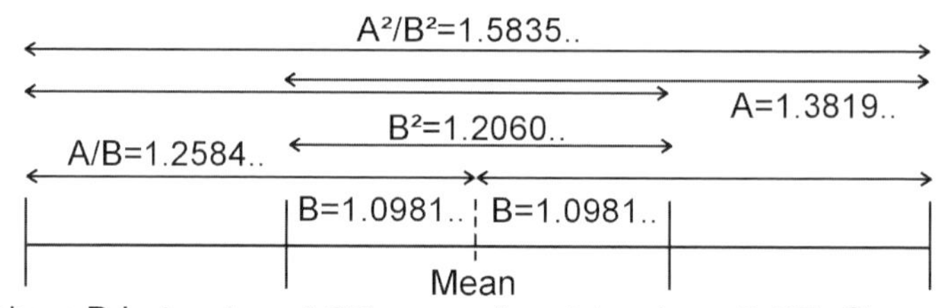

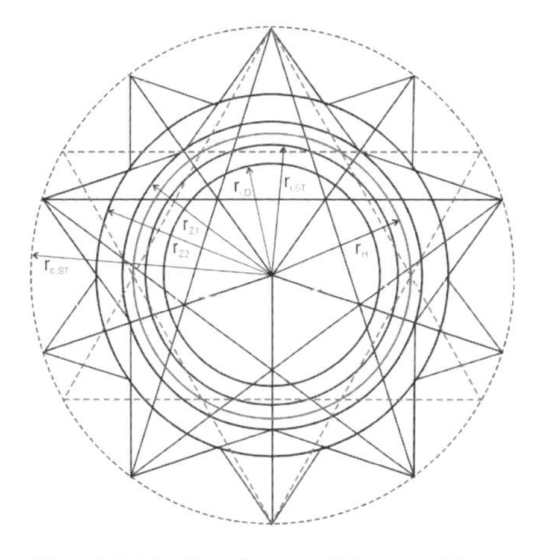

**Fig. 14.19** *Star-figures of the two side-views of the dodecahedron star*

We can now turn to the ratios of the inner planets as they appear in Fig. 13.9. The two different side-views of the 'Hedgehog' with the circles marked out by the star-figures are shown here. Once again we shall consider the area ratios. The inmost circle (with radius $r_{i,D}$) marks the insphere of the dodecahedron, the next shows the insphere of the 'Hedgehog' ($r_{i,ST}$), and the outer dotted one its circumsphere ($r_{c,ST}$). The area of the circle of the side-view of the latter is 5, as before in relation to the value of the basic circle of value 1. The middle circle ($r_H$) is produced by the inner hexagon of the hexagram. This circle—we are still talking of areas—has the ratio 1:3 (see Appendix 1.3, hexagon) to the dotted outer circle. Its size is therefore 5:3. In calculating the other intervals we refer to Appendix 2.2. The circles formed by the intersection points of the decagram with radii $r_{Z1}$ and $r_{Z2}$ are shown as $r_6$ and $r_7$ in Fig. 14.14 where the areas are $F_6$ and $F_7$. Table 14.2 shows the ratios for $F_8/F_7 = 1.894427$ and $F_7/F_6 = 1.381966$, whereas $F_8$ corresponds to the dotted outer circle in Fig. 14.19. Thus we arrive at:

$$F_{Z2} = \frac{5}{1.894427} = 2.63932..(= A^3);$$

$$F_{Z1} = \frac{2.63932..}{1.38196..} = 1.90983..(= A^2)$$

## 3. Astronomical calculations

### 3.1 General

In this section we shall present the formulae or, in some cases, give indications of astronomical methods which will in principle enable anyone to verify all the results we have been putting forward. Four factors must be taken into account in comparative studies of planetary data and movements.

***Frame of reference***

In astronomy, positional data must always be seen in connection with a specific frame of reference. All longer-term geometrical depictions only yield meaningful results, i.e. results which take into account the actual constellations in space and thus also the interactions among the planets, when they relate to a common fixed point of departure. The point in question lies at the intersection of the celestial equator with the ecliptic where the Sun is located at the time of the vernal equinox. Owing to precession, i.e. the conical turning movement of the Earth's axis over the course of about 25,800 years, the equinox must be given with reference to a specific date. At present the equinox of 1 January 2000 is the one which is applied, so all the position data given in this book refer to this.

***Time references***

Statements concerning the resonances of planetary orbital periods must also take into account the precession of the Earth's axis. In some cases, the relevant literature gives synodic periods in tropical years. A tropical year covers 365.24219 days and is thus about 0.014 days or 20 minutes shorter than the sidereal year which relates to the fixed stars. Over the course of one thousand years the difference thus amounts to about 14 days. This is irrelevant where data are mathematically compared solely in tropical years. However, longer-term geometrical considerations that are intended to take into account actual, heliocentric relationships ought to be calculated on the basis of sidereal periods. For this reason the times shown in the diagrams in this book correspond to the sidereal year, i.e. the period of the Earth's orbit of 365.25636 days.

***Conjunctions***

In general the term 'conjunction' signifies a constellation of two celestial bodies lying in a line with the Sun. The point at which this occurs is given heliocentrically in the ecliptic longitude, i.e. as an angle relating to the vernal point. This provides a specific point in time. As seen from the Earth, however, a conjunction can be depicted in various frames of reference which can lead to the relevant points in time varying by several days. In this book the conjunctions are viewed exclusively from the heliocentric viewpoint, so the reader is referred to the relevant literature for more detailed information regarding these differences.[6] The term 'conjunction' or 'true conjunction' can also be used to denote the situation in which two planets are spatially closest to one another (see Fig. 14.20 in Appendix 3.4). The Table in Appendix 6.3 lists the range of temporal differences occurring in the two types of conjunction. The deviations are rarely important in the longer-term movement-images of the planets in which the times of conjunctions are used (see the explanations preceding Fig. 6.19). As a rule, therefore, the heliocentric conjunctions given in the geometrical depictions in this book are based on the ecliptic longitude. With the help of search programs the dates of conjunctions (and other constellations) are for the most part determined to the nearest <0.01 days, while always making sure that any remaining inexactitude does not affect the diagrams.

***Orbital inclination***

In connection with the depictions of the planetary movements in the plane, the inclination, i.e. the small angle of inclination of a planetary orbit to the ecliptic, can be ignored as it is negligible in the scale used here.

## 3.2 Kepler's Third Law

Johannes Kepler's Third Law states the relationship between a planet's orbital period T and its mean distance from the Sun a. It is valid provided that the mass of a planet compared to that of the Sun is considered to be negligible. When related to a specific basic unit, e.g. the relevant values of Earth's orbit, the law states that the square of the orbital periods is equal to the cube of the mean distances. (Strictly speaking the conversion ought to involve a normalizing constant which, however, we can ignore.)

$$T^2 = a^3 \text{ can be transformed into } \frac{T^2}{a^3} = 1$$

$$\left(\text{e.g. for Jupiter: } \frac{11.862..^2[\mathit{years}]}{5.201..^3[\mathit{AU}]} = 1\right)$$

This relation is valid for every planet, i.e. for two planets with parameters $T_1$ and $a_1$ or, respectively, $T_2$ and $a_2$, we have:

$$\frac{T_1{}^2}{a_1{}^3} = \frac{T_2{}^2}{a_2{}^3} \text{ and thus: } \frac{T_1{}^2}{T_2{}^2} = \frac{a_1{}^3}{a_2{}^3}$$

The latter equation is, by the way, the original form of the Third Law as Kepler formulated it (in words) in *The Harmony of the World.*

If we now make $\frac{a_1}{a_2}$ equal to 100, the result is:

$$\frac{T_1{}^2}{T_2{}^2} = 100^3 = 10^6 \Rightarrow \frac{T_1}{T_2} = \sqrt{10^6} = 1000$$

## 3.3 Velocities and distances; angles according to Johannes Kepler

***Velocities***

The following are the formulae needed to calculate the velocities at various points along the elliptical orbit:

Mean velocity:

$$v_m = \sqrt{\frac{G * M}{a}} * 10^{-3} \left[{}^{km}/_{sec}\right]$$

Velocity at perihelion:

$$v_p = \sqrt{\frac{G * M}{a} * \left(\frac{1+\epsilon}{1-\epsilon}\right)} * 10^{-3} \left[{}^{km}/_{sec}\right]$$

Velocity at aphelion:

$$v_a = \sqrt{\frac{G * M}{a} * \left(\frac{1-\epsilon}{1+\epsilon}\right)} * 10^{-3} \left[{}^{km}/_{sec}\right]$$

Velocity at any point at distance r from the Sun:

$$v_r = \sqrt{G * M * \left(\frac{2}{r} - \frac{1}{a}\right)} * 10^{-3} \left[{}^{km}/_{sec}\right]$$

For the results in the two-body problem (see Appendix 3.4, Basic data) the relevant planetary mass m must be added to the solar mass M, i.e. ..G★(M + m) instead of ..G★M.

For the planetary masses given in the table in Appendix 6.1 as reciprocal fractional solar masses, ..G★M★(1 + 1/m′) must therefore be substituted for ..G★M.

with gravitational constant $G = 6.67259 \star 10^{-11} m^3 kg^{-1} s^{-2}$
solar mass $M = 1.98891944534281 \star 10^{30} kg$*
$\epsilon$ = numerical eccentricity of the planet
a = semi-major axis [m] (!)

The difference between ‘v at b’ and the arithmetical mean of perihelion and aphelion velocity given in Table 4.3 can now be determined as follows (whereby I here merely indicate how it is done):

The arithmetical mean of the velocity is given by $v_{am} = \frac{v_p + v_a}{2}$, so substituting for $v_p$ and $v_a$ and simplifying gives:

$$v_{am} = \sqrt{\frac{G * M}{a} * \frac{1}{1-\epsilon^2}} * 10^{-3} \left[{}^{km}/_{sec}\right]$$

For ‘v at b’ (i.e. the velocity when a planet is at a distance from the

* The constants used here and also the data for the astronomical unit (AU) are the values laid down in 1994 by the International Astronomical Union.

Sun equal to the semi-minor axis), we substitute r in the equation for $v_r$, using $b = \sqrt{a^2(1-\epsilon^2)}$ and then simplify to give:

$$v_b = \sqrt{\frac{G*M}{a} * \frac{2-\sqrt{1-\epsilon^2}}{\sqrt{1-\epsilon^2}}} * 10^{-3} \left[{}^{km}/_{sec}\right]$$

The small difference between $v_{am}$ and $v_b$, let us call it x, can now be determined by substitution and simplification in the equation $v_{am} \star x = v_b$, giving:

$$x = \sqrt{2 * \sqrt{1-\epsilon^2} - (1-\epsilon^2)}$$

For small eccentricities the factor x very quickly approaches 1.

***Distances***

A planet's mean distance or semi-major axis a, related to the astronomical unit (AU), arises on the basis of the Kepler's Third Law as:

$$T^2 = a^3 \Rightarrow a = \sqrt[3]{T^2} = \left(\frac{T}{365.25636}\right)^{\frac{2}{3}} [AU]$$

with AU = 149,597,870.691 km, T = orbital period, days

In the two-body problem a is calculated as:

$$a = \sqrt[3]{G*M*\left(1+\frac{1}{m'}\right)*\frac{1}{4\pi^2}*T^2} * \frac{1}{AU}$$

with G,M: as before, AU [m], T [sec] m′: reciprocal fractional solar mass

The semi-minor axis b is calculated via a and the linear eccentricity according to the theorem of Pythagoras; perihelion and aphelion distance pe and ap (other abbreviations are frequently used, but these show clearly what we mean) also follow from the form of the ellipse (See Appendix 1.1):

$$b = \sqrt{a^2 - e^2} \qquad pe = a - e \qquad ap = a + e$$

***Angles according to Kepler***

'Kepler's angles' (my shorthand for the angles which, as shown in Fig. 3.6, Kepler used to determine his planetary harmonies) correspond to the angular velocities. It is expedient to state them as degrees/day. They can be calculated for a point at any distance r from the Sun by means of the formula:

$$\alpha = \frac{1}{r^2} * \sqrt{GM * \left(1 + \frac{1}{m'}\right) * a * (1 - \epsilon^2) * 86400 * \frac{180}{\pi}} \quad [^\circ/_d]$$

with r, a in [m] G, M, m′ see above

## 3.4 Synodic periods, mean orbital periods, basic data

***Synodic periods***

A synodic period is the interval of time between two consecutive conjunctions or oppositions. From the point of view of one of the planets, the other appears to have accomplished one orbit relative to the Sun in this period. The eccentricities of the elliptical orbits cause the synodic period to vary around a specific, exactly calculable mean value, while the conjunction only coincides exactly with the point of shortest distance in very few cases. The following diagram shows two Mercury/Venus conjunctions in the year 2007. In the case of the second conjunction the minimum spatial distance occurring 4 days later is shown dotted in addition.

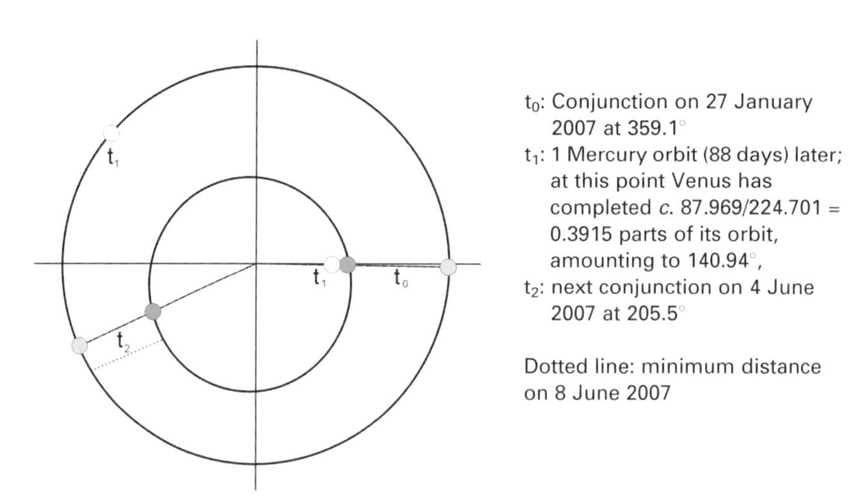


**Fig. 14.20** *Synodic period in the example of Mercury/Venus*

The condition for a synodic orbit is that the inner planet has always accomplished one orbit more than the outer one. The mathematical formulation is as follows:

$$\frac{T_S}{T_1} = \frac{T_S}{T_2} + 1$$

with $T_S$ the synodic period; and $T_1$, $T_2$ the orbital periods of inner or, respectively, outer planet

By rearrangement of the equation one arrives at the formula for calculating the synodic period:

$$T_S = \frac{T_2 * T_1}{T_2 - T_1}$$

in the example: $T_S = \frac{224.701 * 87.969}{224.701 - 87.969} =$ 144.566 *days*

The formula applies equally to the relationship of orbital and rotational periods. This becomes clear when one imagines the tip of the rotating, horizontal axis of a body as a point on an orbit. For the retrograde rotation, e.g. of Venus, the condition is that the simultaneous contrary movements of both bodies together make up one circle, or:

$$\frac{T_S}{T_1} = 1 - \frac{T_S}{T_2} \quad \Rightarrow \quad T_S = \frac{T_1 * T_2}{T_1 + T_2} \quad \text{(retrograde rotation)}$$

Let us now return to the connection (mentioned in Chapter 8) between the periods of time during which—owing to its rotation—Venus views the Sun or, respectively, the Earth, i.e. in which its axis points in the respective directions. The periods arrived at by the formulae for retrograde rotation just introduced are:

$$VSV = \frac{V * R}{V + R} = 116.7506 \textit{ days}; \quad VEV = \frac{E * R}{E + R} = 145.9277 \textit{ days}$$

where VSV is the 'Venus-Sun-View', VEV the 'Venus-Earth-View', V, E the orbital periods of Venus or, respectively, Earth, and R is Venus rotation.

Since both measures of time are determined with the help of the period of rotation, changing it would of course also influence the VSV and the VEV, so there is an obvious mathematical interdependency. In order to find an expression for this one can solve the two equations for R and then equate the respective other sides. By further somewhat more complicated manipulations which we do not need to demonstrate here we then arrive at:

$$\frac{1}{VSV} = \frac{1}{VEV} + \frac{E - V}{E * V}$$

The expression to the right of the plus-sign depicts the reciprocal of the formula for calculating the synodic period of Venus and Earth. So it can be replaced by the value ($T_{VE}$ or, respectively, $1/T_{VE}$). By rearranging the equation once again we then have:

$$VSV = \frac{VEV * T_{VE}}{VEV + T_{VE}} \quad \textit{resp.} \quad T_{VE} = \frac{VEV * VSV}{VEV - VSV}$$

If, now, 4 VEVs occur during one Venus-Earth synodic period, as happens in a very close approximation, then one arrives at the following, by substitution into the second equation:

$$T_{VE} \cong 4\ VEV \cong \frac{VEV * VSV}{VEV - VSV} \Rightarrow \frac{4\ VEV * (VEV - VSV)}{VEV} \cong VSV$$

$$\Rightarrow 4\ VEV \cong 5\ VSV$$

### *Mean orbital periods*

It turns out that the tables of planetary data, e.g. concerning orbital periods, contained in different astronomical textbooks often vary by slight amounts. Yet exact values for the mean orbital periods are a prerequisite for some of the statements made about the order in our solar system. For this reason the mean orbital periods used here (see Table 6.1) have been determined as a mean value for a period of 6000 years, i.e. from 1000 BC to AD 5000. They have been determined by the program for calculating mean orbital data (see Appendix 3.6), i.e. without consideration being given to the perturbations, which anyway balance out sufficiently over a period of this length. It is important to calculate the orbital period for at least 4 points (at 90° intervals) because the average period of time changes somewhat at various points on the orbit. The mean value of 4 points is sufficient, however, with a greater number not having any marked effect on the results.

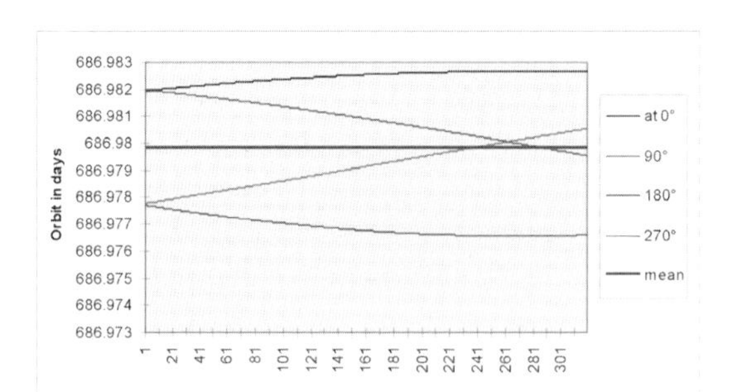


The diagram shows double the period otherwise used for determining the mean, beginning at the year −1000. The genuine mean obtained from data of four different orbital points is as good as constant.

**Fig. 14.21** *Development of the orbital period of Mars over a period of 12,000 years calculated from mean orbital data, in units of 20 Mars years. Precision < 0.00001°**

* This degree of precision, related to the individual value, is considerably higher than that obtainable via the planetary programs used. Since any small deviations do not accumulate (i.e. they cancel out), very exact mean values can be calculated in this way. (For details regarding precision see Appendix 3.7.)

***Basic data***

One can calculate the semi-major axes on the basis of the mean orbital periods with the help of Kepler's Third Law. And one can then use these values together with the eccentricities to work out other parameters such as the semi-minor axes, the velocities, etc. (see Appendix 3.3). By taking account of the planetary masses in the formulae one could obtain somewhat more exact values, but even these are not the optimum since they only provide solutions for the two-body problem (Sun with any one planet). However, especially the planets beyond Jupiter are noticeably influenced by its mass, which is anyway not involved in the calculations of the two-body problem, just as little as are the—anyway not so important—gravitational effects of the other planets. An exact arithmetical solution to the three- or the multi-body problem is anyway not possible.

Planetary calculation methods such as the VSOP Planetary Theory (see Appendix 3.6) enable us, *inter alia*, to state the distance of a planet from the Sun at a specified moment in time. So one might think of obtaining a mean value for the semi-major axis over a longer span of time, determining the distances from the Sun for any number of points within this span (either at regular or irregular intervals) and then deriving the mean value from this. As we know, the semi-major axis a of an elliptical planetary orbit represents the mean distance from the Sun. But alas, the devil is here in the detail. For it turns out that by using this method the mean value (of an unperturbed orbit) does not come out as $a$ but as the value $a * (1 + \epsilon^2/2)$. This perhaps initially surprising finding results from the fact that in the region of its aphelion the planet firstly moves more slowly and secondly is further from the Sun. In other words the larger distances are disproportionately represented in the mean value calculated on the basis of time. And, by the way, if one calculates the mean value of the distances from the Sun on the basis of equal angles (for example by dividing the ellipse into 360 one-degree angles and in each case determining the distance from the Sun) the result—once again very surprisingly—amounts to the exact value of the semi-minor axis b. Only by dividing the elliptical orbit into equal segments and determining the distance from the Sun for each section (i.e. its end point or its starting point) does one arrive at the semi-major axis a as the mean value. However, this method is not practical when using computer programs to calculate the distances as a function of time.

With the help of the VSOP Planetary Theory and relevant relatively easy-to-write search programs, however, one can very precisely obtain the exact aphelion and perihelion passages as, in each case, the largest and smallest distances from the Sun. From these distances it is then easy to determine the semi-major axes a and the linear or, respectively, the numerical eccentricities e and $\epsilon$:

$$a = \frac{ap + pe}{2}; \qquad e = \frac{ap - pe}{2} \qquad resp. \qquad \epsilon = \frac{ap - pe}{ap + pe}$$

The velocities at these points in time can also be calculated accurately in the same way.

Starting with the exact perihelion and aphelion passages and the relevant distances from the Sun in a period of 6000 years (3000 before and 3000 after 1 January 2000), we determined—in respect of each pair of these passages—the values for the semi-major axes, the eccentricities and the perihelion and aphelion velocities for each planet* from which the mean values as shown in the table in Appendix 6.1 were calculated.

However there are two further points to be reckoned with if the most accurate mean values are to be obtained.

Firstly, in the case of the four large gaseous planets one has to take into account the longer-term rhythms in which the orbital parameters oscillate somewhat. See, for example, Fig. 6.14 (right) showing the variations in the Uranus/Neptune relationship; over a period of 4300 years the orbital parameters vary in a manner similar to the positional differences shown in that diagram. If the mean values of the parameters are to be calculated for a period of 6000 years, the rhythms mentioned cannot be exactly covered, so it is obvious that slightly less accurate values would result when not taking the relevant period into account. So the data were ascertained in the subsequent periods—as far as possible centred on 1 January 2000—whereby the periodicity was determined by means of the parameter that showed it most clearly.†

**Table 14.5** *Rhythms in the outer planetary system. The temporal limits are shown by the aphelion and perihelion passages relevant to each planet.*

| | From (year before 1 Jan. 2000) | To (year after 1 Jan. 2000) | Parameter | Number of Periods | Period (years) |
|---|---|---|---|---|---|
| Jupiter | −3233.18 | 3101.46 | Maximum a | 7 | 904.95 |
| Saturn | −2810.95 | 2729.54 | Minimum a | 6 | 923.41 |
| Uranus | −3562.12 | 5008.99 | Maximum e | 2 | 4285.55 |
| Neptune | −2756.31 | 5403.56 | Maximum e | 2 | 4079.94 |

* For Pluto for a period of 2000 years (see Appendix 3.6, Pluto).

† In this rather complicated procedure, all the data which lead to the mean values shown on p. 356 can be found in the Internet at *www.keplerstern.com*, heading Calculations.

Secondly, in the case of Neptune, we have the phenomenon of a secondary aphelion or, respectively, perihelion arising in time quite close to the genuine distance from or closeness to the Sun. Between the actual and the secondary perihelion an apparent maximum then occurs which Jean Meeus terms an apheloid (or, at the opposite extreme point, a periloid). However, we shall have to leave it to the technical literature to deal with this very specialized subject in more detail.[7] It is sufficient here to point out that when determining the extreme orbital points in the case of Neptune one must always check for an 'even more extreme' point. Also, we should note that owing to the above empirical determination of some of the data, small differences from the values given in Table 6.1 can occur when the formulae in Appendix 3.3 are applied.

## 3.5 Cycle-resonance, calculation, evaluation

***Cycle-resonance***

My term cycle-resonance gives us a quantitative measure as to the precision of a geometrical figure traced by a specific planetary configuration. It states the deviation in degrees shown by two sequential occurrences of the star-figure or polygon in question. Thus the cycle-resonance gives us a means of assessing whether, for example, there is indeed a resonant, or rather a resonance-like ratio of small whole numbers, there being, of course, no one-hundred-per-cent resonance such as exactly 3:2, 4:3, 13:8, etc. regarding the orbits or the comparison of synodic periods. The examples below of the constellations Mercury/Earth and Mars at Venus/Saturn conjunctions show the significance of this measure:

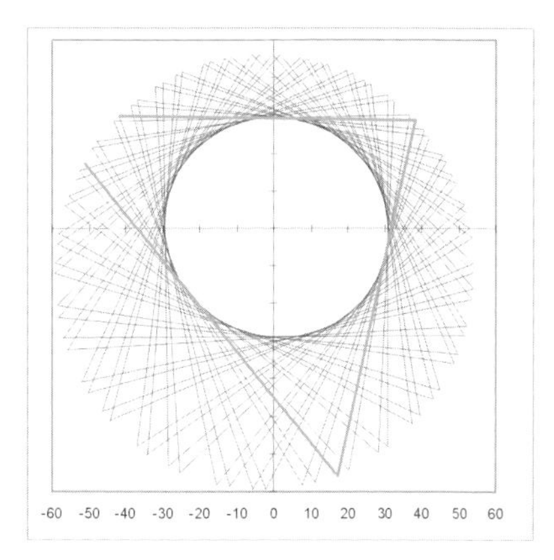


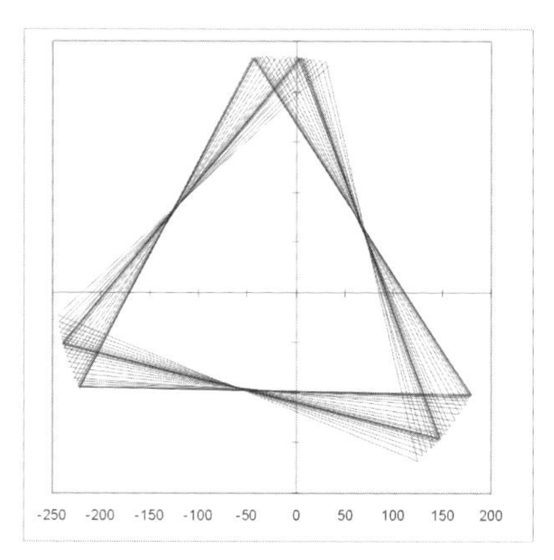


**Fig. 14.22** *Cycle-resonance. Left: Mercury at Mercury/Earth conjunctions. Right: Mars at Venus/Saturn conjunctions. In each case 83 times (82 connecting lines); period c. 26.01 or, respectively, 51.52 years; scale in millions of km*

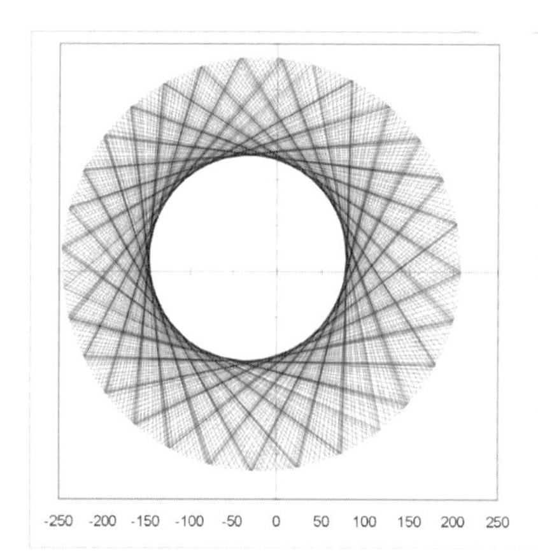


**Fig. 14.23** *Mars at Venus/Saturn conjunctions; 450 times; period c. 282.74 years*

In the first case the triangle emphasized as an example deviates from the subsequent one by an average of −17.37°—with the minus sign indicating that the figure is not yet closed. (The calculation procedure will be shown in the following.) Despite this, in some of the literature, including the scientific literature, we are told of a triangle of stellar conjunctions or else of a 4:1 resonance of the orbital periods of Earth and Mercury. In my view this is an exceedingly generous estimation. After more than three lines a 41-pointed star with a cycle-resonance of 2.606° (corresponding to a ratio of 54:13) soon* takes shape.

But after 4 cycles this, too, is no longer recognizable. The following 191:46 resonance is much stronger with a value of 0.435° for the relevant 145-pointed star. However, these are not necessarily small whole numbers, so we have to conclude that there is no commensurability in the real sense between the orbital periods of Earth and Mercury. The situation is very different in the case of the right-hand triangle which rotates on average by a mere 0.787°. One also notices that there is a further resonance superimposed on this triangle which leads over time to the formation of a 30-pointed star. In this case the triangle is not emphasized in the drawing since it is formed by the connecting lines of the positions themselves, as is apparent in Fig. 14.23. This effect is brought about by hidden correspondences between the Mars orbit, formed as it is by its eccentricity, and the rhythms of the Venus/Saturn conjunctions. (For more on such resonances, see Chapter 6.)

### *Calculation*

In the example of the Mercury/Earth conjunction, the ratio between synodic and sidereal period, which causes the formation of the relevant star-figure, is:

---

* Further whole-number ratios occur between the triangle and the 41-pointed star; in these cases the ratios of the orbital periods are *c.* 25:6 (cycle-resonance 9.99°) or, respectively, 29:7 (−7.38°)

$$\frac{Me/Ea}{Me} = \frac{T_2}{T_1} = \frac{115.877}{87.969} = 1.317 \cong \frac{I}{J} = \frac{4}{3}$$ (for one triangle; in this case $T_2$ is the synodic period)

After 3 conjunctions, Mercury has completed 3.952 orbits. A totally resonant triangle would result if it were to complete exactly 4 orbits during the period of 3 conjunctions. The cycle-resonance Kr arises from the difference through multiplication by 360°. For arbitrary time-periods $T_1$ and $T_2$ the result is:

$$Kr = \left(\frac{J * T_2}{T_1} - I\right) * 360^\circ = \frac{J * T_2 - *T_1}{T_1} * 360^\circ$$

$$for \quad \frac{T_2}{T_1} \cong \frac{I}{J}$$

When Kr is negative, the relevant geometrical figure is not closed

For this example of the calculation we thus arrive at:

$$Kr = \frac{3 * 115.877 - 4 * 87.969}{87.969} * 360^\circ = -0.04825 * 360^\circ = -17.37^\circ$$

By using a slightly modified formula one obtains the same result from the orbital periods of the two planets. In this case we shall call them U in order to distinguish them:

$$Kr = \frac{I * U_1 - J * U_2}{U_2 - U_1} * 360^\circ \quad for \quad \frac{U_2}{U_1} \cong \frac{I}{J}$$ (in this case I = 4; J = 1)

We shall now demonstrate how the cycle-resonance can be used to obtain the period which is needed for one complete revolution of a geometrical figure in the circle. The example in Fig. 2.10 is the starting point. This shows the square which Mars traces in the heavens when its position is entered every time a Venus/Earth pentagram is completed. The mean values are:

| | |
|---|---|
| Venus/Earth conjunction | E/V = 583.92137 days |
| Pentagram | Pe = E/V★5 = 2919.60685 days (= $T_2$) |
| Mars orbital period | Ma = 686.97985 days (= $T_1$) |
| Ratio pentagram/Mars | Pe/Ma = 4.249916 = 16.99966/4 |
| | ≅ 17:4 (= I/J) |

The value of the quotient pentagram to Mars denotes in this case that Mars completes 4.249.. orbits per pentagram, i.e. in 16.99.. orbits it would have completed the square. It would trace a completely resonant figure in exactly 17 orbits. Thus the cycle-resonance arises as:

$$Kr = \frac{4 * 2919.60685 - 17 * 686.97985}{686.97985} * 360^\circ = -0.120563^\circ$$

Since this is the deviation for one square (from the previous one), the figure rotates in the circle after 360/|Kr| = 2985.982 appearances. The period for this, let us call it Z, arises through multiplication by the time needed for one square, i.e. 16.99.. Mars orbits:

$$Z = \frac{2985.982 * 16.99966 * 686.97985}{365.25636\ (\textit{Earth year})} = 95471.5\ \textit{years}$$

Since 16.99.. Mars orbits correspond to 4 pentagrams, a more general formulation is also possible:

$$Z = \frac{360^\circ}{|Kr|} * \frac{J * T_2}{T_1} * \frac{T_1}{365.256..} = \frac{360^\circ * T_2 * J}{|Kr| * 365.25636} (\textit{years},\ T_{1,2}\ \textit{in days})$$

The number obtained, *c.* 95,500 years, can however only be taken as a guide value because exact data for this period of time exceed the possibilities of the computer programs used (see Appendix 3.7).

***Evaluation***

The question as to whether a planetary ratio can be assessed as being resonant depends—as we have seen—on the method used, and above all on a limit which has to be set. Since there are no absolute criteria, it seems to me to be appropriate to use the situation present in the planetary system itself. In this way we come to the following evaluation for the cycle-resonance (Kr):

| | | |
|---|---|---|
| Kr < 5° : | Weak resonance | e.g. Jupiter/Uranus conjunction |
| Kr < 2.5° : | Resonance given | e.g. Venus/Earth conjunction |
| Kr < 1° : | Strong resonance | e.g. Mars at Venus/Saturn conjunction |
| Kr < 0.2° : | Very strong resonance | e.g. Mars at Venus/Earth conjunction |

This grouping is valid for an I up to *c.* 12, corresponding to a 12-pointed star or a dodecagon. For Is with higher values the borderlines would have to be drawn progressively closer together. The above example of the Mercury/Earth conjunction shows a 41-pointed star with a cycle-resonance of about 2.6°. Thus this corresponds closely to the value of the Venus/Earth pentagram of *c.* 2.4° (see explanation to Fig. 2.8). In spite of this, the latter is much more resonant. It would still be recognizable after more than 20 appearances, whereas the 41-pointed star would cease to be discernible after the fourth. However, 41 can no longer be termed a small whole number, so there is no need to fix any further criteria.

## 3.6 Planetary programs used

Planetary positions and data from Mercury to Neptune have been calculated according to the VSOP Planetary Theory (Variations Séculaires des Orbites Planétaires*) developed by P. Bretagnon of the Bureau des Longitudes in Paris and published in 1982. A version revised for practical application followed in 1987 in collaboration with G. Francou. The data and calculation procedures needed for application of the VSOP Theory may be found (in a somewhat abbreviated form, as used here†) in *Astronomical Algorithms* by Jean Meeus.

These excellent publications make it possible in principle for anyone to calculate the planetary orbits for periods of several thousands of years to a fantastic degree of accuracy with the use of a simple PC. This being the case, I wish to put on record here my profound gratitude to P. Bretagnon, G. Francou and Jean Meeus. With regard to the investigations of the harmonic and geometric order in our solar system undertaken here, what they have achieved equals the importance of Tycho Brahe's observations for the relevant researches conducted by Johannes Kepler.

### *Mean orbits*

Jean Meeus has derived the elements for determining mean orbital data from the 1982 VSOP version. This enables us to calculate the mean positions of the planets—i.e. disregarding gravitational interactions—and also the orbital parameters which vary slowly, such as eccentricity, perihelion longitude (position of the perihelion in relation to the vernal point), etc. In addition the velocities can be ascertained for any point in time. The Kepler angle can be accurately obtained from the positions at two points in time shortly before and after aphelion or perihelion of a planet by means of subtraction of the two values. It makes sense to determine this angle (for the purposes discussed in Chapter 4) via the mean orbital data because the influence of the perturbations varies slightly at every perihelion or aphelion passage.

---

* Meaning roughly 'long-term variations of the planetary orbits'. In astronomy, secular perturbations or changes are either those which persist long-term always in the same direction (unlike those which are periodic and cause a variation around a mean value) or those which repeat non-periodically and become apparent only over very long periods of time.

† The terms of the complete version and brief instructions may be found on the FTP-Server of the Bureau des Longitudes at *www.bdl.fr* or *www.imcce.fr*.

***Pluto***

Pluto's orbit was calculated by means of JPL-Ephemerides DE406 (NASA's Jet Propulsion Laboratory). This is a numerical integration of planetary movements from 3000 BC to AD 3000. The orbital parameters were obtained in accordance with the description in Appendix 3.4 (Basic data), but only for a period of 2000 years, i.e. 1000 before and after 1 January 2000. As Pluto's semi-major axis and its eccentricity grow continuously during the integration span, the determination of a mean value over 6 millennia would not be centred on our present time. This would mean firstly that the values would probably be too low and secondly that the reference period would not concur with that of the other planets.

***Moon***

Calculations of the lunar orbit are based on the lunar ephemeris ELP-2000/82 (Éphémerides Lunaires Parisienne) of M. Chapront-Touzé and J. Chapront published in 1983.[8] The necessary terms and procedures and some corrections published by the authors in 1988 are to be found in Jean Meeus' book. The positions to be obtained relate to the equinox of the date* and have been converted to that of 1 January 2000.

## 3.7 Accuracy and validity

***General***

In the first place one has to ask how it is possible to ascertain the degree of accuracy of astronomical calculations. And secondly we should clarify what degree of accuracy is needed in order to secure the validity of the results put forward here.

Calculations of the orbits can and must, of course, be tested against the observed movements of the planets or, conversely, the development of the procedures includes reference to the actual positions. There are many very accurate data that can be applied with reference to periods covering several centuries. And there are also historical references to unusual celestial events which can be included in program testing, such as perhaps the most famous example, already mentioned in Chapter 5, namely the Jupiter/Saturn conjunction in the year 6 BC, which was recorded by the neo-Babylonian school of astrology at Sippar.[9] Jean Meeus gives the exact point in time in the form of two oppositions, visible from Earth, of both planets in relation to the Sun: Jupiter 15 September 6 BC, 07h; Saturn 14 September 6 BC, 09h TDT.[10] (In the terminology used here: an Earth/

---

* i.e. no fixed point of reference, but the vernal point which moves with the precession.

Jupiter and an Earth/Saturn conjunction.) Use of the abbreviated version of the VSOP Theory shows these values with a deviation of less than two hours. In short, we can surely assume that the accuracy of the calculation procedures in question, which will be discussed next, is indeed entirely reliable.

There are two aspects to be taken into account regarding the necessary accuracy: firstly, that of the calculation of the orbital data shown in the table in Appendix 6.1 (as mean values over 6000 years), from which we have derived e.g. the ratios of the velocities and of the semi-minor axes; and secondly, the positions on which the geometrical figures are based. The only decisive factor in both cases is that the error should not accumulate above a specific amount. In the depictions of the geometrical figures a possible deviation would not become optically visible below, say, 0.5° in the case of a position marked, for example, by a point or two planetary positions joined by a line. Similarly an error of this magnitude in the calculation of mean orbital periods and eccentricities over 6000 years would even out to a negligible amount. The prerequisite here in every case is that rather than accumulating, the differences must vary periodically around a mean value.

***Calculating the planetary orbits***

P. Bretagnon and G. Francou give the following data with regard to the accuracy of the VSOP 87 version: the error in every case remains lower than one arcsecond (1/3600 degree of the full 360° circle), for Mercury to Mars 4000 years, for Jupiter and Saturn 2000 years, and for Uranus and Neptune 6000 years, in each case before and after 1 January 2000.[11] These values relate to the full version of the Theory. According to formulae given by P. Bretagnon or, respectively, Jean Meeus, the deviations of the results arising from the abbreviated version used here over against the full version can be calculated with reference to the time factor. Seven thousand years (before and after 1 January 2000) show a difference in ecliptic longitude of maximum 0.33° in the case of Neptune and of less than 0.1° for all the other planets.

So the slightly lower accuracy of the shorter version can be regarded as irrelevant as regards the geometrical figures of the planets. All the movement pictures arising within the above time periods of ± 2 to 6 millennia, in which the exactitude is in the range of one or a few arcseconds, are therefore verified. So all that remains is to look more closely at the data for the formations of the outer planets pictured in Chapter 12, because here the period of 4000 years (2000 years before and after 1 January 2000 as the limit given for Jupiter and Saturn) is exceeded. In order to estimate the degree to which the VSOP Theory provides qualitatively reliable results which are also sufficiently accurate for our purposes beyond the limits set by P. Bretagnon, we shall now show for several planets examples of the differences between the VSOP procedure and the positions calculated on the basis of mean orbital data, i.e. disregarding gravitational interactions.

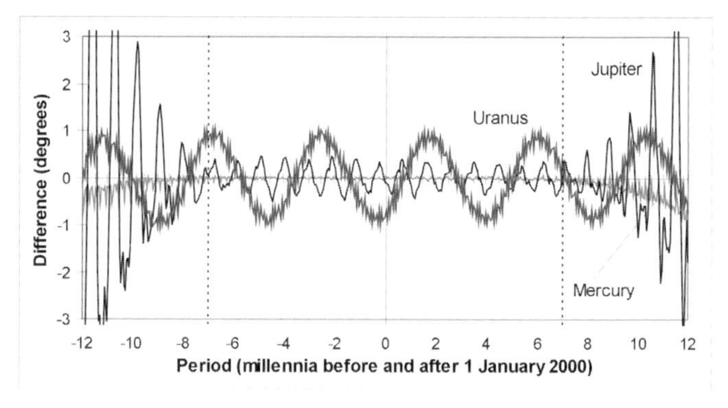


**Fig. 14.24** *Deviations of the calculated positions (in ecliptic longitude) between VSOP 87, shortened version, and calculation based on mean orbital data; stepping interval 50 years. The differences in the case of Mercury are shown magnified 8-fold*

Very similar curves arise for the other planets. We see that for about ± 7 to 8 millennia the differences fluctuate periodically within boundaries that are typical for each planet, and that compared with the mean orbital data the accuracy of the VSOP Theory only begins to deteriorate noticeably after that or for some of the planets later still. For at least the first few millennia the accuracy, as shown above, of the VSOP Theory is very high. So the fluctuations shown within the area marked by the dotted lines depict the periodic error of the calculations with the mean orbital data—i.e. of the regularly evening out influence of the perturbations. On the basis of the periodicity of deviations in these calculations we can therefore assume that we shall obtain sufficient accuracy by calculating the mean values of orbital periods and eccentricities via the mean orbital data.

Furthermore no noticeable difference is detectable between the first millennia, in which the errors can amount at most to a few arcseconds, and the rest of the period within ± 7000 years. This leads us to conclude that the deviations arising in the results of the VSOP calculations at the edges of the marked area of ± 7000 years can also not be very much greater and certainly remain considerably below 0.5°. Furthermore the calculations based on mean orbital data provide results that are sure to be qualitatively true for several tens of millennia. This is shown by the fact that movement pictures arising from conjunction data over about 30,000 years still show figures that are principally the same.

***Calculating Pluto's orbit***
There is no need to discuss this further here since we can rely on NASA's Ephemerides DE406 being among the most accurate calculations of planetary movements ever.

***Calculating the Moon's orbit***
In the ELP 2000-85 version (a further improvement of ELP 2000), a maximum error of *c.* 20 arcseconds is given for historical times, going back to 1500 BC.[12] Even in ELP 2000/82 the error is already very much lower for the initial centuries. In the version he has published—although this is shorter by a few terms but contains elements of the improved version—Jean Meeus states an accuracy of 10 arcseconds without, though, specifying any validity period. However, from the other data given one can assume that this includes the stated accuracy for at least several centuries. (The longest period for the depictions regarding the Moon in Chapter 9 is *c.* 180 years.)

## 4. *Probability calculations*

### 4.1 Ascertaining the harmonic probability

The diagram in Fig. 4.3 requires alteration only to the extent that instead of differences we now have to show quotients because we are dealing here with musical ratios. In consequence the deviation from complete correspondence with a given interval (or tone value) now has to be reached via the geometric mean. Hence the equation:

$$\text{Harmonic probability} = \frac{\text{Test value} - \text{Tone value 1}}{\sqrt{\text{Tone value 2} * \text{Tone value 1}} - \text{Tone value 1}}$$

A simple program can be written for the investigation of large numbers of test values. This will initially determine the tone values 1 and 2 relating to a test value and on this basis then show the harmonic probability in accordance with the above formula.

## 4.2 The t-test

The t-test is the most appropriate procedure for testing small samples where the variance of a population is unknown, on condition that the latter is distributed normally. Here the population is the set of all possible deviations from the 12 musical intervals; the values to be investigated are the sample. The precondition of normal distribution is given.

The probabilities quoted in Chapter 4 are the values arising from a one-tailed test. This signifies that our only concern is whether the planetary intervals come close to harmonic ratios and not whether they are especially far off these. The latter would of course also be a deviation from a random distribution. This explains why maximum probabilities of 0.5 appear in the tables in Chapter 4 even where the mean value of the harmonic probability is almost exactly 0.5. Were the mean value to be 0.5 in relation to a two-tailed analysis there would of course be an almost 100% certainty that the distribution was random.

The value T of the t-test now comes out in the calculation as:

$$T = \sqrt{n} * \frac{x - m}{s}$$

with:

n: Number of test values (of the sample)
x: Mean value of the sample
m: Mean value of the population
s: Standard deviation of the sample

The mean value of the population is 0.5. With the help of tables or with an appropriate computer program the relevant probability p can then be ascertained or, respectively, calculated from the value T in accordance with the t-test. In doing this one must take account of the number of degrees of freedom. This is n-1. Remember that some computer programs offer two standard deviations. Choose the one relating to a sample, not the one relating to a population.

For the example in Chapter 4 of the velocities of Mercury 'at b' compared with the 17 other possible values of the planets 'at b' and at aphelion, we thus arrive at the following (please accept the values for x and s or else check against Tables 4.3 or Appendix 6.4 for the velocities given there):

$$T = \sqrt{17} * \frac{0.23738 - 0.5}{0.20552} = -5.268 \Rightarrow p = 0.0000382\ (0.0038\%)$$

for 16 degrees of freedom (where p = probability)

Purely mathematically it does not matter to which of the 18 possible key-notes one relates the intervals (see Fig. 4.7). In order to take this into account I used a suitable computer simulation with 'random planetary systems' to ascertain the probability that can actually be expected statistically. (I do not know of an exact mathematical procedure for this.) The random numbers were arrived at as follows:

1st number between 40 and 60 to match the approximate velocity of Mercury 'at b';
2nd to 9th number: the previous value is multiplied by a random number between 0.5 and 0.9 to match the range in our planetary system between the velocities of neighbouring planets;
10th to 18th number: division of the 1st to 9th value by a random number from 1 to 1.35 to simulate the possible differences between v 'at b' and the aphelion velocity.

This arrangement is not of course to be described as a realistic simulation of a planetary system; it serves merely to delineate the layout of the planets in case this should turn out to be having an effect on the result. Random numbers could lie very close together, whereas one must presuppose for the planets that they have a minimum distance from one another as well as a maximum eccentricity which limits the difference between a planet's two velocities.

This enables us to find out not only how often the result of the simulation falls below the above-mentioned value T, but also how probable it is for the mean value of Mercury 'at b' to be attained. Since both the mean value and also the standard deviation are included in the calculation of T, it is possible for the two results to differ somewhat. However, in our case the result for the mean value of the harmonic deviations which is so close to the real musical relationships should be the decisive factor. In an analysis of 50,000 series, from which in each case 18 different possibilities (relating to different key-notes) can be formed, the value of $T < -5.268$ was found 147 times and a mean value of $< 0.23738$ was found 62 times. This gives us a probability (p) of

$$p = \frac{62}{50.000 * 18} * 100 = 0.0069\%$$

In order to take account of the fact that the velocity ratios were selected from various investigated possibilities, this value must then be multiplied by 9*:

$$9 * p = 9 * 0.0069 = \underline{\underline{0.062\%}}$$

The extent of this trial can be regarded as being entirely sufficient for the achievement of a safe result. This can be checked against the continuously determined mean values of the mean values of the results of individual test runs of, e.g. 10,000 trials in each case. In this instance the overall result stabilizes after 40,000 planetary systems.

## 4.3 The probability of the star-figures

To calculate the probability of a star-formation arising one must clarify the connection between the orbital periods and the extent of the deviation from a full resonance. The probability can then be determined for any difference. With the Venus/Earth pentagram, for example, the deviation or in other words, the value of the cycle-resonance is Kr = 2.4°. During one synodic orbit Venus completes 2.5986.. orbits and Earth exactly one fewer. In Chapter 6 we explained that for the geometrical formation of a star-figure one needs only the number after the decimal point, which we shall call x. In a random distribution a pentagram would arise with equal probability if the relevant number were to exceed the value of an exact resonance of r = 0.6 in the other direction, as the following sketch shows:

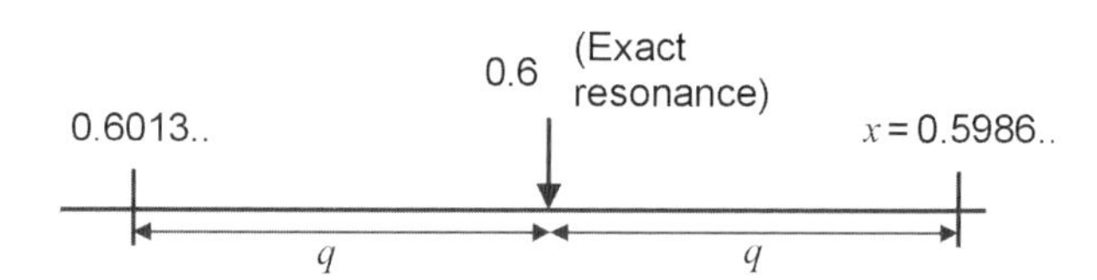


The expression q = r − x denotes the one-tailed probability for the formation of a pentagram. In this example, q = 0.6 − 0.59866 = 0.00134, i.e. in a random distribution an average of 13 out of ten thousand values between 0 and 1 would lie in the region of q. One must take into account, though, that q is present on both sides and that the value r = 0.4 also gives

* The number nine arises from all the possibilities of combining extreme values (at perihelion and aphelion) and mean values. For example one can form the intervals perihelion/aphelion, perihelion/mean value and mean value/aphelion for the velocities. The same goes for the distances and the angular velocities, so that in this way 9 conceivable possible combinations arise which must be taken into account in the probability calculation, since any of them might theoretically have led to exceptional ratios.

rise to a pentagram (just as all other star-figures are formed symmetrically). Therefore q must be multiplied by 4:

$$p = 4 * q = 4 * (r - x)$$

Thus we obtain a probability of p = 4⋆0.00134 = 0.00536 = 0.536% or 1:186 for the formation of a pentagram when the deviation from a full resonance is as stated above.

In addition the probability can be related to the value of the cycle-resonance. Here one must ascertain the connection between this and the deviation q. The latter occurs in each of the conjunctions in the sequence. So for the Venus/Earth pentagram its value would first have to be multiplied by 5 and then converted to the 360° circle. For any polygon or star-figure the following is thus valid:

$$Kr = q * n * 360° \Rightarrow p = \frac{4 * Kr}{n * 360°}$$

(where Kr = cycle resonance and n = number of points of the polygon or star-figure)

Table 14.6 shows the probabilities obtained on this basis for all polygons and star-figures up to the number 12 with a maximum deviation of Kr = 3°. The one-cornered and the two-cornered polygon (dot and line where r = 0 or 1 or, respectively, r = 0.5) can only come about in one way, so that they show in column ① as 0.5 because the otherwise doubled possibility of the formation is included in the calculation of p. Columns ③ and ⑥ show the probabality with which the figures up to the corresponding n ought to be found in our planetary system. Here one must note how many constellations there are in which a formation would be theoretically possible. An octagon, for example, can only come about when the ratios of the orbital periods are maximally 9:1 or 15:7, which corresponds to the values r = 0.125 or, respectively, r = 0.875. The first condition is fulfilled by 15 and the second by 4 real constellations. When one calculates these figures for every polygon up to the dodecagon one arrives at an average of *c.* 10, which—somewhat simplified but for our purposes perfectly adequately—shows up in column ③. In the case of the star-figures the corresponding average is about 9, but here, too, we might as well take this to be 10.*

* A pentagram can arise in two ways: firstly if the orbital periods of two planets differ by a maximum factor of 3.5, since the smallest possible ratio is 7:2 (corresponding to the 4 after the decimal point, as explained above). In our solar system this is the case for 9 pairs. The second possibility is limited by the ratio 8:3 (corresponding to the 6 after the decimal point). In the planetary system 6 intervals of orbital periods come within this measure. There thus arises an average of 7.5 and for the deviation of 3° an overall probability of 7.5⋆0.00667 = 0.05 (or 5% or 1:20). For a cycle-resonance of 2.4° the value is 1:25.

Hence, seen purely statistically, there ought to be 0.78 polygons and 0.46 star-figures up to n = 12 in the conjunction relationships of the planets in the solar system. And in total this does indeed fit very nicely with the actual situation.

**Table 14.6** *Probability of formation of polygons and star-figures at maximum deviation of 3° from full resonance*

| N | Possibilities for polygons | Probability *p* | ② * 10 summed up | Possibilities for star-figures | Probability *p* | ⑤ * 10 summed up | Combined probability |
|---|---|---|---|---|---|---|---|
| | ① | ② | ③ | ④ | ⑤ | ⑥ | ⑦ |
| 1 | 0.5 | 1/60 | 0.1667 | — | — | 0 | 1/60 |
| 2 | 0.5 | 1/120 | 0.2500 | — | — | 0 | 1/120 |
| 3 | 1 | 1/90 | 0.3611 | — | — | 0 | 1/90 |
| 4 | 1 | 1/120 | 0.4444 | — | — | 0 | 1/120 |
| 5 | 1 | 1/150 | 0.5111 | 1 | 1/150 | 0.0667 | 1/75 |
| 6 | 1 | 1/180 | 0.5667 | — | — | 0.0667 | 1/180 |
| 7 | 1 | 1/210 | 0.6143 | 2 | 1/105 | 0.1619 | 1/70 |
| 8 | 1 | 1/240 | 0.6560 | 1 | 1/240 | 0.2036 | 1/120 |
| 9 | 1 | 1/270 | 0.6930 | 2 | 1/135 | 0.2776 | 1/90 |
| 10 | 1 | 1/300 | 0.7263 | 1 | 1/300 | 0.3110 | 1/150 |
| 11 | 1 | 1/330 | 0.7566 | 4 | 2/165 | 0.4322 | 1/66 |
| 12 | 1 | 1/360 | 0.7844 | 1 | 1/360 | 0.4600 | 1/180 |

## 4.4 Probability calculations in Chapter 11

From the 15 values taken into account in Fig. 11.5 one can create 105 intervals (1 + 2 + 3 ... + 14). This is the first step to calculating a harmonic overall probability in accordance with the procedures described in Appendix 4.1 and 4.2. Because the 105 values are not independent of one another they lead to an extremely low figure which still awaits correction. We have here 14 values which are independent of one another and which make a statistical estimation possible. The lower graph shown in Fig. 11.5 gives as examples the 14 intervals which arise for the proton as the 'key-note'. In this case the mean value of the harmonic deviation is 0.2951 and the standard deviation is 0.1945. This leads to a random variable T of −3.942 and a probability in accordance with the t-test of 0.084%, i.e. *c.* 1:1200 that this degree of harmony could have come about by chance. However, to ascertain the overall harmonic probability one must take all possibilities into account (see Fig. 4.6 and the explanation thereto). A computer analysis with random numbers—according to the method described in Appendix

4.2—shows that this is lower still by the factor of 1.7, i.e. approximately 1:2000.*

Another aspect to be investigated is whether the particle masses possess an inner connection which leads inevitably to harmonic ratios. For example the atomic weights of the light chemical elements have ratios close to small whole numbers because they are composed mainly of the masses of protons and neutrons, so that correspondingly formed intervals are naturally very close to musical ratios. What is the situation in this respect as regards the masses of the elementary particles? The mass of some particles was predicted before they had even been discovered. There are particles whose mass cannot be deduced from models developed so far, e.g. the leptons,[13] and others for which the mass can be estimated from already-known values. Here are two formulae by means of which the ratios of masses of certain particle groups can be determined:[14]

$$\frac{M_N + M_X}{2} \cong \frac{3M_L + M_S}{4}; \quad M_K^2 \cong \frac{3M_E^2 + M_P^2}{4}$$

with M: Mass; the indices stand for Neutron, Xi, Lambda, Sigma, Kaon, Eta and Pion

The formulae have no further importance here, except that from them—separately for each equation—the intervals of the particle masses were derived and the mean values of the harmonic deviations calculated. The result is a value of 0.415, which is considerably higher than the mean value of all the 105 intervals. It follows that the above-mentioned relations between the particle masses cannot contribute substantially to the correspondences with harmonic intervals. Since a number of other masses can also not be deduced, one can say that the harmony discussed here cannot be attributed to already known relationships.

## 4.5 Probability calculations in Chapter 13

The hypothesis under investigation states that the configuration of the semi-minor axes shown in Fig. 13.14 is entirely random. Having first

---

* The 105 intervals show a mean value of 0.3498, a standard deviation of 0.2438 and the value T equalling −6.3107. In 100,000 trials with in each case 15 random numbers between 0.5 and 5800 (corresponding to the range of the elementary particle masses) the result falls below the stated mean value 53 times and below the T value 44 times. This gives a probability of 0.053% or, respectively, 0.044% which amounts to a mean of *c.* 1:2000.

determined the probability in general we shall then proceed to introduce the results of a computer simulation which includes data pertaining to the facts of the planetary system.

We begin by listing all the intervals in the following table in which all the ratios are related firstly to Mercury and secondly to the Sun. This gives a clearer picture and, as we shall see, enables us to reach a relatively simple estimation of the probability. The ratios are also listed as exponents, whereby only exponents to the bases 2, 3 and π exist. This means, for example, that the Mars/Mercury ratio $\cong 4 = 2^2$ for base 2 has the exponent 2 and for the others the value zero. The Venus/Mercury interval $\cong 6/\pi$ can be written as $2^1 \star 3^1 \star \pi^{-1}$, which means that it has the exponents 1, 1 and −1. In this way all the intervals relating to any planet can be obtained by means of adding or subtracting the exponents.

**Table 14.7** *Ratios of the semi-minor axes or, respectively, the solar radius: (a) relating to Mercury, (b) relating to the Sun*

| | Ratio (/Me) | In decimals | Exponents rel. to Me | | | Planetary intervals | Differ. (%) | Exponents rel. to Sun | | | Planetary intervals | Differ. (%) |
|---|---|---|---|---|---|---|---|---|---|---|---|---|
| | | | 2 | 3 | π | | | 2 | 3 | π | | |
| So | $\pi/2^8$ | 1/81.4873 | −8 | 0 | 1 | 1/81.3945 | 0.114 | 0 | 0 | 0 | 1 | 0 |
| Me | 1 | 1 | 0 | 0 | 0 | 1 | 0 | 8 | 0 | −1 | 81.3945 | 0.114 |
| Ve | $6/\pi$ | 1.90986 | 1 | 1 | −1 | 1.90935 | 0.027 | 9 | 1 | −2 | 155.4109 | 0.141 |
| Ea | $(6/\pi)^{3/2}$ | 2.63938 | 3/2 | 3/2 | −3/2 | 2.63936 | 0.001 | 9.5 | 1.5 | −2.5 | 214.8295 | 0.115 |
| Ma | 4 | 4 | 2 | 0 | 0 | 4.00455 | 0.114 | 10 | 0 | −1 | 325.9482 | 0.000 |
| Ju | $(18/\pi)^{3/2}$ | 13.71463 | 3/2 | 3 | −3/2 | 13.71788 | 0.024 | 9.5 | 3 | −2.5 | 1116.560 | 0.090 |
| Sa | $8\pi$ | 25.13274 | 3 | 0 | 1 | 25.15189 | 0.076 | 11 | 0 | 0 | 2047.226 | 0.038 |
| Ur | $16\pi$ | 50.26548 | 4 | 0 | 1 | 50.60730 | 0.680 | 12 | 0 | 0 | 4119.157 | 0.565 |
| Ne | $8\pi^2$ | 78.95684 | 3 | 0 | 2 | 79.36936 | 0.522 | 11 | 0 | 1 | 6460.231 | 0.408 |
| Pl | $32\pi$ | 100.5310 | 5 | 0 | 1 | 100.9463 | 0.413 | 13 | 0 | 0 | 8216.475 | 0.299 |

This depiction of the exponents once again serves to show how simple and clear the order is without, however, imposing a rigid pattern on it. As explained in Chapter 13, we can only take account of the simplest basic pattern which in terms of geometry involves circumscription by squares and circles. An additional degree of structuring as was shown in other constructions, e.g. use of the triangle to depict the Jupiter/Earth ratio, could even further reduce the probability that we are looking at a random order, but this cannot be taken into consideration here.

Beginning with a basic circle, circumscription with a square indicates the ratio $4/\pi$. The next circle leads to the value 2, the next square to $4/\pi \star 2 = 8/\pi$, the following circle to 4, the following square to $16/\pi$, and so on. Thus all the values can be expressed by means of two series of numbers: (a) $2^2/\pi$, $2^3/\pi$, $2^4/\pi$ or generally $2^n/\pi$; or, respectively, (b) $2^1$,

$2^2$, $2^3$ or $2^n$. In relation to the solar radius we now see in the depiction of the exponents that the Mercury/Sun and Mars/Sun intervals belong to series (a) and Saturn/-. Uranus/- and Pluto/Sun to series (b). Thus we have 5 out of 9 intervals which obey the basic requirement (with a certain mean difference). The next thing to be clarified is how high the probability is that a value arising from a random distribution will come close to that of one of the members of the two number series with a predetermined deviation which, as we have said, is here shown by the mean difference. With the help of binomial distribution* we shall then be able to calculate how high the overall probability is that in a random selection at least 5 out of 9 values will correspond to the two number series or, respectively, to the geometrical sequence of circle-square-circle etc.

As we have already said, the first thing to clarify is how likely it is that a random interval will come close with a specified degree of accuracy to a member of one of the stated number series. As an example let us take the number four which is to be obtained with the deviation a of maximally 1%. It is a member of the series $2^n$, and its neighbours are the numbers two and eight. The relevant interval range encompassing the random numbers that could come close to the number four arises from the geometric mean of 2 and 4 or, respectively, 4 and 8. So the extent of the range reaches from $2\sqrt{2}$ to $4\sqrt{2}$. Smaller or greater values would be closer to 2 or, respectively 8. The size of the range is thus $4\sqrt{2} - 2\sqrt{2} = 2\sqrt{2}$, i.e. in round numbers 2.83. Let us now assume the random values to be subdivided into discrete steps of the size 0.01. There would then be 283 possibilities of which 8 would fulfil the condition of approaching <1%, namely those of 3.96 to 4.03. The probability which corresponds to this is 8/283 = 0.0283. This corresponds to double the permissible deviation multiplied by the initial value 4 divided by the value of the interval range. As a generalization this can be stated as follows:

- Limits of the interval range for $2^n$:

  $g_1 = \sqrt{2^{n-1} * 2^n} = \sqrt{2} * 2^{n-1}$; $g_2 = \sqrt{2^n * 2^{n+1}} = \sqrt{2} * 2^n$

- Probability that a random interval with deviation a will reach the value $2^n$:

  $$p = \frac{2a * 2^n}{g_2 - g_1} = \frac{2a * 2^n}{\sqrt{2} * 2^n - \sqrt{2} * 2^{n-1}} = \frac{2a * 2^n}{\sqrt{2} * 2^{n-1}} = a * 2\sqrt{2}$$

  (1>a>0, i.e. 0.01 corresponds to 1%)

---

* This is a standard procedure of probability calculation. For more detail see a mathematical textbook.

From this it follows that probability p is the same for every n (since the absolute value of the deviation also increases, e.g. a = 0.01 for $2^4$ leads to 0.16). Furthermore it transpires that p also has the same value for other series that are based on the number two (e.g. $2^n/\pi$), because if we substitute the series x★$2^n$ generally in place of $2^n$ the x in the above calculation cancels out. So for the two number series the probability we are looking for comes to:

$$p = a * 4\sqrt{2}$$

The mean deviation of the 5 listed planetary intervals from the ideal mathematical values is 0.203%. Since deviations are possible both upwards and downwards, the value has to be doubled, so that one arrives at 2a = 0.00406 or, respectively, 0.406%. The probability p that a random value belongs to one of the number series mentioned with a deviation of a is p = a★2★ $\sqrt{2}$ = 0.00574. Accordingly the probability for two series is p = 0.01149, i.e. the single probability is somewhat above 1%. The binomial distribution now tells us that the overall probability for 5 out of 9 intervals (5 successes out of 9 trials) is 1:41 million, i.e. this is approximately the probability that the ordering of the semi-minor axes as shown could have come about purely by chance. A value of 1:270,000 still remains if one reduces the anyway cautious starting point even further in order to take account of certain temporal oscillations of the ratios over the course of the millennia, for example by basing the calculations on the maximum instead of the mean deviation (0.565% for Uranus/Sun). Alternatively one might take account solely of the planetary ratios. In this case the 4 intervals Mars/-, Saturn/-, Uranus/- and Pluto/Mercury belong to the two series with a mean difference of 0.321%, so that a probability of 1:142,000 arises for 4 of the 8 values.

A computer simulation has been used to test the result for the planetary intervals, i.e. without account being taken of the Sun. The initialization was set so that the values of in each case 8 random numbers increased by values between 1.253-fold and 2.088-fold, i.e. every subsequent number was increased within these limits by a random factor over against the previous one. (In the actual planetary system, the values for the ratios of the semi-minor axes extend from 1.271 to 2.012.) The first value, corresponding to the Venus/Mercury interval of 1.909 was limited to the region of 1.5 to 2.338. Then the gap between Mars and Jupiter was taken into account in that at this point a double multiplication by fictional numbers within the set limits was undertaken. It became apparent that the probabilities turn out to be somewhat higher than those given by the binomial distribution. With a permissible deviation of a = 0.003208, corresponding to the mean difference of the four Mercury intervals mentioned, the single probability is p = 0.018145, and using the binomial distribution the overall probability of *c.* 1:142,000 arises (7.1 out of 1 million) as mentioned. The random test showed a value of *c.* 1:50,000 (20 out of 1 million). The table here shows a comparison between the results expected from the binomial distribution and those arising from a test series of 1 million tested fictional 'planetary systems'. The results differ slightly, but not in essence, when the probabilities decrease. Even though in this case the initialization, i.e. the prior structuring, exerts some influence, in principle the theoretical calculations are confirmed.

| Number of successes | Binomial distribution | Random test | Random test/ Binomial distribution |
|---|---|---|---|
| 0 of 8 | 863,731.3 | 852,741 | 0.987 |
| 1 of 8 | 127,696.7 | 136,417 | 1.068 |
| 2 of 8 | 8,259.6 | 10,339 | 1.252 |
| 3 of 8 | 305.3 | 484 | 1.584 |
| 4 of 8 | 7.1 | 20 | 2.836 |

## 5. *Planetary scales*

### 5.1 Musical scale derived from the distances of the planets according to Hans Kayser

In his book *Akroasis,*[15] Hans Kayser introduces his readers to a musical scale which arises by relating logarithms to the base 10 of the mean planetary distances with logarithms to the base 2 of the musical intervals. Kayser takes the planets up to Neptune, Pluto at the time having been only recently discovered; but I am about to complete his system. Certain aspects of the following table may initially appear somewhat confusing, but clarity will soon be restored. We begin by first relating the musical notes to Mercury.

The logarithms to the base 2 of the notes, suitably transposed, may be found in column ⑤, and the relatively good correspondence with the logarithms to the base 10 in column ③ is immediately evident. In order for the calculations to be musically meaningful and so that, furthermore, comparable intervals can be derived from them for the probability analysis, the various logarithms must now be converted into notes. For this purpose they are subjected to the procedure also used for obtaining the notes themselves from logarithms to the base 2. To do this the number

**Table 14.8** *Logarithmic scale according to Hans Kayser. (**For the values in columns ⑥ and ⑦, '2 ^ col. ③' (or col. ⑤) means: '2 raised to the power of col. ③', or col. ⑤. The resulting values have been transposed into the octave that ranges from 1 to 2)*

| | Semi-major axis a (km*10⁶) | Reciprocal value related to Mercury | Log. to base 10, column ② | Note allocated by Kayser | Log. to base 2 of the note in column ④ | 2 ^ col. ③ = planetary interval** | 2 ^ col. ⑤ = musical interval** | Harmonic probability |
|---|---|---|---|---|---|---|---|---|
| | ① | ② | ③ | ④ | ⑤ | ⑥ | ⑦ | ⑧ |
| Mercury | 57.9092 | 1.0000 | 0.0000 | *C* | 0.0000 | 1.0000 | 1.0000 | |
| Venus | 108.2089 | 0.5352 | −0.2715 | *A* | −0.2630 | 1.6569 | 1.6667 | 0.2901 |
| Earth | 149.5980 | 0.3871 | −0.4122 | *G* | −0.4150 | 1.5030 | 1.5000 | 0.0605 |
| Mars | 227.9406 | 0.2541 | −0.5951 | *F* | −0.5850 | 1.3240 | 1.3333 | 0.2199 |
| Jupiter | 778.3221 | 0.0744 | −1.1284 | *B♭* (B) | −1.1699 | 1.8297 | 1.7778 | 0.9201 |
| Saturn | 1427.607 | 0.0406 | −1.3919 | *G* | −1.4150 | 1.5243 | 1.5000 | 0.4938 |
| Uranus | 2871.068 | 0.0202 | −1.6953 | *E* | −1.6781 | 1.2352 | 1.2500 | 0.5874 |
| Neptune | 4498.187 | 0.0129 | −1.8903 | D♭ | −1.9069 | 1.0790 | 1.0667 | 0.4290 |
| Pluto | 5906.846 | 0.0098 | −2.0086 | *C* | −2.0000 | 1.9881 | 1.0000 | 0.1873 |

two must be raised to a higher power by the relevant logarithm value from column ③ or, respectively, column ⑤ and—for clarity—transposed into the octave space from 1 to 2 (i.e. must have been transposed up by one or more octaves by multiplying by 2 until the value lies between 1 and 2). Now the harmonic probabilities of the separate intervals can be determined. By means of the by now familiar procedure shown in Appendix 4.2 we thus arrive at a mean value of 0.399, a standard deviation of 0.273 and a random variable T of − 1.053. From this, with 7 degrees of freedom, we arrive at the probability of 16.4% that we are here dealing with values arising from a random distribution. As already remarked in Chapter 4, this value tells us very little statistically. And when we apply the same procedure to the other planets as the key-note nothing essentially different emerges. The most favourable value arises in the case of Pluto with a mean value of 0.35 and a probability from the t-test of 9%.

A somewhat better but also not exceptional correspondence is reached when one refrains from converting logarithms to the base 10 into note values as described, even though this is the logical thing to do, but instead takes them as they are. (One does not apply the reciprocal values of the planetary distances but the direct ratios.) In the case of Pluto the maximum value from the t-test is then *c.* 3%.

## 5.2 Musical scale from the synodic periods according to Thomas Michael Schmidt

With Schmidt it is the synodic periods, related to the Earth, that give expression to a musical order.[16] He works with the synodic periods of the five planets visible from Earth, with the Sun whose orbit in relation to the Earth is 365.256 days, and with the lunar year which is defined according to 12 synodic lunar orbits equalling 354.367 days. From these he forms direct intervals or intervals between differences. Here are examples of the various possibilities:

| | | |
|---|---|---|
| Mars/Venus | 779.94/583.92 | = 1.336 |
| Venus/Sun | 583.92/365.256 | = 1.599 |
| Sun/(Venus-Sun) | 365.256/218.664 | = 1.670 |
| Venus/(Venus-Mercury) | 583.92/468.04 | = 1.248 |
| (Venus-Mercury)/(Sun-Mercury) | 468.04/249.376 | = 1.877 |
| (Jupiter-lunar year)/(Saturn-lunar year) | 44.51/23.72 | = 1.877 |

Schmidt gives a total of 17 of these combinations all of which lie very close to the values of harmonic intervals. From these 17 ratios a mean value of harmonic probability of 0.0437 (as in Appendix 4.1) arises which in this case would lead to an extremely low overall probability of it being solely the result of chance. However, there are two snags. On the one hand the differences are not independent values, and on the other very many more, unfortunately less harmonic, intervals can be arrived at by this means.

Take the example of Venus/Sun and Sun/(Venus-Sun). The first proportion yields a ratio of almost exactly 8:5, corresponding to the minor sixth. From this there inevitably arises the second interval: Su/(Ve-Su) = 5/(8-5) = 5:3, i.e. the major sixth. This is indeed pleasing, but in view of a probability estimate only one of the two intervals can be applied. For all 12 musical intervals the result in 9 cases, purely from mathematical necessity, is another or also the same harmonic proportion if one uses the scheme described to form ratios with differences of the initial values. The only exceptions are the sevenths and the tritone.

So let us consider separately all the direct ratios and all the ratios arising from differences. We have 7 initial values, so that in total 21 (6 + 5 + 4 + 3 + 2 + 1) direct interval combinations are possible. One can also form 21 differences. One can divide each synodic period by the 21 differences, so that in this way 7 * 21 = 147 possibilities arise. Thirdly one can form intervals from the 21 differences themselves (as in the last two examples above), for which there are 210 possibilities. Of course these, too, are not values which are independent of one another; but we would only need to investigate them more thoroughly if an analysis of all the combinations were to show significant features as discussed in Chapter 4. Taking all the intervals formed by each relevant method, the following 3 mean values of harmonic probability result:

(1) 0.548  (2) 0.483  (3) 0.497

Obviously there is no need to go to any more trouble. The only harmonically notable cases are the synodic periods relating to the Earth if one separates these. So we find that the intervals formed for Sun, Mercury, Venus and Mars on the one hand—and for Jupiter, Saturn and the lunar year on the other—are all close or very close to the musical intervals. For the 6 possible combinations of the first half, the preliminary overall probability according to the t-test is 0.4%, and for the 3 possibilities of the second half it is 3.8%. The values would be somewhat altered if we were to take into account that there are only 3 or, respectively, 2 values that are actually independent of one another, i.e. the direct ratios of, in each case, neighbouring planets. But we shall not embark on an even more detailed consideration. It would not change the fact that the synodic periods do not provide a convincing overall picture of a harmony—in the musical sense—among the planets.

# 6. *Astronomical Data*

## 6.1 Basic data

| | Orbital period T (days) | Semi-major axis a (AU) | a (km*10^6) | Numerical eccentricity ϵ | Linear eccentricity e (km*10^6) | Semi-minor axis b (km*10^6) |
|---|---|---|---|---|---|---|
| Mercury | 87.9692561 | 0.38709911 | 57.9092029 | 0.20562367 | 11.9075026 | 56.6717493 |
| Venus | 224.7008 | 0.72333165 | 108.208875 | 0.00680155 | 0.7359883 | 108.206372 |
| Earth | 365.256363 | 1.00000066 | 149.59797 | 0.01667094 | 2.4939392 | 149.57718 |
| Mars | 686.979852 | 1.52368855 | 227.940563 | 0.09337366 | 21.2836440 | 226.944722 |
| Jupiter | 4332.58926 | 5.20276167 | 778.322068 | 0.04823674 | 37.5437164 | 777.416047 |
| Saturn | 10759.2268 | 9.54296408 | 1427.60711 | 0.05556198 | 79.3206720 | 1425.4018 |
| Uranus | 30688.4775 | 19.1919065 | 2871.06835 | 0.04618996 | 132.614541 | 2868.00398 |
| Neptune | 60182.291 | 30.0685223 | 4498.1869 | 0.00910687 | 40.964412 | 4498.00037 |
| Pluto | 90544.186 | 39.4848273 | 5906.8461 | 0.24899858 | 1470.7963 | 5720.80318 |

| | Perihelion distance (km*10^6) | Aphelion distance (km*10^6) | Mean velocity v (km/sec) | v 'at b' (km/sec) | v at perihelion (km/sec) | v at aphelion (km/sec) |
|---|---|---|---|---|---|---|
| Mercury | 46.0017004 | 69.8167055 | 47.8720537 | 48.9061922 | 58.9758672 | 38.8590408 |
| Venus | 107.472886 | 108.944863 | 35.0207069 | 35.0215170 | 35.2597302 | 34.7833023 |
| Earth | 147.104030 | 152.091909 | 29.7847273 | 29.7888667 | 30.2836296 | 29.2942780 |
| Mars | 206.656919 | 249.224207 | 24.1293198 | 24.2349688 | 26.4979927 | 21.9721685 |
| Jupiter | 740.774787 | 815.876538 | 13.0642167 | 13.0794332 | 13.7102636 | 12.4482336 |
| Saturn | 1348.32161 | 1506.82066 | 9.64303042 | 9.65793811 | 10.2005165 | 9.12835536 |
| Uranus | 2738.50934 | 3003.66561 | 6.79897638 | 6.80623698 | 7.12380473 | 6.49756886 |
| Neptune | 4457.19735 | 4539.23583 | 5.43185608 | 5.43208133 | 5.47379381 | 5.39772797 |
| Pluto | 4436.16405 | 7377.64394 | 4.73999617 | 4.89171472 | 6.11310864 | 3.67752393 |

| | 'Kepler's angles', mean (°/day) | at perihelion (°/day) | at aphelion (°/day) | Orbital inclination (°) | Rotation period (days) | Equatorial radius (km) | Mass m′ as reciprocal fractional solar mass |
|---|---|---|---|---|---|---|---|
| Mercury | 4.092339 | 6.346544 | 2.755284 | 7.0052 | 58.6462 | 2439.7 | 6023600 |
| Venus | 1.602130 | 1.624113 | 1.580522 | 3.3849 | −243.0185 | 6051.8 | 408523.71 |
| Earth | 0.985609 | 1.019169 | 0.953418 | −0.0028 | 0.99726963 | 6378.14 | 328900.56 |
| Mars | 0.524033 | 0.634748 | 0.436437 | 1.8429 | 1.02595675 | 3397 | 3098708 |
| Jupiter | 0.083091 | 0.091622 | 0.075531 | 1.3133 | 0.41354 | 71492 | 1047.3486 |
| Saturn | 0.033460 | 0.037428 | 0.029968 | 2.4741 | 0.44401 | 60268 | 3497.898 |
| Uranus | 0.011731 | 0.012872 | 0.010699 | 0.7743 | −0.71833 | 25559 | 22902.98 |
| Neptune | 0.005982 | 0.006088 | 0.005870 | 1.7700 | 0.67125 | 24764 | 19412.24 |
| Pluto | 0.003976 | 0.006821 | 0.002466 | 17.14 | −6.3872 | 1195 | 135200000 |

The orbital periods and orbital inclinations for Mercury and Neptune were obtained as mean values over 6000 years, i.e. 3000 years before and after 1 January 2000, with orbital elements derived from the VSOP Planetary Theory (see Appendix 3.6, Mean orbits). The semi-major axes, the eccentricities, the distances and the velocities at aphelion and perihelion were determined or, respectively, calculated as mean values for the same period of time from the exact aphelion and perihelion passages using the VSOP Planetary Theory (see details in Appendix 3.4, Basic data). The semi-minor axes, the mean velocities, the velocities 'at b', and the 'Kepler angles' were calculated from the semi-major axes and eccentricities obtained as shown using the formulae given in Appendix 3.3 for the two-body problem. The data obtained in this way do indeed show mean values over a longer period of time—unlike many so-called mean values which, however, only apply to short periods. The values for Pluto were determined according to the JPL Ephemerides DE406 for the period 1000 years before and 1000 after 1 January 2000 (see Appendix 3.6, Pluto). The velocity 'at b' is that of the planet when its distance from the Sun is that of the semi-minor axis (see more in Chapter 4). The rotation periods and equatorial radii are taken from the *Astronomical Almanac 1999*. The planetary masses are those last given in 1994 by the International Astronomical Union (IAU).

## 6.2 Resonances of the orbital periods

The table shows relatively exact approximations of whole number ratios to those of the sidereal periods, and in each case the next most exact approach. The cycle-resonances (for their significance and calculation see Appendix 3.5) and the geometrical formations of the figures are related to the synodic periods because thereby the same value or, respectively, the same figure arises for both planets (whereas otherwise, for example in the

| | Ratio of the orbital periods c. | Star-figure or polygon | Value of the cycle-resonance (°) | Improved approximation | Value of the cycle-resonance (°) |
|---|---|---|---|---|---|
| Venus/Mercury | 23:9 | 14 | 2.59522 | 212:83 | −1.80122 |
| Earth/Mercury | 54:13 | 41 | 2.60582 | 137:33 | −2.17061 |
| Mars/Mercury | 39:5 | 34 | −2.46303 | 164:21 | 0.22904 |
| Jupiter/Mercury | 49:1 | 48 | −1.87401 | 148:3 | 1.83893 |
| Earth/Venus | 13:8 | 5 | −2.40888 | 382:235 | 1.17899 |
| Mars/Venus | 52:17 | 35 | 4.50438 | 107:35 | −1.01955 |
| Jupiter/Venus | 19:1 | 18 | −5.54510 | 58:3 | 3.05663 |
| Saturn/Venus | 48:1 | 47 | 0.90257 | 383:8 | −0.45821 |
| Mars/Earth | 79:42 | 37 | 2.34860 | 284:151 | −1.28745 |
| Jupiter/Earth | 12:1 | 11 | 4.58125 | 83:7 | −1.07498 |
| Saturn/Earth | 29:1 | 28 | −5.77693 | 59:2 | 1.09697 |
| Uranus/Earth | 84:1 | 83 | −0.08243 | 4369:52 | 0.05011 |
| Jupiter/Mars | 63:10 | 53 | −4.55844 | 82:13 | 0.85788 |
| Saturn/Mars | 47:3 | 44 | 0.37073 | 1018:65 | −0.15210 |
| Uranus/Mars | 45:1 | 44 | 2.70726 | 134:3 | −0.12158 |
| Saturn/Jupiter | 72:29 | 43 | −3.98571 | 149:60 | 0.12261 |
| Uranus/Jupiter | 7:1 | 6 | −4.92212 | 85:12 | 0.11415 |
| Neptune/Jupiter | 14:1 | 13 | 3.05508 | 125:9 | −0.43163 |
| Pluto/Jupiter | 21:1 | 20 | 1.83812 | 188:9 | −1.54878 |
| Uranus/Saturn | 20:7 | 13 | 6.59684 | 77:27 | −2.31986 |
| Neptune/Saturn | 28:5 | 23 | 2.52681 | 151:27 | −2.02937 |
| Pluto/Saturn | 42:5 | 37 | −3.76043 | 101:12 | 0.68437 |
| Nept./Uranus | 51:26 | 25 | 4.55020 | 151:77 | −0.93142 |
| Pluto/Uranus | 59:20 | 39 | −1.58515 | 298:101 | 1.22372 |
| Pluto/Neptune | 3:2 | 1 | −6.42054 | 167:111 | 0.44949 |

case of Venus/Mercury, a 23-pointed star and a 9-pointed star with varying cycle-resonances are the result). When the value of the cycle-resonance is high, the star-figure in question very quickly becomes superimposed upon by the one arising from the next pair of numbers (e.g. the 35-pointed star of Mars/Venus becomes a 72-pointed star, see Fig. 7.3). In some cases less exact approximations with smaller whole numbers are possible. (For example the ratio Mars/Earth can be expressed as 32:17, but then the cycle-resonance would already have risen to 10.68°.)

## 6.3 Conjunction periods/synodic periods

All data (except percentages in one case) are in days. The mean values are synodic periods calculated from the sidereal periods according to the table in Appendix 6.1. Maxima and minima were obtained in each case from several hundred conjunctions or, respectively, minimum distances and may thus be regarded as very good reference values for past and future millennia. Over and above this and depending on the period of time under investigation, the extreme values for the outer planets, especially the Saturn/Uranus and Saturn/Neptune constellations, could vary noticeably because in those cases the size of the fluctuations in itself shows larger fluctuations.

| | Mean value | Maximum conjunctions | Minimum | Fluctuation range | In % of the mean value | Max. of min. distances | Minimum | Fluctuation range |
|---|---|---|---|---|---|---|---|---|
| Me/Ve | 144.56622 | 158.14 | 125.89 | 32.25 | 22.31 | 154.29 | 133.81 | 20.48 |
| Me/Ea | 115.87748 | 129.35 | 105.71 | 23.64 | 20.40 | 122.62 | 108.92 | 13.71 |
| Me/Ma | 100.88821 | 109.91 | 95.39 | 14.52 | 14.39 | 105.67 | 96.97 | 8.69 |
| Me/Ju | 89.79241 | 90.62 | 89.18 | 1.44 | 1.60 | 90.16 | 89.41 | 0.75 |
| Me/Sa | 88.69410 | 88.95 | 88.49 | 0.46 | 0.52 | 88.76 | 88.59 | 0.18 |
| Me/Ur | 88.22209 | 88.36 | 88.13 | 0.23 | 0.26 | 88.28 | 88.16 | 0.13 |
| Me/Ne | 88.09803 | 88.16 | 88.05 | 0.11 | 0.12 | 88.13 | 88.07 | 0.06 |
| Me/Pl | 88.055 | | | | | | | |
| Ve/Ea | 583.92137 | 587.70 | 579.83 | 7.87 | 1.35 | 587.14 | 581.26 | 5.88 |
| Ve/Ma | 333.92152 | 354.87 | 313.34 | 41.53 | 12.44 | 349.19 | 318.86 | 30.33 |
| Ve/Ju | 236.99190 | 238.46 | 235.70 | 2.77 | 1.17 | 238.18 | 235.94 | 2.24 |
| Ve/Sa | 229.49365 | 230.03 | 229.03 | 1.01 | 0.44 | 229.99 | 229.07 | 0.92 |
| Ve/Ur | 226.35819 | 226.51 | 226.21 | 0.30 | 0.13 | 226.52 | 226.21 | 0.31 |
| Ve/Ne | 225.54290 | 225.57 | 225.52 | 0.05 | 0.02 | 225.57 | 225.52 | 0.05 |
| Ve/Pl | 225.260 | | | | | | | |
| Ea/Ma | 779.93610 | 811.33 | 763.85 | 47.47 | 6.09 | 798.77 | 767.74 | 31.03 |
| Ea/Ju | 398.88405 | 402.99 | 395.32 | 7.67 | 1.92 | 402.13 | 396.05 | 6.08 |
| Ea/Sa | 378.09190 | 379.20 | 377.08 | 2.12 | 0.56 | 379.19 | 377.08 | 2.11 |
| Ea/Ur | 369.65603 | 370.09 | 369.27 | 0.82 | 0.22 | 370.09 | 369.25 | 0.84 |
| Ea/Ne | 367.48670 | 367.56 | 367.41 | 0.15 | 0.04 | 367.57 | 367.38 | 0.19 |
| Ea/Pl | 366.736 | | | | | | | |
| Ma/Ju | 816.43456 | 835.92 | 798.82 | 37.10 | 4.54 | 826.51 | 806.60 | 19.91 |
| Ma/Sa | 733.83547 | 748.46 | 722.44 | 26.02 | 3.55 | 742.66 | 725.92 | 16.74 |
| Ma/Ur | 702.71044 | 707.96 | 698.70 | 9.26 | 1.32 | 705.83 | 699.87 | 5.96 |
| Ma/Ne | 694.91226 | 696.49 | 693.51 | 2.98 | 0.43 | 695.68 | 694.02 | 1.65 |
| Ma/Pl | 692.232 | | | | | | | |
| Ju/Sa | 7253.4525 | 7511.75 | 6968.56 | 543.19 | 7.49 | 7422.33 | 7079.64 | 342.69 |
| Ju/Ur | 5044.8145 | 5221.56 | 4901.42 | 320.15 | 6.35 | 5157.00 | 4944.11 | 212.89 |
| Ju/Ne | 4668.6936 | 4701.68 | 4637.52 | 64.16 | 1.37 | 4685.49 | 4651.00 | 34.49 |
| Ju/Pl | 4550.32 | | | | | | | |
| Sa/Ur | 16567.823 | 17639.0 | 15421.6 | 2217.38 | 13.38 | 17288.6 | 15805.7 | 1482.95 |
| Sa/Ne | 13101.473 | 13652.1 | 12542.7 | 1109.30 | 8.47 | 13372.4 | 12869.4 | 503.03 |
| Sa/Pl | 12210.1 | | | | | | | |
| Ur/Ne | 62620.009 | 62877.2 | 62401.2 | 476.00 | 0.76 | 62744.0 | 62481.1 | 262.92 |
| Ur/Pl | 46422.7 | | | | | | | |
| Ne/Pl | 179473.5 | | | | | | | |

## 6.4 Velocity intervals

The first table contains all the intervals of the aphelion velocities and the velocities 'at b', i.e. those that play the decisive role in Chapter 4. The second table shows these intervals transposed into the octave of 1 to 2, so that they can be directly compared with musical intervals. The third table gives the harmonic deviations (see Appendix 4.1) for each interval. In addition, two mean values for each planet have been formed. The first column shows the mean values of the rows, i.e. for example the top number arises from the intervals of the aphelion velocities of all the planets, in each case related to the 'v at b' of Mercury. Accordingly the

| | V at → aphelion | Mercury | Venus | Earth | Mars | Jupiter | Saturn | Uranus | Neptune | Pluto |
|---|---|---|---|---|---|---|---|---|---|---|
| | v 'at b'↓ | *38.8590* | *34.7833* | *29.2943* | *21.9722* | *12.4482* | *9.12836* | *6.49757* | *5.39773* | *3.67752* |
| Me | *48.9062* | 1.25855 | 1.40602 | 1.66948 | 2.22582 | 3.92877 | 5.35761 | 7.52684 | 9.06051 | 13.2987 |
| Ve | *35.0215* | 1.10958 | 1.00685 | 1.19551 | 1.59390 | 2.81337 | 3.83656 | 5.38994 | 6.48820 | 9.52312 |
| Ea | *29.7889* | 1.30448 | 1.16766 | 1.01688 | 1.35575 | 2.39302 | 3.26333 | 4.58462 | 5.51878 | 8.10025 |
| Ma | *24.2350* | 1.60343 | 1.43525 | 1.20876 | 1.10298 | 1.94686 | 2.65491 | 3.72985 | 4.48985 | 6.59002 |
| Ju | *13.0794* | 2.97100 | 2.65939 | 2.23972 | 1.67990 | 1.05071 | 1.43284 | 2.01297 | 2.42314 | 3.55659 |
| Sa | *9.65794* | 4.02353 | 3.60152 | 3.03318 | 2.27504 | 1.28891 | 1.05802 | 1.48639 | 1.78926 | 2.62621 |
| Ur | *6.80624* | 5.70933 | 5.11050 | 4.30403 | 3.22824 | 1.82895 | 1.34118 | 1.04751 | 1.26094 | 1.85077 |
| Ne | *5.43208* | 7.15362 | 6.40331 | 5.39283 | 4.04489 | 2.29161 | 1.68045 | 1.19615 | 1.00636 | 1.47710 |
| Pl | *4.89171* | 7.94385 | 7.11066 | 5.98855 | 4.49171 | 2.54476 | 1.86608 | 1.32828 | 1.10344 | 1.33017 |

**Velocity intervals**

| | | Mercury | Venus | Earth | Mars | Jupiter | Saturn | Uranus | Neptune | Pluto |
|---|---|---|---|---|---|---|---|---|---|---|
| Me | Values | 1.25855 | 1.40602 | 1.66948 | 1.11291 | 1.96438 | 1.33940 | 1.88171 | 1.13256 | 1.66233 |
| Ve | trans- | 1.10958 | 1.00685 | 1.19551 | 1.59390 | 1.40669 | 1.91828 | 1.34749 | 1.62205 | 1.19039 |
| Ea | posed | 1.30448 | 1.16766 | 1.01688 | 1.35575 | 1.19651 | 1.63167 | 1.14615 | 1.37969 | 1.01253 |
| Ma | into the | 1.60343 | 1.43525 | 1.20876 | 1.10298 | 1.94686 | 1.32746 | 1.86493 | 1.12246 | 1.64751 |
| Ju | octave | 1.48550 | 1.32969 | 1.11986 | 1.67990 | 1.05071 | 1.43284 | 1.00649 | 1.21157 | 1.77829 |
| Sa | 1 to 2 | 1.00588 | 1.80076 | 1.51659 | 1.13752 | 1.28891 | 1.05802 | 1.48639 | 1.78926 | 1.31310 |
| Ur | | 1.42733 | 1.27763 | 1.07601 | 1.61412 | 1.82895 | 1.34118 | 1.04751 | 1.26094 | 1.85077 |
| Ne | | 1.78840 | 1.60083 | 1.34821 | 1.01122 | 1.14581 | 1.68045 | 1.19615 | 1.00636 | 1.47710 |
| Pl | | 1.98596 | 1.77766 | 1.49714 | 1.12293 | 1.27238 | 1.86608 | 1.32828 | 1.10344 | 1.33017 |

**Transposed values**

| | Mean → | Mercury | Venus | Earth | Mars | Jupiter | Saturn | Uranus | Neptune | Pluto |
|---|---|---|---|---|---|---|---|---|---|---|
| Me | *0.20529* | 0.20866 | 0.00609 | 0.05146 | 0.40899 | 0.56083 | 0.16873 | 0.10914 | 0.20502 | 0.12866 |
| Ve | *0.33287* | 0.52187 | 0.20883 | 0.11791 | 0.11999 | 0.00946 | 0.70387 | 0.39341 | 0.66829 | 0.25218 |
| Ea | *0.59932* | 0.68144 | 0.84868 | 0.51481 | 0.62328 | 0.09159 | 0.95980 | 0.57337 | 0.71881 | 0.38211 |
| Ma | *0.40741* | 0.10392 | 0.62887 | 0.35404 | 0.74489 | 0.83674 | 0.13883 | 0.20452 | 0.08589 | 0.56902 |
| Ju | *0.28112* | 0.30439 | 0.08594 | 0.17390 | 0.24215 | 0.47122 | 0.57646 | 0.19779 | 0.46751 | 0.01075 |
| Sa | *0.39363* | 0.17940 | 0.47920 | 0.33725 | 0.33930 | 0.94920 | 0.25543 | 0.28568 | 0.23939 | 0.47781 |
| Ur | *0.48458* | 0.45713 | 0.67390 | 0.32462 | 0.42797 | 0.93497 | 0.21799 | 0.56572 | 0.26698 | 0.49198 |
| Ne | *0.28826* | 0.22156 | 0.02508 | 0.41346 | 0.34219 | 0.56395 | 0.25222 | 0.10111 | 0.19406 | 0.48072 |
| Pl | *0.22264* | 0.22104 | 0.00201 | 0.06010 | 0.07012 | 0.54591 | 0.18099 | 0.11934 | 0.72939 | 0.07483 |
| | Mean ↓ | *0.32215* | *0.32873* | *0.26084* | *0.36876* | *0.55154* | *0.38381* | *0.28334* | *0.39726* | *0.31867* |

**Harmonic deviations**

bottom row contains the mean values of the columns, so the first value shows the results of all the 'v at b' numbers pertaining to Mercury's aphelion velocity, etc.

## 6.5 Intervals of the semi-minor axes

The table shows the present ratios (b calculated as the mean value of 3000 years before and after 1 January 2000) and those in 10,000 years, determined by applying mean orbital data. It contains the measure of the semi-

| Semi-minor axis | In 10,000 years → | Mercury | Venus | Earth | Mars | Jupiter | Saturn | Uranus | Neptune |
|---|---|---|---|---|---|---|---|---|---|
| Mean value over 6,000 years ↓ | | *56.6522* | *108.208* | *149.589* | *226.767* | *776.985* | *1429.14* | *2872.18* | *4504.24* |
| Me | 56.671749 | 1.00035 | *1.91004* | *2.64047* | *4.00279* | *13.7150* | *25.2265* | *50.6985* | *79.5070* |
| Ve | 108.20637 | 1.90935 | 1.00001 | *1.38242* | 2.09566 | 7.18048 | 13.2073 | 26.5431 | 41.6258 |
| Ea | 149.57718 | 2.63936 | 1.38233 | 1.00008 | *1.51594* | *5.19415* | *9.55379* | *19.2005* | *30.1109* |
| Ma | 226.94472 | 4.00455 | 2.09733 | 1.51724 | 1.00078 | *3.42636* | *6.30223* | *12.6658* | *19.8629* |
| Ju | 777 41605 | 13.7179 | 7.18457 | 5.19742 | 3.42557 | 1.00055 | *1.83934* | *3.69657* | *5.79708* |
| Sa | 1425.4018 | 25.1519 | 13.1730 | 9.52954 | 6.28083 | 1.83351 | 1.00262 | *2.00973* | *3.15172* |
| Ur | 2868.0040 | 50.6073 | 26.5049 | 19.1741 | 12.6375 | 3.68915 | 2.01207 | 1.00146 | *1.56823* |
| Ne | 4498.0004 | 79.3694 | 41.5687 | 30.0714 | 19.8198 | 5.78583 | 3.15560 | 1.56834 | 1.00139 |
| Pl | 5720.8032 | 100.946 | 52.8694 | 38.2465 | 25.2079 | 7.35874 | 4.01347 | 1.99470 | 1.27186 |

minor axes b at km*$10^6$ and the corresponding ratios for both periods. The future distances and their intervals are printed in italics; the grey shading shows the quotient of present and subsequent values for each planet. Of course no absolute guarantee can be given for data spanning 10 millennia. Regarding the validity range of the calculation procedures used, see Appendix 3.7.

## 6.6 Intervals of the aphelion and perihelion distances

| | Aph. → | Mercury | Venus | Earth | Mars | Jupiter | Saturn | Uranus | Neptune | Pluto |
|---|---|---|---|---|---|---|---|---|---|---|
| | Perih. ↓ | 69.8167 | 108.945 | 152.092 | 249.224 | 815.877 | 1506.82 | 3003.67 | 4539.24 | 7377.64 |
| Me | 46.0017 | 1.51770 | 2.36828 | 3.30622 | 5.41772 | 17.7358 | 32.7558 | 65.2947 | 98.6754 | 160.378 |
| Ve | 107.473 | 1.53936 | 1.01370 | 1.41517 | 2.31895 | 7.59146 | 14.0205 | 27.9481 | 42.2361 | 68.6466 |
| Ea | 147.104 | 2.10700 | 1.35026 | 1.03391 | 1.69420 | 5.54626 | 10.2432 | 20.4186 | 30.8573 | 50.1526 |
| Ma | 206.657 | 2.95999 | 1.89689 | 1.35876 | 1.20598 | 3.94798 | 7.29141 | 14.5346 | 21.9651 | 35.7000 |
| Ju | 740.775 | 10.6103 | 6.7995 | 4.8706 | 2.97232 | 1.10138 | 2.03411 | 4.05476 | 6.12769 | 9.95936 |
| Sa | 1348.32 | 19.3123 | 12.3762 | 8.8652 | 5.41007 | 1.65260 | 1.11755 | 2.22771 | 3.36658 | 5.47172 |
| Ur | 2738.51 | 39.2243 | 25.1367 | 18.0056 | 10.9881 | 3.35652 | 1.81741 | 1.09683 | 1.65756 | 2.69404 |
| Ne | 4457.20 | 63.8414 | 40.9124 | 29.3059 | 17.8843 | 5.46308 | 2.95801 | 1.48392 | 1.01841 | 1.65522 |
| Pl | 4436.16 | 63.5402 | 40.7194 | 29.1677 | 17.7999 | 5.43730 | 2.94406 | 1.47692 | 1.02323 | 1.66307 |

Aphelion and perihelion distances in km*$10^6$, calculated from the mean values over a period of 6000 years (see above) and corresponding ratios.

## 6.7 Positions for Venus, Earth and Venus rotation at Venus-Earth-View

This table contains the positions needed for Figs 8.4 and 8.13 (the latter being the rotation figure of Venus which, as mentioned in Chapter 8, p. 155, one could draw oneself). Row 0 shows in addition the data for 1 January 2000, zero hours TDT, which would not be included in the drawing.

| No. | Days (0 for 1 January 2000, zero hours) | Interval (days) | Position Venus (°) | Position Earth (°) | Venus rotation (°) retrograde |
|---|---|---|---|---|---|
| *0* | *0* | | *181.79* | *99.87* | *0* |
| 1 | 110.27 | | 357.32 | 210.49 | 163.36 |
| 2 | 243.07 | 132.80 | 211.42 | 337.96 | 0.08 |
| 3 | 378.06 | 134.99 | 66.20 | 112.92 | 200.05 |
| 4 | 560.54 | 182.48 | 358.69 | 292.14 | 110.37 |
| 5 | 694.49 | 133.94 | 214.66 | 63.25 | 308.79 |
| 6 | 826.22 | 131.73 | 64.19 | 196.21 | 143.93 |
| 7 | 962.88 | 136.66 | 284.15 | 327.64 | 346.37 |
| 8 | 1144.01 | 181.13 | 214.84 | 148.88 | 254.69 |
| 9 | 1278.07 | 134.06 | 68.14 | 279.76 | 93.29 |
| 10 | 1410.46 | 132.39 | 281.27 | 48.58 | 289.40 |
| 11 | 1545.29 | 134.83 | 136.97 | 184.92 | 129.14 |
| 12 | 1728.54 | 183.25 | 69.85 | 1.75 | 40.60 |
| 13 | 1861.57 | 133.03 | 283.98 | 135.78 | 237.66 |
| 14 | 1993.96 | 132.39 | 135.77 | 265.80 | 73.78 |
| 15 | 2130.02 | 136.06 | 353.24 | 37.61 | 275.34 |
| 16 | 2311.20 | 181.19 | 284.34 | 219.64 | 183.74 |
| 17 | 2445.84 | 134.64 | 139.81 | 348.86 | 23.19 |
| 18 | 2577.55 | 131.71 | 350.27 | 121.02 | 218.30 |
| 19 | 2713.35 | 135.80 | 209.13 | 255.17 | 59.46 |
| 20 | 2895.86 | 182.51 | 140.80 | 73.21 | 329.82 |
| 21 | 3028.92 | 133.06 | 353.40 | 207.17 | 166.94 |
| 22 | 3161.71 | 132.80 | 207.46 | 334.67 | 3.66 |
| 23 | 3296.66 | 134.95 | 62.16 | 109.40 | 203.56 |
| 24 | 3479.13 | 182.46 | 354.67 | 288.83 | 113.86 |
| 25 | 3613.16 | 134.03 | 210.74 | 59.84 | 312.40 |
| 26 | 3744.86 | 131.71 | 60.23 | 192.85 | 147.51 |
| 27 | 3881.42 | 136.56 | 280.09 | 324.26 | 349.80 |
| 28 | 4062.65 | 181.23 | 210.88 | 145.44 | 258.27 |
| 29 | 4196.70 | 134.05 | 64.15 | 276.50 | 96.85 |
| 30 | 4329.12 | 132.42 | 277.40 | 45.18 | 293.01 |
| 31 | 4463.85 | 134.73 | 132.83 | 181.45 | 132.60 |
| 32 | 4647.14 | 183.29 | 65.81 | 358.38 | 44.11 |
| 33 | 4780.24 | 133.10 | 280.12 | 132.34 | 241.28 |
| 34 | 4912.59 | 132.35 | 131.75 | 262.54 | 77.34 |
| 35 | 5048.61 | 136.02 | 349.25 | 34.15 | 278.84 |
| 36 | 5229.81 | 181.20 | 280.38 | 216.28 | 187.26 |
| 37 | 5364.49 | 134.68 | 135.81 | 345.56 | 26.77 |
| 38 | 5496.22 | 131.73 | 346.39 | 117.57 | 221.91 |
| 39 | 5631.87 | 135.65 | 204.96 | 251.80 | 62.86 |
| 40 | 5814.50 | 182.63 | 136.79 | 69.76 | 333.40 |
| 41 | 5947.57 | 133.07 | 349.49 | 203.83 | 170.52 |

The values for two cycles are listed. Those referring to the Venus rotation would have to be plotted clockwise in a circle and then joined up continuously. (The positions for Venus and Earth would run anticlockwise.) Value 2, 0.08°, referring to the Venus rotation does not mean, for example, that there is an angle of 0.08° between positions 1 and 2 and that they therefore almost coincide, but that the second point must be plotted very close to the zero point, whereas the first item lies 163.36° from the zero point.

The data in this table can be easily checked, at least roughly, with the help of the angles covered in a single day. The orbits of Venus and Earth are almost circular, so that for an approximate calculation the assumption of constant angular velocities $\omega$ (omega) is sufficient. In degrees per day these amount to:

$$\omega_{Ea} = \frac{360}{365.256} = 0.98561 \qquad \omega_{Ve} = \frac{360}{224.701} = 1.60213$$

$$\omega_{VeR} = \frac{360}{243.019} = 1.48137$$

## 6.8 Data pertaining to Sun/Mercury/Venus rotation

| | Axis parallelism, angle < 10° | | | | | Axis parallelism, angle < 7.5° | | | | |
|---|---|---|---|---|---|---|---|---|---|---|
| No. | Days after 1.1. 2000 | *Sun* axis (°) | *Mercury* axis (°) | *Venus* axis (°) | Max. distance (°) | Days after 1.1.2000 | *Sun* axis (°) | *Mercury* axis (°) | *Venus* axis (°) | Max. distance (°) |
| 1 | 45.96 | 291.915 | 282.126 | 291.916 | 9.791 | 850.26 | 180.426 | 179.328 | 180.454 | 1.126 |
| 2 | 804.3 | 248.511 | 257.202 | 248.538 | 8.692 | 1700.52 | 0.851 | 358.656 | 0.908 | 2.252 |
| 3 | 850.26 | 180.426 | 179.328 | 180.454 | 1.126 | 2504.82 | 249.362 | 255.858 | 249.446 | 6.496 |
| 4 | 1654.56 | 68.936 | 76.530 | 68.992 | 7.594 | 2550.78 | 181.277 | 177.984 | 181.362 | 3.378 |
| 5 | 1700.52 | 0.851 | 358.656 | 0.908 | 2.252 | 3355.08 | 69.787 | 75.186 | 69.900 | 5.399 |
| 6 | 2504.82 | 249.362 | 255.858 | 249.446 | 6.496 | 3401.04 | 1.702 | 357.312 | 1.816 | 4.504 |
| 7 | 2550.78 | 181.277 | 177.984 | 181.362 | 3.378 | 4205.34 | 250.213 | 254.514 | 250.354 | 4.301 |
| 8 | 3355.08 | 69.787 | 75.186 | 69.900 | 5.399 | 4251.3 | 182.128 | 176.640 | 182.270 | 5.630 |
| 9 | 3401.04 | 1.702 | 357.312 | 1.816 | 4.504 | 5055.62 | 70.922 | 73.965 | 70.778 | 3.187 |
| 10 | 4205.34 | 250.213 | 254.514 | 250.354 | 4.301 | 5101.58 | 2.837 | 356.091 | 2.695 | 6.746 |
| 11 | 4251.3 | 182.128 | 176.640 | 182.270 | 5.630 | 5905.88 | 251.348 | 253.293 | 251.232 | 2.061 |
| 12 | 5055.62 | 70.922 | 73.965 | 70.778 | 3.187 | 6756.14 | 71.773 | 72.621 | 71.686 | 0.935 |
| 13 | 5101.58 | 2.837 | 356.091 | 2.695 | 6.746 | 7606.4 | 252.199 | 251.949 | 252.140 | 0.250 |
| 14 | 5905.88 | 251.348 | 253.293 | 251.232 | 2.061 | 8456.66 | 72.624 | 71.277 | 72.594 | 1.347 |
| 15 | 5951.84 | 183.262 | 175.419 | 183.149 | 7.844 | 9260.96 | 321.135 | 328.479 | 321.132 | 7.347 |
| 16 | 6756.14 | 71.773 | 72.621 | 71.686 | 0.935 | 9306.92 | 253.050 | 250.605 | 253.048 | 2.445 |
| 17 | 6802.1 | 3.688 | 354.747 | 3.603 | 8.941 | 10111.22 | 141.560 | 147.807 | 141.586 | 6.247 |
| 18 | 7560.44 | 320.284 | 329.823 | 320.224 | 9.599 | 10157.18 | 73.475 | 69.933 | 73.502 | 3.570 |
| 19 | 7606.4 | 252.199 | 251.949 | 252.140 | 0.250 | 10961.48 | 321.986 | 327.135 | 322.040 | 5.149 |
| 20 | 8410.7 | 140.709 | 149.151 | 140.678 | 8.473 | 11007.44 | 253.901 | 249.261 | 253.956 | 4.696 |
| 21 | 8456.66 | 72.624 | 71.277 | 72.594 | 1.347 | 11811.74 | 142.411 | 146.463 | 142.494 | 4.052 |
| 22 | 9260.96 | 321.135 | 328.479 | 321.132 | 7.347 | 11857.7 | 74.326 | 68.589 | 74.410 | 5.822 |
| 23 | 9306.92 | 253.050 | 250.605 | 253.048 | 2.445 | 12662 | 322.837 | 325.791 | 322.948 | 2.954 |
| 24 | 10111.22 | 141.560 | 147.807 | 141.586 | 6.247 | 12707.96 | 254.752 | 247.917 | 254.865 | 6.948 |
| 25 | 10157.18 | 73.475 | 69.933 | 73.502 | 3.570 | 13512.26 | 143.262 | 145.119 | 143.402 | 1.856 |

*Cont'd*

*Cont'd*

| | Axis parallelism, angle < 10° | | | | | Axis parallelism, angle < 7.5° | | | | |
|---|---|---|---|---|---|---|---|---|---|---|
| No. | Days after 1.1. 2000 | *Sun* axis (°) | *Mercury* axis (°) | *Venus* axis (°) | Max. distance (°) | Days after 1.1.2000 | *Sun* axis (°) | *Mercury* axis (°) | *Venus* axis (°) | Max. distance (°) |
| 26 | 10961.48 | 321.986 | 327.135 | 322.040 | 5.149 | 14362.54 | 323.972 | 324.570 | 323.827 | 0.743 |
| 27 | 11007.44 | 253.901 | 249.261 | 253.956 | 4.696 | 15212.8 | 144.397 | 143.898 | 144.281 | 0.500 |
| 28 | 11811.74 | 142.411 | 146.463 | 142.494 | 4.052 | 16063.06 | 324.823 | 323.226 | 324.735 | 1.597 |
| 29 | 11857.7 | 74.326 | 68.589 | 74.410 | 5.822 | 16867.36 | 213.333 | 220.428 | 213.272 | 7.156 |
| 30 | 12662 | 322.837 | 325.791 | 322.948 | 2.954 | 16913.32 | 145.248 | 142.554 | 145.189 | 2.695 |
| 31 | 12707.96 | 254.752 | 247.917 | 254.865 | 6.948 | 17717.62 | 33.759 | 39.756 | 33.726 | 6.029 |
| 32 | 13512.26 | 143.262 | 145.119 | 143.402 | 1.856 | 17763.58 | 325.674 | 321.882 | 325.643 | 3.792 |
| 33 | 13558.22 | 75.177 | 67.245 | 75.319 | 8.074 | 18567.88 | 214.184 | 219.084 | 214.180 | 4.903 |
| 34 | 14362.54 | 323.972 | 324.570 | 323.827 | 0.743 | 18613.84 | 146.099 | 141.210 | 146.097 | 4.890 |
| 35 | 14408.5 | 255.887 | 246.695 | 255.743 | 9.191 | 19418.14 | 34.610 | 38.412 | 34.634 | 3.802 |
| 36 | 15166.84 | 212.482 | 221.772 | 212.364 | 9.408 | 19464.1 | 326.525 | 320.538 | 326.551 | 6.013 |
| 37 | 15212.8 | 144.397 | 143.898 | 144.281 | 0.500 | 20268.4 | 215.035 | 217.740 | 215.088 | 2.704 |
| 38 | 16017.1 | 32.908 | 41.100 | 32.818 | 8.282 | 20314.36 | 146.950 | 139.866 | 147.005 | 7.139 |
| 39 | 16063.06 | 324.823 | 323.226 | 324.735 | 1.597 | 21118.66 | 35.461 | 37.068 | 35.542 | 1.607 |
| 40 | 16867.36 | 213.333 | 220.428 | 213.272 | 7.156 | 21968.92 | 215.887 | 216.396 | 215.996 | 0.509 |
| 41 | 16913.32 | 145.248 | 142.554 | 145.189 | 2.695 | 22819.18 | 36.312 | 35.724 | 36.450 | 0.727 |
| 42 | 17717.62 | 33.759 | 39.756 | 33.726 | 6.029 | 23669.46 | 217.021 | 215.174 | 216.875 | 1.847 |
| 43 | 17763.58 | 325.674 | 321.882 | 325.643 | 3.792 | 24473.76 | 105.532 | 112.377 | 105.412 | 6.964 |
| 44 | 18567.88 | 214.184 | 219.084 | 214.180 | 4.903 | 24519.72 | 37.447 | 34.502 | 37.329 | 2.944 |
| 45 | 18613.84 | 146.099 | 141.210 | 146.097 | 4.890 | 25324.02 | 285.957 | 291.705 | 285.867 | 5.838 |
| 46 | 19418.14 | 34.610 | 38.412 | 34.634 | 3.802 | 25369.98 | 217.872 | 213.830 | 217.783 | 4.042 |
| 47 | 19464.1 | 326.525 | 320.538 | 326.551 | 6.013 | 26174.28 | 106.383 | 111.032 | 106.321 | 4.712 |
| 48 | 20268.4 | 215.035 | 217.740 | 215.088 | 2.704 | 26220.24 | 38.298 | 33.158 | 38.237 | 5.140 |
| 49 | 20314.36 | 146.950 | 139.866 | 147.005 | 7.139 | 27024.54 | 286.809 | 290.360 | 286.775 | 3.586 |

This table shows as examples the first 49 data in each case which lead to the two diagrams in Fig. 9.8 that involve accuracies of < 10° or, respectively, < 7.5°. The calculations are based on rotation periods of 25.38 days (Sun), 58.64617 days (Mercury) and 243.019 days (Venus). The axes of the three bodies are defined to point towards the vernal point (0°) at the starting moment of zero hours on 1 January 2000.

## 6.9 Elementary particles

This table first of all gives the data of the 41 elementary particles shown in Fig. 11.4 in the order of their lives. The last two columns only show the values of the 15 longest-lived particles for Fig. 11.5, whereby related particles of similar weights are not taken into consideration. The mean values of the harmonic probabilities refer to the 14 intervals of each of the particles in relation to each of the others. Where necessary, the ratios in relation to the proton have been transposed into the octave from 1 to 2. In the column headed 'Type', B means baryon, L means lepton, M means meson and Bo means boson. The masses and lives are taken from: *Reviews of Modern Physics*, No. 68, January 1996, pp. 615–52.

| Name | Mass (MeV) | Life (sec) | Type | Mean harm. prob. | Ratio to proton |
|---|---|---|---|---|---|
| Proton | 938.27231 | > 1.6E + 25 | B | 0.2951 | 1 |
| Electron | 0.51099907 | > 4.3E + 23 | L | 0.2721 | 1.7931 |
| Neutron | 939.56563 | 887 | B | | |
| Muon | 105.658389 | 2.197E-06 | L | 0.5402 | 1.1100 |
| Pion ± | 139.56955 | 2.6033E 08 | M | 0.2601 | 1.6807 |
| Kaon ± | 493.677 | 1.2386E-08 | M | 0.3054 | 1.9006 |
| Xi 0 | 1314.9 | 2.9E-10 | B | 0.3578 | 1.4014 |
| Lambda 0 | 1115.684 | 2.632E-10 | B | 0.2573 | 1.1891 |
| Xi − | 1321.32 | 1.639E-10 | B | | |
| Sigma − | 1197.436 | 1.479E-10 | B | 0.5383 | 1.2762 |
| Kaon 0 | 497.672 | 8.927E-11 | M | | |
| Omega − | 1672.45 | 8.22E-11 | B | 0.2952 | 1.7825 |
| Sigma + | 1189.37 | 7.99E-11 | B | | |
| B ± | 5278.9 | 1.62E-12 | M | 0.3006 | 1.4065 |
| B 0 s | 5369.3 | 1.61E-12 | M | | |
| B 0 | 5279.2 | 1.56E-12 | M | | |
| Lambda 0 b | 5641 | 1.14E-12 | B | 0.3118 | 1.5030 |
| D ± | 1869.3 | 1.057E-12 | M | 0.3121 | 1.9923 |
| D ± s | 1968.5 | 4.67E-13 | M | 0.3801 | 1.0490 |
| D 0 | 1864.5 | 4.15E-13 | M | | |
| Xi + c | 2465.6 | 3.5E-13 | B | 0.4533 | 1.3139 |
| Tau | 1777 | 2.91E-13 | L | 0.3681 | 1.8939 |
| Lambda c + | 2284.9 | 2.06E-13 | B | | |
| Xi 0c | 2470.3 | 9.8E-14 | B | | |
| Omega 0c | 2704 | 6.4E-14 | B | | |
| Pion 0 | 134.9764 | 8.4E-17 | M | | |
| Eta-Meson | 547.45 | 5.578E-19 | M | | |
| Sigma 0 | 1192.55 | 7.4E-20 | B | | |
| Gamma | 9460.37 | 1.254E-20 | M | | |
| J/Psi | 3096.88 | 7.566E-21 | M | | |
| Eta' | 957.77 | 3.275E-21 | M | | |
| Phi | 1019.413 | 1.486E-22 | M | | |
| Omega | 781.94 | 7.808E-23 | M | | |
| Eta c | 2979.8 | 4.987E-23 | M | | |
| Kappa ± | 891.59 | 1.322E-23 | M | | |
| Kappa 0 | 896.1 | 1.303E-23 | M | | |
| Delta | 1232 | 5.485E-24 | B | | |
| D ⋆ ± s | 2112.4 | 4.577E-24 | M | | |
| Rho | 768.5 | 4.368E-24 | M | | |
| W-Boson | 80330 | 3.170E-25 | Bo | | |
| Z-Boson | 91187 | 2.643E-25 | Bo | | |

# Glossary

(Terms coined by the author in German for which English equivalents have therefore also been coined are marked with an asterisk.)

**Angular momentum:** The product of mass, velocity and radius; or for a planet the product of mass, orbital velocity and radial distance.

**Aphelion:** Point furthest from the Sun on a planet's elliptical orbit one focus of which is occupied by the Sun. Opposite: → Perihelion

**Astronomical unit:** The mean distance of the Earth from the Sun, defined as 149,597,870.691 km (IAU 1994).

**Barycentre:** Centre of mass or gravity of two or more bodies, e.g. in the Earth/Moon system or the solar system.

**Commensurability:** A relationship between two quantities which is closely similar to that of two small whole numbers. With reference to the bodies of the solar system one usually talks of commensurability when referring to such a ratio of their orbital periods. → Resonance

**Conjunction:** In this book this term always refers to the heliocentric view of the universe and therefore exclusively denotes the situation where two planets are on the same side of the Sun and in line with it. (For more detail on conjunction in ecliptic longitude see also Appendix 3.1). For the opposite of conjunction → Opposition.

**Conjunction, Double*:** A method of portraying geometrically the positions of a planet in its conjunctions with two others. Chronologically sequential positions are linked regardless of which of the two planets takes up the conjunction position.

**Cosmology:** The science of the creation, evolution and current state of the universe.

**Cycle-resonance*:** The measure (in degrees) by which a star-figure or a polygon traced by sequential planetary positions, e.g. the conjunctions, diverges from a perfect geometrical figure.

**Eccentricity:** The measure of deviation of an ellipse from a circle. Linear eccentricity is the distance between the centre of an ellipse and a focus. It is shown as a distance; and numerical eccentricity is shown as a ratio of the linear eccentricity and the semi-major axis of the ellipse.

**Ecliptic:** The plane of the Sun's apparent annual orbit round the Earth, synonymous with the plane of the Earth's actual orbit round the Sun. The orbits of the planets lie close to the ecliptic with the exception of Mercury and Pluto, which have inclinations of *c.* 7° and 17° respectively.

**Ecliptic longitude:** The angle of a body in the ecliptic relative to the vernal point. For example the vernal point itself has the ecliptic longitude of 0° while a planet exactly opposite at a specific point in time would have the longitude of 180°. → Conjunction

**Epicycloid:** A curve described by a point on an outer circle rolling over an inner circle. In the geocentric view the epicycloids serve to explain the loop figures traced by the planets in the firmament.

**Equinox:** When day and night are of equal length. The term denotes the two points at which the Sun's apparent annual path on the celestial sphere (Ecliptic) crosses the celestial equator (the projection of the Earth's equator onto the celestial sphere). This occurs about 21 March (vernal equinox) and 23 September (autumnal equinox) respectively. In locating the position of celestial bodies, the vernal equinox serves as the initial reference point for the measurement of right ascensions along the celestial equator and for the measurement of celestial longitudes along the ecliptic (see also Appendix 3.1, Frame of reference).

**Fractional resonance*:** A relationship of the orbital periods according to small whole numbers which can come about between three planets. To work out whether a fractional resonance is present the first step is to calculate the equalization of their conjunction periods, i.e. the interval of time in which they coincide approximately in whole numbers. The next step is to work out what portions of their orbits the three planets involved have covered during that period. Here only the fractional part of the number is relevant. For example if one planet has covered fairly accurately $5\frac{1}{7}$ orbits, the next $2\frac{1}{7}$ orbits and the third $\frac{1}{7}$ of an orbit, i.e. all with the same fractional part, then one can speak of a 'fractional resonance'. A depiction in this case would show a sevenfold regular geometrical figure.

**Gravity, Newton's Law of:** The law discovered by Sir Isaac Newton according to which the force of attraction between two bodies is the product of their masses divided by the square of their distance apart and, finally, multiplied by the gravitational constant.

***Harmony of the World, The*:** Johannes Kepler's main work (*De harmonice mundi*) published in 1619 in which, among much else, he made known what later came to be called Kepler's Third Law and also his ideas about the musical harmony of planetary movements.

**Hexagram:** A six-pointed star, a geometrical figure composed of two equilateral triangles which according to various schools of thought symbolizes the polar order of the cosmos.

**Kepler's Angle*:** The angle between the perihelion and the position of a planet in its orbit is generally termed the 'true anomaly'. In connection with his harmony of the spheres, Kepler used the angle that a planet traverses at aphelion, or at perihelion, during the course of one day. He was thus speaking of the change in true anomaly during the course of one day. In terms of physics these angles correspond to the angular velocity over one day.

**Kepler's stellar solids*:** In his work *The Harmony of the World*, Johannes Kepler described two regular star-figures which can be constructed by extending the edges of the dodecahedron or the icosahedron (see Fig. 3.9). In accordance with their origin we have named these the *Dodecahedron star* and the *Icosahedron star*. Other terms are sometimes used in the literature.

**Line of apsides:** Line joining aphelion and perihelion. During periods ranging from tens of thousands to hundreds of thousands of years this line rotates about the Sun. The terms are: apsidal precession and perihelion precession.

**Linkline*:** (in German *Raumgerade*) The imagined line between two planets or celestial bodies at a specific point in time. Geometric figures depicted in terms of linklines are usually heliocentric. (Exceptions are specifically indicated.)

**Motion, Forward:** → Retrograde motion

**Node:** Term describing the two points of intersection of an orbit with a plane, usually the ecliptic. The ascending node is, for the ecliptic, the one in which a celestial body crosses it from south to north. The longitude of a node is the angle between this point and the reference point, usually the vernal point.

**Numerical integration:** With a few exceptions, the equations of motion resulting from the gravitational interaction of at least three bodies (the three-body problem) cannot be solved mathematically. However, very good approximations can be obtained by (mathematical) series expansion or by numerical integration using computers.

**Octavation*:** The process by which any numerical ratio is brought by repeated multiplication or division by 2 into the space of a reference octave (usually the numerical region between 1 and 2) so that even widely separated ratios can be compared directly.

**Opposition:** In this book this term refers exclusively to the situation of two planets on opposite sides of the Sun and in line with it. The opposite → Conjunction

**Orbital Sphere*:** A circular ring defined by the rotation or precession of the aphelion and perihelion of a planet (→ Line of Apsides), within which that planet is always situated.

**Pentagram:** A five-pointed star. Since the Pythagoreans, the pentagram and the number five have repeatedly been seen as symbols standing for the human being.

***Pentagramma Veneris*:** The five-pointed star-figure formed from the lines joining sequential conjunctions or oppositions of Earth and Venus has long been known by this term. It is discernible both in the geocentric and the heliocentric view of the universe.

**Perihelion:** Point closest to the Sun on a planet's elliptical orbit one focus of which is occupied by the Sun. Opposite: → Aphelion

**Perihelion, Longitude of the:** The position of the perihelion in relation to the ecliptic, usually taken as the sum of the longitude of the ascending node and the argument of the perihelion. The former is the angle between the vernal point and the ascending node, and the latter the angle between that and the perihelion. The perihelion precesses over long aeons of time; the ellipse of a planetary path as it were turns in space. The terms are perihelion precession and apsidal precession. → Line of apsides

**Perturbations:** An unperturbed orbit is the path followed by a body free from extraneous interference, e.g. in the motion of a planet round the Sun. Perturbations are caused by the gravitational influence of the other planets which exercise a relatively small but clearly perceptible influence on the otherwise unperturbed orbit.

**Precession:** The gyroscopic rotation of the Earth's axis which is inclined at approximately 23° to the ecliptic. It is caused by the gravitational forces of the Moon, of the Sun and, to a lesser degree, of the planets. This causes the position of the vernal point to change in the same measure, namely by *c.* 1.4°, in 100 years. The precession is not entirely uniform, so that no totally exact period of time can be given for a rotation of 360°. It amounts to *c.* 25,800 years, termed a Platonic year.

**Quadratic Time*:** An imaginary temporal plane. In contrast to the familiar characteristics of time moving exclusively in one direction, it would be possible in quadratic time to reach any point from any other. In other words this is a form of eternity or divine time. In Kepler's Third Law, time is included in the calculation as a quadratic term. This Law can therefore be interpreted as pointing to a transcendental temporal plane.

**Resonance:** In physics this is an amplifying effect that can arise when the ratios of two oscillations, or two quantities perceived as oscillations, are small whole numbers. When referring to planetary orbital periods this is the same as → Commensurability. See also → Fractional resonance*.

**Retrograde motion:** Motion against the general direction of the rotation and orbiting of the planets in the solar system. The normal direction is forward motion, which runs anticlockwise.

**Sidereal period:** Actual orbital period of a body in the solar system in relation to the fixed stars. → Synodic period, Tropical year

**Significance:** In probability calculations this is a term describing a deviation that exceeds an expected mean value by more than merely a chance amount. The level of significance has to be decided in each case since probability calculations can only ever make relative statements.

**Silver section*:** A value that occurs repeatedly in connection with the golden section and in a number of geometric constructions. For example if the number 1 is divided into the golden section, the smaller part has the value of 0.3819... In the present work the number 1.3819.. is termed the silver section.

**Synodic period:** The synodic period refers to the movement of a planet as seen in relation to the Sun from the viewpoint of another planet. For example the synodic period of two planets is the average time between two consecutive conjunctions or oppositions. → Sidereal period

**Three-body problem:** → Numerical integration

**Time, Terrestrial Dynamic (TDT):** Absolutely uniform time defined in 1967 in accordance with the period of oscillation of the caesium atom. Astronomical calculations require a uniformly flowing time which deviates somewhat from → Universal Time which is based on the slightly variable rotation of the Earth.

**Time, Universal (UT):** The measure of time given by the Earth's rotation as related to the Greenwich meridian. In astronomical reckoning a new day traditionally begins at 12 noon. Since the Earth's rotation is not absolutely uniform, terrestrial dynamic time (TDT) has been introduced for astronomical calculations. Depending on the purpose of the calculation or else the degree of accuracy required it may be necessary to convert between these two measurements.

**Tropical year:** This is the period between one passage of the Sun through the vernal point and the next. On account of precession this is *c.* 1/25800 days shorter than the sidereal year. Its numerical value is at present 365.24219 days.

**Vernal Point** → Equinox

# Bibliography


Abell, Arthur M., *Gespräche mit berühmten Komponisten*, Schroeder, Garmisch-Partenkirchen 1962

Arp, Halton C., 'Der kontinuierliche Kosmos' in *Neue Horizonte 1992/93*, Piper, Munich 1993

*The Astronomical Almanac 1999*, Washington, US Government Printing Office

Barth, H. (Ed.), *Allgewalt Musik*, Langewiesche-Brandt, Ebenhausen 1953

Bethge, Klaus, *Elementarteilchen und ihre Wechselwirkungen*, Wissenschaftliche Buchgesellschaft, Darmstadt 1991

Breuer, Reinhard (Ed.), *Immer Ärger mit dem Urknall*, Rowohlt Taschenbuch, Reinbek 1993

*Bruno, Giordano—Ausgewählt und vorgestellt von Elisabeth von Samsonow*, Diederichs, Munich 1995

Bühler, Walther, *Das Pentagramm und der Goldene Schnitt als Schöpfungsprinzip*, Verlag Freies Geistesleben, Stuttgart 1996

Caspar, Max, *Johannes Kepler*, W. Kohlhammer Verlag, Stuttgart 1948

Cousto, Hans, *Die kosmische Oktave*, Synthesis-Verlag, Essen 1984

Cronin, Vincent, *The View from Planet Earth. Man Looks at the Cosmos*, Collins, London 1981

Croswell, Ken, *Wir sind Kinder der Milchstrasse*, Scherz Verlag, Berne 1997

Davies, P.C.W. & J.R. Brown (Ed.), *The Ghost in the Atom. A Discussion of the Mysteries of Quantum Physics*, Cambridge University Press, 1986

Devlin, Keith J., *Mathematics: The Science of Patterns. The Search for Order in Life, Mind, and the Universe*, Scientific American Library, New York 1994

Donne, John, *The Complete Poetry of John Donne*, Ed. John T. Shawcross, Anchor Books, Garden City, New York 1967

Draeger, Hans-Heinz, 'Die Verbindlichkeit der mathematischen Intervalldefinitionen' in *Musikalische Zeitfragen*, Vol. 10, Bärenreiter, Kassel 1962

Dürr, Hans-Peter (Ed.), *Physik und Transzendenz*, Scherz Verlag, Berne, 4th edition 1987

Eddington, Arthur, *Space, Time and Gravitation*, Cambridge University Press, reprinted 1990

*Encyclopedia of Planetary Sciences*, Eds. James H. Shirley & Rhodes W. Fairbridge, Chapman & Hall, London, New York 1997

Euler, Manfred, 'Biophysik des Gehörs' in *Biologie in unserer Zeit*, 3/96 and 5/96

Fahr, Hans-Jörg, *Der Urknall kommt zu Fall*, Franck-Kosmos, Stuttgart 1992

Fahr, Hans-Jörg, *Universum ohne Urknall*, Spektrum Akademischer Verlag, Heidelberg, Berlin 1995

Fahr, Hans-Jörg & Eugen A. Willerding, *Die Entstehung von Sonnensystemen—Eine Einführung in das Problem der Planetenentstehung*, Spektrum Akademischer Verlag, Heidelberg, Berlin 1998

Fischer, Daniel, *Das Hubble-Universum: Neue Bilder und Erkenntnisse*, Birkhäuser, Basle 1998

Giovanelli, Ronald G., *Secrets of the Sun*, Cambridge University Press 1984

Goethe, Wolfgang von, *Faust*. Translations used are by David Luke, Oxford World's Classics, Oxford University Press, Oxford 1998 and Sir Theodor Martin, Everyman's Library, Oxford World Classics. And the 'Prologue in Heaven' by Percy Bysshe Shelley

Haase, Rudolf, *Johannes Keplers Weltharmonik*, Diederichs, Munich 1998

Hamel, Jürgen, *Geschichte der Astronomie*, Birkhäuser, Basle 1998

Harrison, Edward R., *Cosmology, the science of the universe*, Cambridge University Press, 1981

Heitler, Walter, *Man and Science*, tr. Robert Schlapp, Oliver & Boyd, Edinburgh, London 1963

Hesse, Hermann, *The Glass Bead Game*, tr. Richard & Clara Winston, Jonathan Cape, London 1969

Hildebrandt, Stefan & Anthony Tromba, *Kugel, Kreis und Seifenblasen*, Birkhäuser, Basle 1996

Husmann, Heinrich, *Vom Wesen der Konsonanz*, Müller-Thiergarten, Heidelberg 1953

Iamblichus, *Iamblichus' Life of Pythagoras*, tr. from Greek by Thomas Taylor, John Watkins, London 1965

Jeans, James H., *The Mysterious Universe*, Cambridge University Press, 1930

Jenny, Hans, *Cymatics. A Study of Wave Phenomena and Vibration*, Tr. D.Q. Stephenson, MACROmedia Publishing, Newmarket, NH 03857, USA 2004

Kayser, Hans, *Akroasis: The Doctrine of the World Harmonic*, tr. Suzanne Doucet, The Hans Kayser Translation Project 2009

Keller, Werner, *Und die Bibel hat doch recht*, Econ, Düsseldorf 1955

Kepler, Johannes, *The Harmony of the World*, tr. E.J. Aiton, A.M. Duncan, J.V. Field, *Memoires of the American Philosophical Society*, Philadelphia 1997

Kepler, Johannes, *Mysterium Cosmographicum. The Secret of the Universe* (in Latin and English), tr. A.M. Duncan, Abaris Books, 1981 and 1999, Norwalk CT06850, USA

Kepler, Johannes, *New Astronomy*, tr. William H. Donahue, Cambridge University Press, 1992

*Kosmos Himmelsjahr 1999*, Ed. Hans-Ulrich Keller, Franckh-Kosmos, Stuttgart 1998

Kraul, Walter, *Erscheinungen am Sternenhimmel*, Verlag Freies Geistesleben, Stuttgart 2002

Kreyszig, Erwin, *Statistische Methoden und ihre Anwendung*, Vandenhoeck & Ruprecht, Göttingen 1968

Küppers, Bernd-Olaf (Ed.), *Ordnung aus dem Chaos*, Piper, Munich 1987

Lang, Kenneth R., *Sun, Earth and Sky*, Springer, New York 1995

Lang, Kenneth R. & Charles A. Whitney, *Wanderers in Space, Exploration and Discovery in the Solar System*, Cambridge University Press, 1991

Laskar, Jacques, Thomas Quinn and Scott Tremaine, 'Confirmation of Resonant Structures in the Solar System' in *Icarus* 95 (1992)

Ledermann, Leon L. & David N. Schramm, *From Quarks to the Cosmos. Tools of Discovery*, The Scientific American Library, New York 1989

Lemcke, Mechthild, *Johannes Kepler*, Rowohlt Taschenbuch, Reinbek 1995
Lermer, Reinhardt, *Grundkurs Astronomie*, 2nd Edition, Bayerischer Schulbuch Verlag, Munich 1989
*Lexikon der Astronomie*, Spektrum Akademischer Verlag, Heidelberg 1995
Mann, Thomas, *Confessions of Felix Krull, Confidence Man*, tr. Denver Lindley, Penguin Books, 1958
Mansfeld, Jaap, *Die Vorsokratiker*, Philipp Reclam Junior, Stuttgart 1987
Martineau, John, *A Little Book of Coincidence*, Wooden Books, Walkmill, Cascob, Wales 2001
Meeks, John, *Planetensphären*, Philosophisch-Anthroposophischer Verlag, Dornach 1990
Meeus, Jean, *Astronomical Algorithms*, William Bell Inc., Editions in 1991 and 1998
*Meyers Handbuch Weltall*, 7th Edition, Meyers Lexikonverlag, Mannheim 1994
Michelsen, Neil F., *Tables of Planetary Phenomena*
Miyazaki, Koji, *Polyeder und Kosmos*, Vieweg & Sohn, Brunswick 1987
Monod, Jacques, *Chance and Necessity: An Essay on the Natural Philosophy of Modern Biology*, tr. from French by Austryn Wainhouse, Collins, London 1972
Montenbruck, Oliver, *Grundlagen der Ephemeridenrechnung*, 6th Edition, Verlag Sterne und Weltraum, Heidelberg 2001
Montenbruck, Oliver & Thomas Pfleger, *Astronomie mit dem Personalcomputer*, Springer, Berlin 1994
Moritz, Karl Philipp, *Götterlehre der Griechen und Römer*, Reclam Junior, Leipzig, no date
Paturi, Felix R., *Harenberg Schlüsseldaten Astronomie*, Harenberg Lexikon Verlag, Dortmund 1996
Peterson, Ivars, *Newton's Clock. Chaos in the Solar System*, W.H. Freeman and Company, New York 1993
Pfrogner, Hermann, *Musik—Geschichte ihrer Deutung*, Karl Alber, Freiburg, Munich 1954
Pichler, Franz, (Ed.), *Der Harmoniegedanke Gestern und Heute—Peuerbach Symposium 2002*, Universitätsverlag Rudolf Trauner, Linz 2003
Plato, *Timaeus and Critias*, tr. Desmond Lee, Penguin Classics, London 1971
Plato, *The Republic*, tr. A.D. Lindsay, J.M. Debt & Sons, London 1935
Plato, *The Laws*, tr. R.G. Bury, William Heinemann, London 1926 (669D-670C)
Remane, Storch & Welsch, *Evolution—Tatsachen und Probleme der Abstammungslehre*, 5th Edition, Dtv Wissenschaft, Munich 1980
Rilke, Rainer Maria, *Rilke's Book of Hours, Love Poems to God*, tr. A. Barrows and J. Macy, Riverhead Books, New York 1996
Ring, Thomas, *Das Sonnensystem—Ein lebender Organismus*, Deutsche Verlagsanstalt, Stuttgart 1939
Roob, Alexander, *Alchemie und Mystik*, Benedikt Taschen, Cologne 1996
Sagan, Carl, *Cosmos*, Macdonald Futura Publishers, London 1980
Scheffler, Helmut & Hans Elsässer, *Bau und Physik der Galaxis*, BI-Wissenschaftsverlag, Mannheim 1992
Schelling, Friedrich Wilhelm Joseph von, *Philosophie und Kunst*, Wissenschaftliche Buchgesellschaft, Darmstadt 1960

Schmidt, Thomas Michael, *Musik und Kosmos als Schöpfungswunder*, Schmidt, Frankfurt 1974

Schneider, Manfred, *Himmelsmeckanic, Vol. 3*, Spektrum Akademischer Verlag, Heidelberg 1996

Schultz, Joachim, *Movement and Rhythms of the Stars: A Guide to Naked-Eye Observation of Sun, Moon and Planets*, Floris Books, Edinburgh 2008

Schultz, Ludolf, *Planetologie: eine Einführung*, Birkhäuser, Basle 1993

Schwentek, Heinrich, *Die Quadratur des Kreises in Mathematik, Kunst und Natur*, private publication, Katlenburg-Lindau 1982

Stephenson, Bruce, *Kepler's Physical Astronomy*, Springer, New York 1987

*Streitfall Evolution*, Ed. Jörg Mey, Robert Schmidt & Stefan Zibulla, Hirzel Wissenschaftliche Verlagsgesellschaft, Stuttgart 1995

Wackenroder, Wilhelm Heinrich, *Phantasien über die Kunst*, Reclam Junior, Stuttgart, edition with supplement, 1983

Waerden, B.L. van der, *Die Pythagoreer—Religiöse Bruderschaft und Schule der Wissenschaft*, Artemis, Zurich, Munich 1979

Waldmeier, M., *Ergebnisse und Probleme der Sonnenforschung*, Geest & Portig, Leipzig 1955

Walker, Daniel P., 'Keplers Himmelsmusik' in *Geschichte und Musiktheorie*, Vol. 6, *Hören, Messen und Rechnen in der frühen Neuzeit*, Joint Editor Carl Dahlhaus, Wissenschaftliche Buchgesellschaft, Darmstadt 1987

Wiora, Walter, 'Die Natur der Musik und die Musik der Naturvölker' in *Musikalische Zeitfragen*, Vol. 10, Bärenreiter, Kassel 1962

Zipp, Friedrich, *Vom Urklang zur Weltharmonie*, Merseburger, Kassel 1985

Zipp, Friedrich, *Vom Wesen der Musik*, Willy Müller, Heidelberg 1974

# Notes

## *Chapter 1*

1. Brahe was of course not alone in his work. As the sixteenth century progressed the ideas of antiquity about the cosmos came to be increasingly questioned. A number of astronomers began to observe celestial appearances such as comets with ever greater accuracy. For more on this subject see, *inter alia*, Felix R. Paturi, *Harenberg Schlüsseldaten Astronomie*, p. 50 and Jürgen Hamel, *Geschichte der Astronomie*, pp. 161ff.
2. William Shakespeare, *The Merchant of Venice*, Act V, Sc. 1; written around 1595.
3. From *The Complete Poetry of John Donne*.
4. Jacques Monod, *Chance and Necessity. An Essay on the Natural Philosophy of Modern Biology*, p. 167.
5. Carl Sagan, *Cosmos*, Chapter VIII, p. 196.
6. Kenneth R. Lang & Charles A. Whitney, *Wanderers in Space, Exploration and Discovery in the Solar System*, p. 236.
7. Johannes Kepler, *The Harmony of the World*, p. 411. The 'observations of Brahe' refers to the data on the motions of the planets compiled over several decades by the Danish astronomer Tycho Brahe.
8. Bruce Stephenson, *Kepler's Physical Astronomy*, quoted by Ivars Peterson in *Newton's Clock. Chaos in the Solar System*, p. 72.
9. Johannes Kepler, towards the end of *The Harmony of the World*, p. 480, following his development of a complex system of interlocking reasons for the eccentricity of the various planetary orbits which have 'their origin in the care taken concerning the harmonies between their movements'.
10. Regarding the orbital periods, I subsequently found a similar depiction in Thomas Ring, *Das Sonnensystem—Ein lebender Organismus*, p. 87.

## *Chapter 2*

1. See, for example, Friedrich Zipp, *Vom Wesen der Musik*, pp. 21ff, or Walter Wiora, 'Die Natur der Musik und die Musik der Naturvölker' in *Musikalische Zeitfragen*, Vol. 10, pp. 112ff.
2. See, for example, Manfred Euler, 'Biophysik des Gehörs' in *Biologie in unserer Zeit*, 3/96, pp. 163ff, and 5/96, pp. 304ff, or Heinrich Husmann, *Vom Wesen der Konsonanz*.
3. This image is also shown in Mechthild Lemcke's biography of Kepler, p. 35.
4. Johannes Kepler, *The Harmony of the World*, p. 139.
5. For more on this see, e.g., Daniel P. Walker, 'Keplers Himmelsmusik' in *Geschichte der Musiktheorie*, Vol. 6, pp. 84ff, or Hans-Heinz Draeger, 'Die Verbindlichkeit der mathematischen Intervalldefinitionen' in *Musikalische Zeitfragen*,

Vol. 10, p. 31. According to Draeger, musical thirds are felt to sound differently when played either sequentially or as chords.
6. Figures similar to the one on the left are shown by Joachim Schultz in *Movement and Rhythms of the Stars*, pp. 121ff, and by Walther Bühler in *Das Pentagramm und der Goldene Schnitt als Schöpfungsprinzip*, pp. 114ff. Walther Bühler refers here to Joachim Schultz, but his own book contains the only reference I have found to the medieval *Pentagramma Veneris*. The pentagram is also mentioned by Thomas Michael Schmidt in *Musik und Kosmos als Schöpfungswunder*, pp. 79ff. He comments that this must already have been known to the Babylonians, since 'the emblem of their goddess Ishtar (Venus, the Lady of Heaven) was the five-pointed star, the pentagram'.
7. From Carl Sagan, *Cosmos*, p. 258.
8. English versions of the quotations from Goethe's *Faust* in this chapter are taken from translations by Sir Theodor Martin, Everyman's Library, and David Luke, Oxford World Classics.

## *Chapter 3*

1. Quoted after Hermann Pfrogner, *Musik—Geschichte ihrer Deutung*, pp. 25 and 27. The texts are from Lu Bu We's 'Spring and Autumn' (third century BC) and the 'Samavidhana-Brahmana' (*c.* 1000 BC). There are many further examples in these and other works.
2. Plato, *Timaeus and Critias*, p. 49
3. Plato, ibid., pp. 53–4.
4. More on this and a depiction similar to that shown in Fig. 3.2 may be found in Friedrich Zipp, *Vom Urklang zur Weltharmonie*, pp. 24ff.
5. Plato, op. cit., p. 41.
6. Hans J. Fahr & Eugen A. Willerding, *Die Entstehung von Sonnensystemen*, p. 14.
7. Ludolf Schultz, *Planetologie: eine Einführung*, p. 52.
8. Ludolf Schultz, *Planetologie: eine Einführung*, p. 52 or, in more detail, S.F. Dermott, 'On the origin of commensurabilities in the solar system—II' in *Monthly Notices of the Royal Astronomical Society*, 141 (1968), pp. 363–7.
9. See *Encyclopedia of Planetary Sciences*, p. 834.
10. Keith J. Devlin, *Mathematics: The Science of Patterns*, p. 114.
11. Johannes Kepler, *The Harmony of the World*, p. 427.
12. Vincent Cronin, *The View from Planet Earth*, p. 111.
13. Walter Heitler, *Man and Science*, p. 9.
14. Kepler's values are calculated according to his data on the apparent daily movements shown on p. 424 of *The Harmony of the World*.
15. Reinhardt Lerner, *Grundkurs Astronomie*, p. 55.
16. Johannes Kepler, *The Harmony of the World*, p. 203.
17. See Daniel P. Walker, 'Keplers Himmelsmusik' in *Geschichte der Musiktheorie*, Vol. 6.
18. Johannes Kepler, *The Harmony of the World*, p. 440.

## Chapter 4

1. Rudolf Haase, *Johannes Keplers Weltharmonik*, p. 56.
2. Ibid., pp. 105ff.
3. Ivars Peterson, *Newton's Clock*, p. 266.
4. Wilhelm Heinrich Wackenroder, *Phantasien über die Kunst*, p. 83.
5. Plato, *The Republic*, p. 321.
6. Plato, *The Laws* (669D–670C), p. 147.
7. Hans Cousto, *Die kosmische Oktave*, p. 27.
8. Johannes Kepler, *The Harmony of the World,* p. 432.
9. Hans Cousto, *Die kosmische Oktave*, p. 154.
10. After Alexander Roob, *Alchemie und Musik*, p. 91.
11. *Iamblichus' Life of Pythagoras*, pp. 31–4.
12. Johannes Kepler, *The Harmony of the World*, p. 432. Kepler here gives the values transposed into an octave in minutes and seconds.
13. Johannes Kepler, *The Harmony of the World*, p. 488.
14. Ludwig van Beethoven to Bettina von Brentano, 28 May 1810; and a diary entry. Gustav Mahler to Anna Bahr-Mildenburg, 18 July 1896; and to Willem Mengelberg, 18 August 1906.
15. John Rodgers & Willie Ruff, 'Kepler's Harmony of the World: A realization for the Ear', in *American Scientist*, Vol. 67 (May–June 1979), p. 286.
16. Friedrich H. Dahlberg (1760–1812) in *Blicke eines Tonkünstlers in die Welt der Geister*, quoted after Hermann Pfrogner, *Musik—Geschichte ihrer Deutung*, p. 253.
17. Heinrich von Kleist (1777–1811) to his fiancée Wilhelmine von Zenge, quoted in *Allgewalt Musik*, ed. H. Barth, p. 28.

## Chapter 5

1. *Sterne und Weltraum*, 5/1995, p. 343.
2. See, for example, *Bild der Wissenschaft*, 1/1998, pp. 69ff.
3. An overview of what actually occurs when a star is born may be found, for example, in special issue 'Schöpfung ohne Ende' of the journal *Sterne und Weltraum*, November 1997.
4. According to Daniel Fischer in *Das Hubble-Universum*, p. 151, only two exceptions were known as of 1998. But these appeared to be so atypical as to be irrelevant with regard to the 10 million year limit. Since then further discs have been discovered which are thought to be up to 400 million years old. It is assumed that they consist of debris formed by some of the collisions between larger clumps. See *Science*, Vol. 286, 1 October 1999, pp. 66ff.
5. Thomas Mann, *Confessions of Felix Krull,* Part 3, Chapter 5, p. 246.
6. Geoffrey W. Marcy & R. Paul Butler, 'Giant Planets Orbiting Faraway Stars' in *Scientific American presents: Magnificent Cosmos*, 1997. The article also gives a more detailed description of the method that has led to the discovery of planets.
7. These pictures are taken from a book that must be unique in this respect: Hans Jenny, *Cymatics. A Study of Wave Phenomena and Vibration*, pp. 22, 23 and 65.

8. In Ludolf Schultz, *Planetologie: eine Einführung*, p. 26, there is a review of the varying ideas about the age of Earth and of the solar system.
9. This is discussed in more detail in Peter H. Richter & Hans-Joachim Scholz, 'Der Goldene Schnitt in der Natur', pp. 199ff: in *Ordnung aus dem Chaos*, ed. Bernd-Olaf Küppers.
10. See for example: *Encyclopedia of Planetary Sciences*, pp. 94ff and pp. 762ff.
11. *Sterne und Weltraum*, 10/1995, p. 672.

## Chapter 6

1. This mode of depicting the Venus/Earth relationship made its appearance independently almost at the same time in: John Martineau, *A Little Book of Coincidence*. As I then discovered, a similar drawing had already been published in Neil F. Michelsen, *Tables of Planetary Phenomena*.
2. See, for example, Jacques Laskar et al., 'Confirmation of Resonant Structures in the Solar System', Icarus 95 (1992), pp. 148–52.
3. Ibid.
4. *Encyclopedia of Planetary Sciences*, p. 765.

## Chapter 7

1. See B.L. van der Waerden, *Die Pythagoreer*, pp. 223ff.
2. Johannes Kepler, *Mysterium Cosmographicum. The Secret of the Universe*, pp. 69 and 73 (Kepler's note in his new 1621 edition of the book).
3. *Encyclopedia of Planetary Sciences*, p. 564.
4. Johannes Kepler, *Mysterium Cosmographicum. The Secret of the Universe*, op. cit., pp. 107 and 221.

## Chapter 8

1. From Stefan Hildebrandt & Anthony Tromba, *Kugel, Kreis und Seifenblasen*, p. 50.
2. Felix R. Paturi, *Harenberg Schlüsseldaten Astronomie*, p. 10.
3. Anthony F. Aveni, 'Venus and the Maya' in *American Scientist*, Vol. 76, pp. 274ff.
4. Carl Sagan, *Cosmos*, p. 95.
5. *Encyclopedia of Planetary Sciences*, p. 887.
6. See e.g. Carl Sagan, op. cit., or John Meeks, *Planetensphären*, p. 22 (where this is dealt with in more detail).
7. Proclus (410–85) in his commentary on the first book of Euclid's *Elements* quoted by Kepler in *The Harmony of the World*, p. 304.
8. Walther Bühler, *Das Pentagramm und der Goldene Schnitt als Schöpfungsprinzip*, p. 98ff.

9. Hermann Hesse, *The Glass Bead Game*, final page of the introduction.
10. Johannes Kepler, *The Harmony of the World*, p. 304.

## *Chapter 9*

1. Arthur Eddington, *Space, Time and Gravitation.*
2. Kenneth R. Lang, *Sun, Earth and Sky*, pp. 53 and 60.
3. Ronald G. Giovanelli, *Secrets of the Sun*, in Chapter 12. He also cites other examples of Earth's climate being influenced by solar activity.
4. Ibid, Chapter 12.
5. Rhodes W. Fairbridge, 'Orbital Commensurability and Resonance' in *Encyclopedia of Planetary Sciences*, pp. 564ff. See also p. 494 and p. 748 in this encyclopedia. The movement of the Sun around the barycentre of the planetary system was described as early as 1939 by Thomas Ring, *Das Sonnensystem—Ein lebender Organismus*, p. 55.
6. Goethe, *Faust*, Part I, 'Prologue in Heaven', translated by Percy Bysshe Shelley.
7. Goethe, *Faust*, Part II, Prologue, Ariel announcing the sunrise.
8. Goethe, *Faust* I, Scene 1, night-time. Faust contemplating the sign of the macrocosm.
9. Proclus (410–85) in his commentary on Plato's *Timaeus*, in Vol. I, Book IV, p. 466 of Thomas Taylor's translation published by The Prometheus Trust.
10. Jack J. Lissauer, 'Planetary Rotation' in *Encyclopedia of Planetary Sciences*, pp. 608ff. Our discussion of current knowledge concerning the origin of planetary rotations is also based on this article.
11. See, for example, *Sterne und Weltraum*, 10/1995, p. 672.
12. Hans Jenny, *Cymatics. A Study of Wave Phenomena and Vibration*, p. 25.

## *Chapter 10*

1. Friedrich Wilhelm Joseph von Schelling, *Philosophie der Kunst*, tr. J. Collis.
2. *Rilke's Book of Hours, Love Poems to God* (translation by Anita Barrows and Joanna Macy).

## *Chapter 11*

1. See, for example, *Meyers Handbuch Weltall*, pp. 435ff.
2. Plato, *Timaeus and Critias*, p. 121.
3. Martin A. Bucher & David N. Spergel, 'Inflation in a Low-Density Universe' in *Scientific American*, January 1999, p. 43.
4. Daniel Fischer, *Das Hubble-Universum*, p. 61 (as at 1998).
5. Edward R. Harrison, *Cosmology, the science of the universe*, p. 228.

6. See the formula and its derivation in Hartmut Schulz, 'Je grösser die Rotverschiebung, desto näher der Quasar' in *Sterne und Weltraum*, 2/97; or—formulated somewhat differently—in Edward R. Harrison, *Cosmology, the science of the universe*, p. 247.
7. Leon M. Lederman & David N. Schramm, *From Quarks to the Cosmos. Tools of Discovery*, p. 169.
8. These figures are given, e.g., in Gerhard Börner, 'Die erste Sekunde' in *Sterne und Weltraum*, special number entitled *Schöpfung ohne Ende, Nr. 2,* 11/97, p. 120. Figures in some other publications differ somewhat.
9. Gerhard Börner, ibid., p. 120.
10. See, e.g., *Meyers Handbuch Weltall*, p. 521.
11. Craig J. Hogan, i.a., 'Surveying Space-time with Supernovae' in *Scientific American*, January 1999, pp. 28ff.
12. Lawrence M. Krauss, 'Cosmological Antigravity' in *Scientific American*, January 1999, p. 41.
13. Plato, op. cit., p. 42.
14. Halton C. Arp, 'Der kontinuierliche Kosmos' in *Neue Horizonte*, 92/93, pp. 113ff.
15. Hans-Jörg Fahr, *Universum ohne Urknall*, pp. 65–6.
16. Stephen D. Landy, 'Mapping the Universe' in *Scientific American*, June 1999, p. 37.
17. See, for example, P.C.W. Davies & R. Brown (Ed.), *The Ghost in the Atom. A Discussion of the Mysteries of Quantum Physics*. This book introduces interpretations by various physicists of the strange characteristics of the subatomic world.
18. Plato, op. cit., p. 43.
19. See, for example Edward R. Harrison, *Cosmology, the science of the universe*, p. 308.

## *Chapter 12*

1. In his book *Musik und Kosmos als Schöpfungswunder*, pp. 83ff, Thomas Michael Schmidt depicts the golden section ratio of the Earth/Jupiter and Earth/Saturn synodic periods and Earth's sidereal period in another way that is mathematically identical.
2. Johannes Kepler, *The Harmony of the World*, p. 391.
3. Johannes Kepler, *New Astronomy*, p. 65.
4. The Wisdom of Solomon, in the Apocrypha, 11:20.

## *Chapter 13*

1. James H. Jeans, *The Mysterious Universe*, pp.148–9. Sir James Jeans was one of the foremost astrophysicists of the first half of the twentieth century.
2. Heinrich Schwentek, *Die Quadratur des Kreises in Mathematik, Kunst und Natur*, pp. 67ff.

3. David Colombara (b. 1922), *Im trigonalen Ton. Der neuen Gedichte zweiter Teil*, p. 48 (tr. J. Collis).
4. Galileo, *The Assayer*, quoted after Ivars Peterson in *Newton's Clock*, p. 284.
5. See for example *Meyers Handbuch Weltall*, pp. 396ff.

## Appendix

1. See B.L. van der Waerden, *Die Pythagoreer*, pp. 240ff and pp. 427ff.
2. Thomas Ring, *Das Sonnensystem—Ein lebender Organismus*, p. 82 and the final sentence on p. 136.
3. In essence this construction is taken from the excellent book by Walther Bühler, *Das Pentagramm und der Goldene Schnitt als Schöpfungsprinzip*, p. 39.
4. Johannes Kepler, *The Harmony of the World*, p. 117. Theophrastus Paracelsus (1493 [or 1494]–1541) is regarded as an important pioneer in the field of medicine who propounded, among much else, the idea of the 'inner physician' as a decisive principle of healing. As so many other ideas, that of the pentagram being connected with health originated with the Pythagoreans.
5. Quoted in Arthur M. Abell, *Gespräche mit berühmten Komponisten*, p. 62.
6. See for example Joachim Schultz, *Movement and Rhythms of the Stars*, pp. 121ff.
7. Jean Meeus, *Astronomical Algorithms*.
8. M. Chapront-Touzé & J. Chapront, 'The lunar ephemeris ELP 2000' in *Astronomy and Astrophysics*, 124 (1983), pp. 50–62.
9. Werner Keller, *Und die Bibel hat doch recht*, p. 338.
10. Jean Meeus, *Astronomical Algorithms*.
11. P. Bretagnon & G. Francou, 'Planetary theories in rectangular and spherical variables. VSOP 87 solutions' in *Astronomy and Astrophysics*, 202 (1988), p. 311.
12. M. Chapront-Touzé & J. Chapront, ELP 2000-85, 'A semi-analytical lunar ephemeris adequate for historical times' in *Astronomy and Astrophysics*, 190 (1988), pp. 342–52.
13. Reinhard Breuer (Ed.), *Immer Ärger mit dem Urknall*, p. 116.
14. Klaus Bethge, *Elementarteilchen und ihre Wechselwirkungen*, p. 327.
15. Hans Kayser, *Akroasis—Die Lehre von der Harmonik der Welt*, p. 132.
16. Thomas Michael Schmidt, *Musik und Kosmos als Schöpfungswunder*, pp. 177ff.

# Picture sources

Page 10, Fig. 1.1: Photo Nicholson/Karkoschka and NASA
Pages 23 and 61, Figs 1.7 and 4.2: Mainz City Library
Page 33, Fig. 2.9: Peter Furst, Delmar, New York
Pages 44 and 49, Figs 3.4 and 3.8: Johannes Kepler, *Weltharmonik,* reprographic print from the 1939 edition, © 1973 by R. Oldenburg Verlag, Munich
Page 45, Fig. 3.5: with kind permission of the Kepler-Kommission der Bayrischen Akademie der Wissenschaften, Munich
Page 83, Fig. 5.1: M.J. McCaughrean, C.R. O'Dell and NASA
Pages 87 and 89, Figs 5.2 and 5.3: Photos Hans Peter Widmer in Hans Jenny, *Cymatics*
Pages 90 and 179, Figs 5.4 and 9.10: Photos Christiaan Stuten in Hans Jenny, *Cymatics*
Page 142, Fig. 8.1: Engraving in Euclid's *Opera Omnia*, David Gregory edition, Oxford 1703
Page 143, Fig. 8.2: Photo NASA
Page 218, Fig. 11.1 top: Photo Bill Schoening/NOAO/AURA/NSF, Tucson AZ
Page 227, Fig. 11.2: Photo NOAO/AURA/NSF, Tucson AZ
Page 230, Fig. 11.3: Photo E. Schreier and NASA
Page 234, Fig. 11.6: Prof. Dr E. Müller, Pennsylvania State University, Osmond Laboratory
Pages 238 and 313, Figs 12.1 and 14.12, University Library, Heidelberg
Page 244, Fig. 12.4: Fondation St Thomas, Strasburg
Page 276, Fig. 13.4: © British Museum, London

Despite intensive searches we have been unable to trace the copyright owners in every case. Please notify us if necessary. All the photos and figures not listed here are by the author.

# Index

# The Computer Program
# 'Signature of the Celestial Spheres'

The program 'The Signature of the Celestial Spheres' invites you to embark on a journey of discovery through our cosmic home with its extraordinary phenomena of movement. It offers a powerful virtual planetarium with many performance and output options some of which are not available elsewhere, based on exceedingly accurate calculations covering several millennia into both the past and the future.

You will experience the planetary system in motion, how the long-term planetary movement figures arise, and how the Sun moves around the barycentre of the planetary system as a whole. You will also discover Johannes Kepler's *Mysterium Cosmographicum. The Secret of the Universe* as a computer simulation, the Platonic solids, Kepler's stellar solids, and much else.

At *www.keplerstern.com* you will find:

- the computer program 'Signature of the Celestial Spheres' of which a trial version can be downloaded
- further information on the subject of the 'harmony of the spheres'
- detailed information about the astronomical calculations